Chemical Oxidation Applications for Industrial Wastewaters

Chemical Oxidation Applications for Industrial Wastewaters

Olcay Tünay, Işık Kabdaşlı, Idil Arslan-Alaton and Tuğba Ölmez-Hancı

Publishing
London • New York

Published by **IWA Publishing**
Alliance House
12 Caxton Street
London SW1H 0QS, UK
Telephone: +44 (0)20 7654 5500
Fax: +44 (0)20 654 5555
Email: publications@iwap.co.uk
Web: www.iwapublishing.com

First published 2010

Originated by The Manila Typesetting Company
Cover by designforpublishing.co.uk

British Library Cataloguing in Publication Data
A CIP catalogue record for this book is available from the British Library

Library of Congress Cataloging-in-Publication Data
A catalog record for this book is available from the Library of Congress

ISBN: 9781843393078
ISBN 10: 1843393077

Contents

Foreword *xi*

Chapter 1: Introduction to redox reactions 1
1.1 Introduction 1
1.1.1 Redox processes 1
1.1.2 Thermodynamics of redox reactions 2
1.1.3 Application of redox processes 10
1.2 Chemicals Used in Wastewater Treatment 11
1.3 Electrochemical Processes 13
1.4 Advanced Oxidation Processes 15
1.4.1 Chemical advanced oxidation processes 16
1.4.1.1 Ozonation at high pH and ozonation in the presence of H_2O_2 16
1.4.1.2 Catalytic ozonation 17
1.4.1.3 Fenton process 18
1.4.1.4 Wet air oxidation 18
1.4.1.5 Supercritical water oxidation 19
1.4.1.6 Sonolysis and enhanced sonolysis (US/O_3; $US/O_3/H_2O_2$; $US/UV/TiO_2$) 20
1.4.2 Photochemical advanced oxidation processes 22
1.4.2.1 Photolysis of H_2O_2 22
1.4.2.2 Photolysis of O_3 and the $O_3/H_2O_2/UV$ system 22
1.4.2.3 Photo-Fenton processes 23
1.4.2.4 Semiconductor-mediated heterogeneous photocatalysis 24
REFERENCES 25

Chapter 2: Textile industry 31
2.1 Introduction 31
2.2 Pollutants Emerging from Textile Preparation, Dyeing and Finishing Activities 33
2.3 Water Consumption 40
2.4 Conventional End-of-Pipe Treatment 41
2.5 Advanced and Emerging Treatment Processes 44
2.5.1 Electrocoagulation 44
2.5.2 Electrochemical treatment (anodic oxidation) 49
2.5.3 Photoelectrocatalysis 53

2.5.4 TiO_2-mediated heterogeneous photocatalysis 54
2.5.5 Ozonation 55
2.5.6 Homogenous and heterogeneous Fenton's processes 57
2.5.7 AOP combinations 59
2.5.8 Wet air oxidation 60
2.5.9 Supercritical water oxidation 61
2.5.10 Electron beam irradiation 62
2.5.11 AOP involving power ultrasound 63
2.6 Concluding Remarks 65
REFERENCES 66

Chapter 3: Leather tanning industry 71
3.1 Leather Tanning Industry 71
3.1.1 Overview 71
3.1.2 Production processes 72
3.1.3 Wastewater generation and characterisation 74
3.1.4 In-plant control and clean technologies 79
3.1.5 Wastewater treatment 82
3.2 Oxidation Processes 85
3.2.1 Sulphide oxidation 85
3.2.1.1 Oxidation of sulphide by air oxygen 85
3.2.1.2 Sulphide oxidation by hydrogen peroxide 92
3.2.1.3 Electrochemical oxidation 93
3.2.1.4 Other sulphide oxidation methods 94
3.2.2 Oxidation applications to raw wastewater 95
3.2.2.1 Electrochemical oxidation and electrocoagulation 95
3.2.2.2 Ozone oxidation 96
3.2.3 Oxidation applications to pre-treated wastewater 97
3.2.3.1 Ozone oxidation 97
3.2.3.2 Other chemical oxidation methods 97
3.2.4 Oxidation applications to biologically treated effluents 98
3.2.4.1 Electrochemical oxidation and electrocoagulation 98
3.2.4.2 Ozonation 99
3.2.5 Colour removal by chemical oxidation 100
3.3 Concluding Remarks 101
REFERENCES 101

Chapter 4: Metal finishing industry 107
4.1 Metal Finishing Industry 107
4.1.1 Overview 107
4.1.2 Production processes 108
4.1.3 Wastewater sources 110
4.1.4 Wastewater characterisation 112
4.1.5 In-plant control and clean technologies 116
4.1.6 Conventional end-of-pipe wastewater treatment 118
4.2 Oxidation and Reduction Processes 121
4.2.1 Cyanide destruction 121

4.2.1.1 Alkaline chlorination 121
4.2.1.2 Ozonation 124
4.2.1.3 Oxidation by H_2O_2 126
4.2.1.4 Sulphur dioxide/air process 128
4.2.1.5 Oxidation by chlorine dioxide 128
4.2.1.6 Hexavalent ferrate oxidation 129
4.2.1.7 Photochemical and photocatalytic oxidation methods 130
4.2.1.8 Other methods 131
4.2.2 Hexavalent chromium reduction 132
4.2.2.1 Chemical reduction by the reduced sulphur compounds 132
4.2.2.2 Chemical reduction by iron and iron compounds 134
4.2.2.3 Electrochemical oxidation and generation 136
4.2.3 Complexed metal bearing wastewater treatment 137
4.2.3.1 Chemical reduction 138
4.2.3.2 Other methods 140
4.2.4 Recovery of precious and common metals 140
4.3 Concluding Remarks 142
REFERENCES 143

Chapter 5: Pharmaceutical industry 149
5.1 Introduction 149
5.2 Pharmaceutical Products and Processes 150
5.2.1 Pharmaceutical products 150
5.2.1.1 Medicals and botanicals (SIC 2833) 151
5.2.1.2 Pharmaceutical preparations (SIC 2834) 151
5.2.1.3 In vivo and in vitro diagnostic substances (SIC 2835) 152
5.2.1.4 Biological products, diagnostics excluded (SIC 2836) 152
5.2.2 Pharmaceutical processes 152
5.2.2.1 Research and development 152
5.2.2.2 The conversion of organic and natural substances into bulk pharmaceutical substances or ingredients through fermentation, extraction, and/or chemical synthesis 153
5.2.2.3 The formulation, mixing and compounding of final pharmaceutical products 154
5.3 Raw Material Inputs and Pollutant Outputs 154
5.3.1 Chemical synthesis 154
5.3.2 Natural product extraction 154
5.3.3 Fermentation 154
5.3.4 Formulation 155
5.4 In-plant Control Strategies 155
5.5 Conventional Treatment Methods 156
5.6 Advanced and Emerging Treatment Technologies 158
5.6.1 Ozone and combinations 158
5.6.2 The H_2O_2/UV-C process 169
5.6.3 Fenton's reagent and combinations 170
5.6.4 Photo-Fenton processes 176
5.6.5 Heterogenous photocatalytic processes 181

5.6.6 Wet peroxide oxidation 183
5.6.7 Power ultrasound 186
5.6.8 Process combinations (hybrid systems) 188
5.6.9 Comparative studies 190
5.7 Concluding Remarks 192
REFERENCES 193

Chapter 6: Pulp and paper industry 199
6.1 Pulp and Paper Industry 199
6.1.1 Overview 199
6.1.2 Pulp and paper manufacturing processes 200
6.1.2.1 Fibre furnish and fibre furnish preparation and handling 202
6.1.2.2 Pulping process 203
6.1.2.3 Chemical recovery processes 205
6.1.2.4 Pulp processing 205
6.1.2.5 Bleaching 206
6.1.2.6 Stock preparation 207
6.1.2.7 Paper and paperboard making processes 207
6.1.3 Wastewater sources 208
6.1.4 Wastewater characterisation 209
6.1.5 Pollution prevention and control 211
6.1.6 Treatment technologies 216
6.2 Chemical Oxidation Processes 217
6.2.1 Pretreatment applications 217
6.2.2 Post-treatment applications 227
6.3 Concluding Remarks 229
REFERENCES 230

Chapter 7: Pesticide industry 235
7.1 Pesticide Industry 235
7.1.1 Overview 235
7.1.2 Pesticide manufacturing processes 237
7.1.2.1 Halogenated organic pesticides 238
7.1.2.2 Nitrogen-containing pesticides 240
7.1.2.3 Metallo-organic pesticides 240
7.1.3 Pesticide formulating, packaging & repackaging processes 240
7.1.4 Wastewater sources 242
7.1.5 Wastewater characterisation 243
7.1.6 Pollution prevention and control 247
7.1.7 Treatment technologies 248
7.2 Chemical Oxidation Processes 250
7.2.1 Cyanide oxidation 252
7.2.2 Priority pollutant pesticide oxidation 253
7.3 Concluding Remarks 261
REFERENCES 261

Chapter 8: Gold and silver mining 267
8.1 Gold and Silver Mining 267
8.1.1 Overview 267
8.1.2 Production processes 268
8.1.3 Waste sources 270
8.1.4 Waste characterisation 271
8.1.5 In-process control technologies 272
8.1.6 Conventional end-of-pipe wastewater treatment 274
8.2 Cyanide Oxidation Processes 274
8.2.1 Alkaline chlorination 274
8.2.2 Copper catalysed hydrogen peroxide oxidation 275
8.2.3 INCO sulphur dioxide/air process 277
8.2.4 Ozonation 280
8.3 Concluding Remarks 280
REFERENCES 281

Chapter 9: Chemicals industry 285
9.1 Chemicals Industry 285
9.1.1 Overview 285
9.1.2 Production processes 286
9.1.3 Wastewater generation and characteristics 287
9.1.4 In-plant control and clean technologies 287
9.1.5 Wastewater treatment 288
9.2 Chemical Oxidation Processes 289
9.2.1 Wet air oxidation 289
9.2.2 Electrochemical processes 290
9.2.3 Ozonation 293
9.2.4 Advanced oxidation methods 293
9.2.5 Sonochemical degradation 296
9.2.6 Developing methods 297
9.3 Concluding Remarks 297
REFERENCES 298

Chapter 10: Other industries and municipal landfill leachate 305
10.1 Alcohol Distilleries 305
10.1.1 Industry overview, processes and wastewater character 305
10.1.2 Oxidation processes 307
10.2 Olive Oil Industry 311
10.2.1 Industry overview, processes and wastewater character 311
10.2.2 Oxidation processes 313
10.3 Municipal Landfill Leachate 316
10.3.1 Overview 316
10.3.2 Oxidation methods applied to landfill leachate 317
REFERENCES 320

Index 327

Foreword

This book is intended to cover chemical oxidation applications applied to industrial wastewater treatment. At the first glance, the subject may seem rather specific, however as anybody who is involved in industrial waste treatment thinks about the place of chemical oxidation in industrial wastewater control as well as the recent developments in the field realises the scope of the subject and may be the need for a monograph focused on this subject. The authors devoted at least last two decades to chemical wastewater treatment methods and the applications to industrial wastewaters and they mostly collaborate in research and application. The idea of writing this book was almost spontaneous and unanimous as we discuss the merits of chemical oxidation and its commonplace in industrial wastewater treatment. Then, the process of writing the book has begun.

The book aims to cover to an up-to-date review, discussion and evaluation of the chemical oxidation applications to industrial wastewaters. Within this context, all existing methods and technologies as well as new and developing methods have been evaluated with specific reference to industrial wastewaters and from the stand points of both theoretical background and applicational aspects such as cost and operation. The wide spectrum of the methods employed in the field call for a review of theoretical basis of the chemical oxidation and the applied methods in the field which have been the subject of first chapter. Chapters 2–9 have been devoted to applications to specific industries which have been selected considering the environmental importance and worldwide common use of the industry as well as the role of chemical oxidation in the treatment of wastewaters. In the last chapter, several industrial process wastewaters and municipal landfill leachate have been the subject of review due to increasing use of chemical oxidation application for their wastewater control.

The book is written for graduate level students, researchers and practitioners. However, design aspects, cost and operation information about the processes address a wider audience.

İstanbul, 2010

Olcay Tünay
Işık Kabdaşlı
Idil Arslan-Alaton
Tuğba Ölmez-Hancı

To our families

Chapter 1

Introduction to redox reactions

1.1 INTRODUCTION

1.1.1 Redox processes

Chemical oxidation is a process in which the oxidation state of an atom is increased. The atom being oxidised may be in the elemental form or in a substance like a molecule or ion. The term *"oxidised"* is also used for the substance containing the oxidised atom. If the oxidation takes place within biological processes the terms biological or biologically-mediated oxidation are used. Chemical reduction is the process by which the oxidation state, the valence, of an atom is reduced. Every oxidation reaction is accompanied by a reduction reaction and these reactions are termed redox reactions.

For inorganic redox reactions, oxidation and reduction are brought about by electron transfer. Oxidation is the loss of electrons and reduction is the gain of electrons by an atom. In the below example:

$$S^{2-} + I_2 \rightarrow 2I^- + S^0 \tag{1.1}$$

Sulphide ion is oxidised to elemental sulphur by losing two electrons, while elemental iodine is reduced to iodide ion by gaining two electrons. In the organic reactions, the mechanism is more complex. An organic reaction oxidation is carried out by replacement of one of the electrons making up the covalent bond between two atoms, by changing one of the atoms in a way for reversing the order of electronegativities of the atoms. If atoms A and B are tied up with a covalent bond and atom A is more electronegative than atom B, replacement of atom B by atom C which is more electronegative than atom A, through breaking the A-B bond and formation of an A-C bond, results in the oxidation of atom A. The basis of this process is expressed by the Pauling rule stating that *"In a covalent bond the charge of an atom is found by assigning each shared pair of electrons completely to the more electronegative of the two atoms sharing them. An electron pair shared by two atoms of the same electronegativity is split between them"* (Stumm and Morgan 1996).

1.1.2 Thermodynamics of redox reactions

A redox reaction is represented by two half-reactions; oxidation and reduction. This representation is quite useful in analysing the redox reactions and defining thermodynamic aspects of the reactions. The half-reactions are *"coupled"*. The redox reaction:

$$I_2(aq) + 2S_2O_3^{2-} \rightarrow 2I^- + S_4O_6^{2-} \tag{1.2}$$

is represented by the following half-reactions:

$$\text{reduction: } I_2\ (aq) + 2e^- \rightarrow 2I^- \tag{1.3}$$

$$\text{oxidation } 2S_2O_3^{2-} \rightarrow S_4O_6^{2-} + 2e^- \tag{1.4}$$

In the reduction half-reaction the reactant that accepts electrons is termed *"oxidant"*. In the oxidation half-reaction the reactant that donates electrons is termed *"reductant"*. An important definition in redox stoichiometry is *"equivalent weight"*. Equivalent weight is calculated by dividing the formula weights of reductant and oxidant to number of electrons in the half-reactions. In thermodynamics spontaneity of a reaction occurring under constant pressure is determined using Gibbs free energy. The negative free energy change indicates that the redox reaction is spontaneous in the direction it is written. The same approach is applied to half-reactions.

As indicated above half-reactions are hypothetical presentations. A half-reaction cannot occur unless it is combined with another half-reaction to yield a redox reaction provided that the redox reaction is thermodynamically possible. So the two half-reactions are "*coupled*". They are coupled because the free electrons cannot exist in solution and the electrons given off in a half-reaction must be received by a substance taking part in the second half-reaction. On the other hand, they are energetically coupled in that the spontaneity of the reactions is judged by the value of the free energy change of final redox reaction. Free energy change of any half-reaction may be positive indicating it is not spontaneous, but the half-reaction proceeds as long as free energy change of the redox reaction is negative, since the half-reactions are energetically coupled. Electron transfer between the substances is readily carried out in solution. Now assume that the half-reactions (Eq. 1.3) and (Eq. 1.4) are attempted to be realised in two separate containers marked I and II in Figure 1.1. None of the half-reactions, even the one having negative free energy can occur. If electron transfer is provided through electrodes dipped in both containers and they are connected with a conductor wire, the half-reactions still do not occur, because electroneutrality condition cannot be satisfied in both containers. In Container I the lack of electrons causes positive charge accumulation and in Container II negative charges will accumulate. If we provide ion transfer between the containers electroneutrality is satisfied and half-reactions begin to take place and electrons flow through the connecting wire. Electron flow creates electric current whose energy corresponds to free energy change of the total redox reaction. Such a system is named as an electrochemical cell.

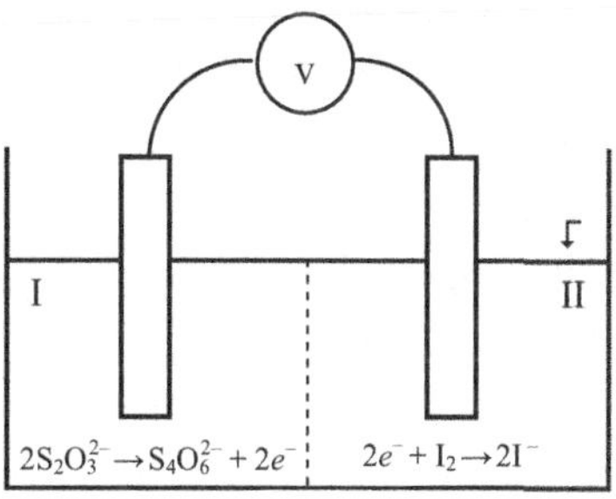

Figure 1.1 Iodine-thiosulphate reaction.

An electrochemical cell (or galvanic cell) is a device for producing an electric current (potential difference) as a result of electrochemical (redox) reactions. A cell consists of two electrodes, an electrolyte (electrolytic conductor) in which electrodes are immersed and a metal conductor by which electrodes are connected. Metal conductor and electrolyte are also called as external circuit and internal circuit, respectively. The potential difference of the electric current obtained by the operation of the cell is (more precisely the maximum potential difference during reversible operation of the cell) called the electromotive force (*e.m.f.*) The *e.m.f.* of a cell is equal to the algebraic sum of the potential differences (jumps) at the interfacial boundaries (surfaces of electrodes) of the cell. *e.m.f.* of a cell is denoted as E_{cell}, and is defined, in a similar manner to reaction free energy, as:

$$E_{cell} = E_{ox} + E_{red} \tag{1.5}$$

where E_{ox} and E_{red} denote potentials of oxidation and reduction half-reactions, respectively. These potentials are termed electrode potentials of the half-reactions. Electrode potential of a half-reaction is a characteristic of the reaction and related to the free energy of this half-reaction.

A cell produces electricity on its external circuit. This corresponds to the work by the cell on the surroundings. Electrical work is defined as:

$$-w_{el} = Q \times E \tag{1.6}$$

where w_{el} is the electrical work and Q is the charge which is moved across a potential difference E. Negative sign of w_{el} is due to the convention that the work (energy) transferred to surroundings by the system is taken negative. The charge Q is carried by the electrons. Faraday constant (F) expresses the charge of a mole of electrons and equals 96,490 Coulomb per mole of electrons. n denotes the number of moles of electrons per mole of reaction with the unit (mol/mol). Then total charge, Q, passing through the external circuit is:

$$Q = n \times F \tag{1.7}$$

The electrical work is, therefore:

$$-w_{el} = n \times F \times E \tag{1.8}$$

On the other hand, the non-expansion work of a reversible electrochemical cell, in other words ΔG equals w_{el}, then:

$$\Delta G = -n \times F \times E \tag{1.9}$$

is the equation that relates free energy to the *e.m.f.* of the cell, or electrode potential of half-reactions. E is measured in volts. Since:

$$\text{Joule} = \text{Coulomb} \times \text{volt} \tag{1.10}$$

the units of ΔG are J/mol, or if two sides of Eq. (1.9) are divided by 4.18 (this is made by assuming that the Faraday constant as 23,060) in cal/mol.

Since individual half-cell reactions do not occur, absolute value of a half-cell potential cannot be determined. However, a half-cell potential can be measured in relation to the potentials of other half-cells. This leads to adopting a reference half-cell reaction and determining the potentials of all other half-reactions against this reference half-reaction. For this purpose, the hydrogen half-cell has been adopted universally as the reference cell. The half-cell reaction:

$$H^+(aq) + e^- \rightarrow \frac{1}{2}H_2(g) \qquad (1.11)$$

is assigned the standard potential of 0.000 V under standard conditions (1atm H_2 pressure, 25°C and pH = 0). This assignment is consistent with $\Delta G_f^o = 0$ for $H_2(g)$ and $H^+(aq)$. The hydrogen reference electrode under standard conditions is named standard hydrogen electrode and abbreviated SHE. For the cell formed by the below reactions where E^o indicates standard conditions:

$$\frac{1}{2}I_2(aq) + e^- \rightarrow I^- \qquad E^o_{I_2,I^-} \qquad (1.12)$$

$$\frac{1}{2}H_2\,(g) \rightarrow H^+ + e^- \qquad E^o_{H_2,H^+} \qquad (1.13)$$

E^o_{cell} is determined by summation of the half-reactions as:

$$\frac{1}{2}I_2(aq) + \frac{1}{2}H_2(g) \rightarrow H^+ + I^- \quad E^o_{cell} \qquad (1.14)$$

and:

$$E^o_{cell} = E^o_{I_2,I^-} + E^o_{H_2,H^+} \qquad (1.15)$$

by definition $E^o_{H_2,H^+} = 0$ and:

$$E^o_{cell} = E^o_{I_2,I^-} \qquad (1.16)$$

Electrode potential of a half-reaction determined against hydrogen reference electrode, under standard conditions, is termed standard electrode potential. In the above given example E^o_{cell} is the standard electrode potential of half-reaction E^o_{I2,I^-}. Direction of the reaction is important for the expression of the electrode potential. Standard electrode potential is universally adopted to be the potential of a half-reaction written in the direction of reduction. Since E^o_{I2,I^-} half-reaction is written in the direction of reduction the value of E^o_{I2,I^-} is its standard electrode potential. Electrode potentials of some reactions are tabulated in Table 1.1. The superscript in the symbol E^o denotes that all the reactions are at standard state. Standard state indicates that reactions are at 1 atm pressure, 25°C temperature while all reactants and products in the reactions in unit activity or 1 molar where it is applicable.

Table 1.1 Standard electrode potentials (Hendricks 2006)

Half-Reaction	E^o (Volt)
$H^+ + e^- \rightarrow 1/2\ H_2(g)$	0
$Cr_2O_7^{2-} + 14H^+ + 6e^- \rightarrow 2Cr^{3+} + 7H_2O$	+1.33
$O_2(aq) + 4H^+ + 4e^- \rightarrow 2H_2O$	+1.27
$O_2(aq) + 2H_2O + 4e^- \rightarrow 4OH^-$	+0.40
$Cl_2 + 2e^- \rightarrow 2Cl^-$	+1.39
$HOCl + 2e^- + H^+ \rightarrow 2Cl^- + H_2O$	+1.49
$OCl^- + 2e^- + H^+ \rightarrow Cl^- + H_2O$	+1.72
$O_3(g) + 2H^+ + 2e^- \rightarrow O_2 + H_2O$	+2.08
$H_2O_2 + 2H^+ + 2e^- \rightarrow 2H_2O$	+1.78
$ClO_2 + 2H_2O + 5e^- \rightarrow Cl^- + 4OH^-$	+0.80
$HSO_5^- + 2H^+ + 2e^- \rightarrow HSO_4^- + H_2O$	+1.85
$FeO_4^{2-} + 8H^+ + 3e^- \rightarrow Fe^{3+} + 4H_2O$	+2.20

The E value of a reaction is independent of reaction stoichiometry, however, as Eq. (1.9) indicates; the ΔG value is dependent on the stoichiometry. ΔG is an extensive property and its value depends on the amount of the matter. Therefore, ΔG values of reactions are always additives. E value of an oxidation reduction reaction cannot be found by the addition of the E values of the half-reactions yielding the resulting oxidation reduction reaction. E value is an intensive property so is independent of the amount of matter. The simple addition of E values of the half-reaction is justified only for the case where number of electrons in the half-reactions is equal and cancels each other resulting in a cell reaction.

In Eq. (1.9), F and n are positive entities; therefore ΔG and E have reverse signs. A reaction is spontaneous when ΔG is a negative number. Then, redox reactions are spontaneous as their E values are positive, in other words, as they make up a cell and produce electricity. According to this thermodynamic property in actual conditions, the half-reactions determine their direction in order to yield a redox reaction with positive E value. The half-reaction with higher E value has the tendency to be reduced. Then, in a redox reaction, the half-reaction with higher E is a reduction while the one with smaller E is oxidation. That is why the substance associated with higher E value is a stronger oxidant with respect to one with smaller E value. The value of ΔG for any condition other than standard conditions is known to be found using:

$$\Delta G = \Delta G^o + R \times T \times \ln Q \qquad (1.17)$$

where R is ideal gas constant (8.31 J/mol.K), T is temperature (in K), and lnQ is the reaction ratio. For the general equation:

$$aA + bB \rightarrow cC + dD \qquad (1.18)$$

Q is given by the equation:

$$Q = \frac{\{C\}^c\{D\}^d}{\{A\}^a\{B\}^b} \qquad (1.19)$$

The activities in Eq. (1.19) may be replaced with molar concentrations, if the solution is diluted. If all terms of Eq. (1.17) is divided by (-n×F) the equation:

$$E = E^o - \frac{RT}{nF} \times \ln Q \qquad (1.20)$$

is obtained. This equation is known as Nernst equation. At 25°C Nernst equation converts to:

$$E = E^o - \frac{0.059}{n} \times \log Q \qquad (1.21)$$

and at equilibrium ΔG = 0 and E = 0:

$$E^o = \frac{RT}{nF} \times \ln K \qquad (1.22)$$

is obtained. In Eq. (1.22) K is equilibrium constant.

Another useful parameter pε is defined as:

$$p\varepsilon = -\log\{e^-\} \qquad (1.23)$$

and expresses the electron activity of a solution at equilibrium. pε is the hypothetical tendency of a solution to accept or to donate electrons, analogous to pH. In a reducing solution electron activity is high, and pε is low, indicating high tendency to donate electrons. In an oxidising solution, electron activity is low and pε is high indicating high tendency to accept electrons.

The equations developed above help to understand and bring solutions to redox reactions. Another way of evaluating the redox data is to use diagrams derived through the basic equations and related thermodynamic data. These diagrams serve to assimilate the system behaviour at a glance. There are several types of diagrams designed for specific purposes. Latimer diagrams show half-cell potentials of an element and provide ease of evaluation for complex systems such as the elements acquiring several oxidation numbers. Free energy-oxidation state diagrams (Frost diagrams) are similar to Latimer diagrams, but are based on free energy changes relative to a common reference point. Potential-pH diagrams are prepared using the Nernst equation and are quite useful for redox evaluations for varying pH. Thermodynamically most stable form of an element at a given pH and electrode potential can be assessed using predominance-area (Pourbaix) diagrams.

An electrochemical cell is a device that produces electric current as a result of a redox reaction. Therefore, an electrochemical cell converts chemical energy to electrical energy. Electrolysis is a reverse process in which redox reactions proceed under the action of an electric current. Since the work obtained from an isothermal reaction is maximum when the reaction is carried out under reversible conditions, the electric work that is obtained from an electrochemical cell is maximum when the cell is operating under reversible conditions. Reversible cells provide maximum work and all thermodynamic expressions developed in this section apply to reversible cells. An electrochemical cell is regarded as reversible if a small amount of current can be passed in either direction without appreciable effect on the measured potential. In a cell as given in Figure 1.1 the electrode to which anions migrate to satisfy the electroneutrality condition is named as anode (positive electrode). The anode releases electrons into external circuit, therefore, it is the place where oxidation half-reaction takes place. The cathode (negative electrode) on the contrary, receives electrons from external circuit by which reduction half-reaction occurs. There are several types of electrodes. The metal electrode reacts reversibly with its cations in the electrolyte. These electrodes are reversible relative to the cations and called electrodes of the first kind. The hydrogen electrode belongs to this kind and it is also a gas electrode. Electrodes of the second kind are the metal electrodes covered by a layer of sparingly soluble salt of this metal and they are in a solution saturated with this salt and containing a highly soluble salt with the same anion. This kind of electrodes is reversible with respect to cation and anion but only the concentration of anion can be controlled. The oxidation-reduction electrodes are metals (noble metals) which exchange only electrons with the electrolyte.

Electrochemical cells are represented by shorthand notation. In this notation phase boundaries are indicated with vertical lines. The notation starts from left to right and composition of each phase is noted. Since hydrogen electrode is not convenient for frequent measurements due to operational problems such as surface poisoning, other reference electrodes are used for measurements. The two most common reference electrodes are calomel:

$$Cl.\ (aq)|HgCl_2(s)|Hg(l) \quad (1.24)$$

and silver-silver chloride:

$$Cl.\ (aq)|AgCl(s)|Ag(s) \quad (1.25)$$

electrodes. A concentration cell is composed of two electrodes of the same kind (metal or gas) dipped into different concentrations of the same electrolyte or two metal electrodes with different concentrations (*e.g.* different alloy ratios) dipped into the same electrolyte. The current in these cells is due to the reactions to equalise the concentration differences according to:

$$E = E^o - \frac{2.3RT}{nF}\log\frac{\{C_1\}}{\{C_2\}} \quad (1.26)$$

where C_1 and C_2 are concentrations of the same electrolyte or electrode.

The electrolysis process is widely used in industrial applications. Chemical transformations in electrolysis vary widely depending on the electrodes, electrolyte and purpose of electrolysis. For instance, metal plating is an electrolysis process where metal ions in the electrolyte are deposited from the solution on the metal electrode. In the electrolysis process the relations between the quantity of electricity flowing through the electrolyte and the amount of substance liberated during electrolysis are quantified by Faraday's laws. Faraday's first law states that for any electrode reaction the amount of substance decomposed is directly proportional to the current density and the duration of flow of electricity. According to Faraday's second law, when equal amounts of electricity pass through solutions of different electrolytes, the amounts of substances undergoing conversion are proportional to their chemical equivalents. The evolution of 1 gram-equivalent of any substance requires the passage of F = 96,480 Coulomb per mole of electrons. In electrolysis a certain potential difference occurs between the electrodes that are oppositely directed to the potential difference applied externally. This effect is called polarisation and is of significant practical importance. Polarisation phenomenon is evaluated as chemical and concentration polarisation. Chemical polarisation is due to the formation of a galvanic cell by the electrolysis products. Concentration polarisation is due to the difference in the concentration of the electrolyte at the anode and cathode arising during the electrolysis.

1.1.3 Application of redox processes

Spontaneity of a process or assessment of the energy requirement for realisation of a process is of primary concern as far as its application is concerned. However, in the practice the rate of the processes, in other words, kinetics of the reactions is as much important as their thermodynamics. Some spontaneous reactions may take place so slowly that they may not be used in the practice unless their rate is accelerated by lowering the activation energy through the use of catalysts or providing proper environmental conditions. In the practice, assessment of the rate of oxidation-reduction processes is of immense importance. The reaction rates of redox processes can be determined experimentally and considering all the factors that may affect the rate of reactions. Among the most important factors affecting the rate of reactions are: initial concentration of reactants, concentration of products building up in the reaction mixture, temperature, composition of the system and its variability through the reaction, ionic strength, use of catalysts, and presence of inhibitors. On the other hand, in many cases, the pH is a major factor determining the rate of redox reactions. Use of catalysts as either heterogeneous or homogeneous catalysis is also a common application in the practice.

Chemical oxidation has long been applied for the treatment of water and wastewaters. The aim of chemical oxidation is conversion of pollutants to acceptable compounds or to the forms suitable for subsequent treatment operations. In the case of organic matter oxidation, mineralisation may not be the ultimate purpose, however, oxidation products as well as intermediates are ensured to be less toxic or less hazardous than the initial substrate. Chemical oxidation applications were relatively limited to processes such as taste and odour control, colour removal, chromium reduction, cyanide oxidation, iron oxidation, manganese oxidation, sulphide oxidation, and disinfection up to 1970's. However, following the recognition of new concepts and definitions such as toxic organics, micropollutants and hazardous wastes and demand for new and powerful treatment techniques, chemical oxidation applications have significantly increased. Development of advanced oxidation processes provided a further acceleration to the use of redox processes. Within this context, a number of new oxidants and oxidation methods have been developed within the last two decades. Conventional and newly developed treatment methods that are based on redox reactions are briefly introduced in the following subsections.

1.2 CHEMICALS USED IN WASTEWATER TREATMENT

The most common oxidants used in industrial wastewater treatment to remove organic and inorganic pollutants can be listed as chlorine, air oxygen, ozone, and hydrogen peroxide. In addition to these traditional oxidants, chlorine dioxide, peroxymonosulphate, and ferrate have been recently recognized as promising oxidising agents for special oxidation applications. Furthermore, ozone and hydrogen peroxide are sometimes used in a combination or in conjunction with ultraviolet light irradiation, or together with a homogenous or heterogeneous catalyst to produce radicals with powerful oxidative properties. The treatment techniques that rely on in-situ formation of radicals will be introduced in the following subsections. Traditional and emerging oxidants are discussed as follows.

Chlorine has been the oxidant traditionally used in industrial wastewater treatment all over the world. However, special concerns about its potential of formation of chlorinated by-products have recently brought about its replacement with other oxidants such as ozone and hydrogen peroxide. Chlorine is available in gaseous form (Cl_2) in pressure metal tanks, or in liquid form as sodium hypochlorite (NaOCl), or in solid form as calcium hypochlorite ($Ca(OCl)_2$). When pure chlorine gas is fed into aqueous solution, it rapidly disproportionates to yield hypochlorous acid (HOCl) being a weak acid ($pK_a = 7.5$ at 25°C), and then hypochlorous acid simultaneously dissociates to hypochlorite ion (OCl^-). When sodium or calcium hypochlorite is added to aqueous solution, the species produced are also hypochlorous acid and hypochlorite ion. Hence, the chemistry for all chlorine sources is basically the same for the aqueous solution. The electrode potentials of all forms are tabulated in Table 1.1 including those of the other oxidants used in the wastewater treatment applications.

Oxygen in the air is the most common and economical oxidant used in the oxidation of some specific pollutants such as the reduced sulphur compounds as well as in the elimination of odour. Oxygen being a colourless, odourless, and tasteless gas dissolves about 9 mg/L in water at 20°C. Its solubility decreases with increasing temperature and total dissolved solids *e.g.* salinity. Oxidation by oxygen in air is a slow reaction since reactivity of molecular oxygen is very low. Therefore, the oxidation rate of aeration process is significantly low. The rate of aeration process can be accelerated via breakdown of oxygen-oxygen bond in molecular oxygen being apparently so strong (119.2 kcal/mol) (Darwent 1970). In order to break this bond, much thermal energy or a catalyst such as manganese sulphate or chloride is necessary. Once the oxygen-oxygen bond is broken, the oxidation proceeds quite rapid.

Ozone (O_3) is one of the most versatile and effective oxidants used in the destruction or abatement of inorganic and specific organic pollutants. It is generated on-site using ozone generators utilising dry compressed air or pure oxygen because it is an unstable gas. Ozone generators can be classified into three groups as low-frequency systems (50 to 60 Hz); medium-frequency systems (400 to 1000 Hz); and high-frequency systems (2000 to 3000 Hz). When generated, the ozone-enriched air or oxygen gas is transferred to the aqueous solution by means of a counter current multistage bubble contactor, an in-line gas injection system, or other gas transfer devices. As can be seen in Table 1.1, ozone with the electrode potential of 2.08 V is a more powerful oxidant than the traditional oxidising agents. Inorganic and organic compounds can be oxidised either by direct reaction with the molecular ozone or by indirect reaction producing free radical species as a consequence of the decomposition of ozone in the water. The rate of ozone decomposition increases with increasing pH, elevating temperature and presence of metals. Molecular ozone is predominant species at lower pH values than 7.0 whereas very low concentrations of molecular ozone can be detected in solution compared to amount dosed at and above pH 9.0 as ozone decomposition resulting in the formation of hydroxyl radicals accelerates with increasing pH (Rice and Wilkes 1994). Molecular ozone reactions are extremely selective while hydroxyl radical is capable of reacting with almost any organic and inorganic substance as is stronger oxidant than molecular ozone.

Hydrogen peroxide (H_2O_2) is one of the most powerful oxidants used in the treatment of some specific pollutants such as the reduced sulphur species, cyanides, and organic compounds. It is commercially available as aqueous solutions grading from 35 to 70 % on the basis of weight to weight ratio. Hydrogen peroxide dissolves in water in all proportions. It decomposes exothermically into water and oxygen gas depending on temperature, solution pH, and presence of impurities and stabilisers. As introduced in the following subsections, H_2O_2 also decomposes in the presence of ferrous ion or ozone, and in conjunction with ultraviolet irradiation to yield radicals being more powerful oxidants.

Chlorine dioxide (ClO_2) which is a neutral compound of chlorine in the (IV) oxidation state, has a great oxidation capacity. It cannot be commercially compressed or stored as a gas since it is explosive under pressure. Hence, chlorine dioxide gas is never shipped. It is always produced using generators on-site. It is soluble in water, particularly in chilled water. In contrast to the hydrolysis of chlorine gas in water, chlorine dioxide does not hydrolyse to any appreciable extent in water, but remains in solution as a dissolved gas (Aieta and Berg 1986). Its aqueous solutions can be prepared by sodium chlorite reacting with gaseous chlorine, hypochlorous acid, or hydrochloric acid under acidic conditions. In presence of a base-catalyst, chlorine dioxide disproportionates to chlorite and chlorate. Therefore, alkaline conditions are to be avoided in practice.

Potassium peroxymonosulphate ($2KHSO_5{\cdot}KHSO_4{\cdot}K_2SO_4$) has proved to be an effective oxidant in the treatment of some organic and inorganic pollutants such as cyanide, the reduced sulphur compounds, ketones, and aldehydes. It is commercially available. Its activation using cobalt as the catalyst or UV irradiation as the activator leads to the formation of highly reactive species, mainly the sulphate radicals (Anipsitakis *et al.* 2008). The main disadvantage of this oxidant is that the oxidation process may yield the treated effluents with high sulphate concentration which may require an additional treatment step.

Ferrate (Fe (VI)) has been recently proposed as an alternative oxidant to traditional oxidising agents. It is capable of oxidising a large number of inorganic and organic pollutants as well as emerging contaminants such as pharmaceuticals, hormones, and endocrine disrupting agents (Sharma *et al.* 2008). Ferrate solution can be prepared by adding K_2FeO_4 to the solution containing a mixture of phosphate and borate at pH 9.0.

The properties of the reductants such as the reduced sulphur compounds are given in the book where they are mentioned.

1.3 ELECTROCHEMICAL PROCESSES

Electrochemical reactions are carried out in electrochemical cells. Basic properties of an electrochemical cell and electrolysis process have been explained in Subsection 1.1. The electrochemical cell can be run in different ways depending on the target reaction and the processes are named according to the purpose of reaction or the method by which treatment of water and wastewater are realised. The electrolysis process has long been used for waste treatment, recovery and in industrial processes. In general it is utilised to convert a compound or ion to its elements such as metal plating and production of chlorine from sodium chloride solution. Electroflotation is used to produce a gas, mostly hydrogen from breakdown of water, in finely bubbled solution for phase separation such as flotation of suspended matter or oils from emulsions. Electroflotation, although, used widely in industry, commonly for enrichment, is also utilised in many electrochemical applications as a side benefit *e.g.* in electrocoagulation (EC).

Electrooxidation is an electrolysis process designed to enhance the oxidation of organic and inorganic compounds in the electrolyte. The process is used as a waste treatment method. Electrooxidation of pollutants is realised by two mechanisms; indirect and direct oxidation. Indirect oxidation is electrochemical generation of an oxidant and releasing it into solution where it realises oxidation reactions.

The oxidation agents generated anodically are chlorine or hypochlorite, hydrogen peroxide, ozone, metal mediators like Ag^{2+} and even hydroxyl radicals through Fenton reaction of electrochemically produced hydrogen peroxide with ferrous ions. Direct anodic reactions are realised on the surface of anode and are achieved through two different pathways. One pathway is through the chemisorbed active oxygen (oxygen in the oxide lattice MO_{x+1}) which is responsible for oxidation of substrate. In the second, physisorbed active oxygen (adsorbed hydroxyl radicals) is responsible for mineralisation (electrochemical combustion) of organic matter. Oxide type anode materials such as Sn-Pd-Ru oxide coated titanium, binary oxide-coated titanium Ru-Ti oxide, PbO_2-coated titanium as well as boron-doped diamond and other inert materials are used for electrooxidation. The process was applied to textile wastewater, tannery wastewater, coke-plant wastewater, cyanide containing wastewater, and phenol containing wastewater (Deng and Englehardt 2007).

Electrocoagulation is an electrolysis process involving in-situ generation of coagulants such as aluminium and iron ions through anodic dissolution. The process has many advantages over conventional coagulation process. These advantages originate mainly from additional mechanisms that enhance the removal of pollutants. Direct and indirect oxidations affect the modified chemical characteristics while flotation facilitates physical separation. EC process also modifies the physical properties of the coagulants formed and increases their destabilisation and adsorption characteristics *e.g.* increase in the surface area of the flocs and modification of surface charges. EC is an inherently versatile process involving several design and operation parameters such as current density, electrode material, electrode spacing, electrolyte concentration, pH, substrate concentration which provide effective optimisation of the process. EC and other electrochemical processes can be applied in various reactor forms as well as electrode arrangements. Electrode polarities can also be changed during operation. Application area of the EC process is quite wide; textile, petroleum industry, oil, turbid effluents, suspended particles of all sorts, organic chemicals, specific organics, fluoride, nitrate, heavy metals, arsenic are examples.

Electroreduction process is another application, particularly, for the treatment of chlorinated hydrocarbons. The mechanism of electrochemical reductive dechlorination is known as electrocatalytic hydrogenolysis. The mechanism at inert electrodes such as glassy carbon involves two successive electron transfers according to following reactions:

$$RX + e^{-} \leftrightarrows RX^{\bullet -} \tag{1.27}$$

$$RX^{\bullet-} \rightarrow R^{\bullet} + X^{-} \tag{1.28}$$

$$RX + e^{-} \rightarrow R^{\bullet} + X^{-} \tag{1.29}$$

$$R^{\bullet} + e^{-} \leftrightarrows R^{-} \tag{1.30}$$

The reduction process involves both radical and carbanion intermediates. The reduction potential of the radical $R^{\bullet}$ is generally more positive than that of the starting halide RX, so that it is readily reduced to the corresponding carbanion. The process works well in non-aqueous solvents such as dimethyl sulphoxide and acetonitrile. Mediated electroreduction can also be used for dechlorination. 1-naphthonitrile, anthracene and many other compounds can be used as mediator whose role is to first accept an electron at cathode and attack the chlorinated compound to yield chloride ion and HC^{-} which converts to HC by combining with H^{+}. Mediated reductive chlorination can also be carried out in non-aqueous solvents. Surfactant aggregates can be used in mediated electrolysis where solubilities of lipophilic organochlorine contaminants are low inhibiting bimolecular reaction between reduced mediator and chlorinated hydrocarbon (Bunce *et al.* 1997; Isse *et al.* 2006).

1.4 ADVANCED OXIDATION PROCESSES

Ozonation and advanced oxidation processes (AOPs) have recently emerged as an important class of technologies for the oxidation and destruction of a wide range of organic pollutants in water and wastewater (Legrini *et al.* 1993; Alvares *et al.* 2001; Zhou and Smith 2001; Oppenländer 2003). The AOPs are characterised by a variety of radical reactions that involve combinations of chemical agents (*e.g.* O_3, H_2O_2, transition metals, and metal oxides) and auxiliary energy sources (*e.g.* ultraviolet-visible (UV-Vis) irradiation, electronic current, γ-radiation, and ultrasound). These treatment processes can either eliminate organic pollutants completely through mineralisation or convert them to the products that are less harmful to human health and the aquatic environment. AOPs are processes involving in-situ generation of highly reactive species such as the hydroxyl radical ($HO^{\bullet}$), which is the primary oxidant in AOPs, while the other radical and active oxygen species are superoxide radical anions ($O_2^{\bullet-}$), hydroperoxyl radicals ($HO_2^{\bullet-}$), triplet oxygen ($3O_2$), and organic peroxyl radicals (ROO^{-}). Unlike many other radicals, $HO^{\bullet}$ is non-selective and thus readily attacks a large group of organic chemicals to convert them to less harmful intermediate products.

1.4.1 Chemical advanced oxidation processes

1.4.1.1 Ozonation at high pH and ozonation in the presence of H_2O_2

Ozonation of organic pollutants involves two types of oxidation reactions, either direct reactions with molecular ozone (O_3; ozonolysis) or indirect oxidation with free radicals that are formed during ozone decomposition. O_3 decomposition is a function of several factors including pH, the presence of inorganic and organic compounds (Bader and Hoigné 1981). Molecular O_3 reactions are extremely selective and limited to unsaturated aromatic and aliphatic compounds. There is a wide spectrum of molecular O_3 reactions such as oxygen atom transfer to anionic, uncharged and cationic species, electron transfer, formation of an oxyl radical, O_3 molecule addition and ring formation (Von Gunten 2003). On the other hand, in the presence of hydroxyl ions (HO^-) at alkaline pH (> 9), O_3 readily decomposes in water to form $HO^•$ and hence ozonation at high pH values is regarded as an AOP. The produced $HO^•$ have a higher oxidation potential (+ 2.8 V) than molecular O_3 (+ 2.07 V) and can attack organic and inorganic molecules almost indiscriminately at diffusion-limited reaction rates (Andreozzi *et al.* 1999; Ikehata and Gamal El-Din 2004; 2005). O_3 decomposition proceeds through the following five-step chain of reactions (Masschelein 1992):

$$O_3 + H_2O \rightarrow 2HO^• + O_2 \tag{1.31}$$

$$O_3 + OH^- \rightarrow O_2^{•-} + HO_2^• \tag{1.32}$$

$$O_3 + HO^• \rightarrow O_2 + HO_2^• \leftrightarrow O_2^{•-} + H^+ \tag{1.33}$$

$$O_3 + HO_2^• \leftrightarrows 2O_2 + HO^• \tag{1.34}$$

$$2HO_2^• \rightarrow O_2 + H_2O_2 \tag{1.35}$$

Similar to O_3, H_2O_2 reacts with organic pollutants present in water through direct and indirect pathways. Indirect reactions are due to the oxidising action of free radicals formed during photolytic or chemically/catalytically induced H_2O_2 decomposition. Although, H_2O_2 is a relatively strong oxidising agent and being used to oxidise aldehydes, alcohols, amines, azobenzene, phenols, cyanides, the reduced sulphur compounds, and metal ions, however, ultimate mineralisation of recalcitrant organic compounds and/or toxicity removal cannot be achieved by mere H_2O_2 oxidation. In the "*peroxone*" process (O_3/H_2O_2), $HO^\bullet$ radicals are generated by a free radical-chain mechanism by the interaction between O_3 and H_2O_2 (Staehelin and Hoigné 1983; Hoigné 1998). The overall (net) reaction is given below:

$$H_2O_2 + 2O_3 \rightarrow 2HO^\bullet + 3O_2 \tag{1.36}$$

As is evident from Eq. (1.36), the molar ratio for the reaction between H_2O_2 and O_3 is 1 to 2 that has been established as the optimum molar ratio in many applications (Glaze *et al.* 1987). At neutral pH values, low H_2O_2 concentrations (10^{-5} to 10^{-4} M) yield a significant O_3 decomposition and hence a high concentration of $HO^\bullet$ (Beltrán 2003). The use of the O_3/H_2O_2 treatment system has proven to ensure a high degree of mineralisation (Yang *et al.* 2005). Peroxone is one of the most studied and applied AOPs in the laboratory, pilot plant works, and even in practical water and wastewater treatment (US EPA 1999).

1.4.1.2 Catalytic ozonation

Recently, alternative ozonation processes catalysed by transition metals have been investigated for degradation of organics. Catalytic ozonation studies can be divided mainly into homogeneous and heterogeneous processes. In homogeneous catalytic ozonation, transition metal ions such as Fe(II), Mn(II), Ni(II), Co(II), Cd(II), Cu(II), Ag(II), Cr(III), and Zn(II) are used together with bulk O_3 (Hordern *et al.* 2003). In homogenous reactions, transition metal ions present in the reaction solution initiate O_3 decomposition by the generation of superoxide radicals ($O_2^{\bullet-}$). The transfer of one electron from the $O_2^{\bullet-}$ molecule to O_3 results in the formation of ozonide, *e.g.* $O_3^{\bullet-}$, and subsequently $HO^\bullet$ (Gracia *et al.* 1995). The mechanism of homogeneous catalytic ozonation can be briefly expressed as Eq. (1.37) (Wu *et al.* 2008):

$$Me^{(n-1)+} + O_3 + H^+ \rightarrow Me^{n+} + HO^\bullet + O_2 \tag{1.37}$$

Heterogeneous catalytic ozonation involves the use of metal oxides, and metals or noble metals supported on metal oxides. TiO_2, Al_2O_3, SiO_2, MnO_2, and Fe_2O_3, or Cu, Ru, Pt, Pb, Pd, Co, metals mounted on these catalyst supports by several ways are used in catalytic ozonation (Hordern *et al.* 2003). The activity of these catalysts is based on the catalytic decomposition of O_3 and the enhanced generation of $HO^\bullet$ (Hordern *et al.* 2003). The efficiency of catalytic ozonation depends to a great extent on the catalyst and its surface properties as well as solution pH (Hordern *et al.* 2003).

The role of metal on the surface of metal oxides in the process of heterogeneous catalytic ozonation in water was explained by Legube and Karpel Vel Leitner (1999). The mechanism is proposed as: On the surface of the reduced Me-catalyst (Me_{red}) O_3 oxidises the metal with the generation of $HO^•$. Organic molecules after adsorption on the catalyst surface are subsequently oxidised by an electron-transfer reaction to give again a reduced catalyst ($Me_{red}R^•$). The organic radical species ($R^•$) are subsequently desorbed from the catalyst and oxidised by $HO^•$ or O_3 either in bulk solution, or more probably, within the thickness of an electric double layer.

1.4.1.3 Fenton process

More than a century ago, H. J. Fenton described the enhanced oxidation power of H_2O_2 on certain organic molecules from H_2O_2 decomposition upon addition of Fe^{2+} as a catalyst at acidic pH values (Fenton 1894; Haber and Weiss 1934). Later on, it was discovered that this was due to the generation of $HO^•$ depending on the reaction given below (Walling 1975):

$$Fe^{2+} + H_2O_2 \rightarrow Fe^{3+} + OH^- + HO^• \tag{1.38}$$

The Fenton process is efficient in the pH range 2–5 and is usually most effective at pH 2.8–3.0 (Pignatello 1992). This AOP has the advantage that H_2O_2 being used as the oxidant, is relatively cheap, abundant and infinitely water soluble. Fenton's reaction has been well studied for its prospective applications in industrial wastewater treatment (Legrini *et al.* 1993; Ollis and Al-Ekabi 1993; Prousek 1996).

1.4.1.4 Wet air oxidation

Wet air or thermal liquid phase oxidation (WAO) in which the generation of active oxidants including $HO^•$ takes place at elevated temperatures (473–593 K) and pressures (20–200 bar), is known to have a great potential for the partial or full treatment of industrial wastewaters having a high toxicity and/or bio-recalcitrance, and a significantly high organic matter content (Schwartz *et al.* 1977). The COD range of interest is usually 10–100 g/L. A high temperature application enhances the reaction kinetics, whereas a pressurised medium is required to increase oxidant (air, oxygen) solubility in the aqueous reaction medium. Residence times may range from 15 to 120 min, and the extent of COD removal may typically be about 75–90 % (DiCorcia *et al.* 1998). Oxidation products usually remain in the aqueous phase. Besides, organic nitrogen is converted to ammonia, nitrate, or nitrogen gas; sulphur is converted to sulphate; halogens to halides and organic phosphorus to phosphates (DiCorcia *et al.* 1998). The major advantage over incineration is the relatively low reaction temperature that prevents the formation of dioxins, carbofuran, and NO_x gases. One of the main drawbacks of WAO is its inability to achieve complete mineralisation of highly polluted effluents, since some low molecular weight oxidation products (especially acetic and propionic acids as well as methanol, ethanol, and acetaldehyde) accumulate in the reaction solution. Besides, the reactor material has to be resistant to corrosion and cracking when chloride ions and other corrosives are present in the waste stream, increasing the capital costs of WAO. Upon addition of catalysts (supported transition metals, metal oxides, noble metals *etc.*) milder reaction conditions can be applied and refractory degradation products (acetic acid and ammonia) can be further oxidised at much lower temperatures by catalytic WAO (CWAO). However, several crucial issues have to be solved concerning the stability of heterogeneous oxidation catalysts at hydrothermal operating conditions including catalyst leaching and sintering loss of surface area of the catalyst support, catalyst poisoning due to carbon monoxide evolution, deposition of organic or inorganic compounds (coking) on the catalyst surface (Ruppert *et al.* 1993; White *et al.* 1994; Staples *et al.* 2001). Oxides of Al, Hf, Zr, and Ti known to be stable are used as catalyst supports. Additional information about studies performed in the field of CWAO is available in recent review papers (Schwartz *et al.* 1977; Sun and Pignatello 1993; Chen and Pignatello 1997; DiCorcia *et al.* 1998; Sagawe *et al.* 2001; Malato *et al.* 2002; Rodríguez 2003; Augugliaro *et al.* 2006).

1.4.1.5 Supercritical water oxidation

Supercritical water oxidation (SCWO) shows substantial promise for the efficient decontamination of many aqueous organic wastes. SCWO is referred to as oxidation in an aqueous medium above the critical temperature and pressure of water, which are 374°C and 22.1 MPa, respectively (Mizuno *et al.* 2000; Shaw and Dahmen 2000). Consequently, SCWO behaves like a low-polarity, largely unassociated solvent medium, in which molecular oxygen and many organic compounds are totally miscible. SCWO can rapidly and efficiently destroy organic substances completely into H_2O and CO_2 with conversion rates as high as 99–99.99%, at 500–650°C in 1–100 s, which are significantly short residence times (Veriansyah *et al.* 2005). Unlike wet air oxidation as well as other AOPs that usually result in partial/incomplete oxidation and hence accumulation of oxidation intermediates, this oxidation process enables the reuse of treated water, thus cutting off the water consumption costs and discharge fees. The major drawback of SCWO are the relatively high capital costs associated with reactor materials and equipment, the requirement of educated staff in charge of running the process and explosion risks (Park *et al.* 2003).

1.4.1.6 *Sonolysis and enhanced sonolysis (US/O_3; $US/O_3/H_2O_2$; $US/UV/TiO_2$)*

Sonochemical treatment (or: sonolysis, sonication) is based on the use of power ultrasound (US) at low to medium frequency (20–1000 kHz) and high-energy to catalyse the destruction of organic pollutants (Augugliaro *et al.* 2006). The treatment effects are essentially the result of ultrasonically induced acoustic cavitation; *e.g.* the formation, growth, and impulsive collapse of bubbles in liquids. The cavitation of a bubble has a variety of effects within homogeneous liquid, heterogeneous solid/liquid, and heterogeneous liquid/liquid phases. Each of these classifications is of interest in water treatment (Mason and Pétrier 2004). The collapse of bubbles induces localised supercritical conditions, *i.e.* high temperatures (more than 5000 K) and pressures (around 1000 atm), electrical discharge and plasma effects (Suslik 1998). The consequences of these extreme environmental conditions are the cleavage of oxygen and water molecules to produce active species, such as hydrogen radicals ($H^•$) and $HO^•$ capable of attacking the organic compounds in water (Kotronarou *et al.* 1991):

$$H_2O \rightarrow H^• + HO^• \tag{1.39}$$

$H^•$ produced from the above reaction can also interact with molecular oxygen present in the system to form $HO_2^•$ and H_2O_2:

$$H^• + O_2 \rightarrow HO_2^• \tag{1.40}$$

$$2HO_2^• \rightarrow H_2O_2 + O_2 \tag{1.41}$$

In general, the sonochemical degradation in aqueous phase involves several reaction pathways and zones such as pyrolysis inside the cavitation bubble and/or at the gas-liquid interfacial region of the cavitation bubbles and $HO^•$ mediated reactions at the gas-liquid interfacial region of the cavitation bubbles, and/or in the liquid bulk (Hoffmann *et al.* 1996). Today, there are several publications reporting the successful removal of a wide range of organic pollutants from various water matrices. However, the ultrasonic energy conversion into specific desirable effects (organic matter removal) is very important from the industrial point of view. Due to the inefficient conversion of energy in producing ultrasonic cavitation and possible difficulties in its scale-up, no industrial installation for wastewater treatment has been reported in the literature. Researchers are trying to circumvent these difficulties by using hybrid techniques for wastewater treatment using ultrasound.

Mere sonolysis is rather ineffective in the destruction of recalcitrant compounds; thus chemical oxidants, such as O_3 or UV irradiation can be used to increase the efficiency and oxidising power of sonochemical treatment processes. The combination of sonolysis and ozonation provides three sources of $HO^•$ which can be listed as (i) from sonochemical decomposition of water, (ii) from O_3 reactions in water and (iii) from the thermolytic decomposition of O_3, in the acoustic cavitation bubble (Weavers and Hoffmann 1998). The simultaneous use of sonolysis and O_3 has shown higher removal efficiencies for a range of organic pollutants than their sequential combination. When O_3 is introduced into a sonolytic system, the ultrasonic irradiation enhances mass transfer, dispersion, and solubility of O_3, thus increasing $HO^•$ production. By combining power US and H_2O_2, it was possible to achieve the formation of free radicals within the cavitation bubbles (Shemer and Narkis 2005):

$$H_2O_2 \rightarrow 2HO^• \tag{1.42}$$

$$H_2O_2 + O_2 \rightarrow 2HO_2^• \tag{1.43}$$

$$H_2O_2 + HO^• \rightarrow HO_2^• + H_2O \tag{1.44}$$

The presence of H_2O_2 simultaneously may enhance the contribution of O_3 in the overall generation of free radicals by the following reaction mechanism (Gogate and Pandit 2004):

$$H_2O_2 \leftrightarrow HO_2^- + H^+ \tag{1.45}$$

$$O_3 + HO_2^- \rightarrow HO^• + O_2 + O_2^{•-} \tag{1.46}$$

The so-called sonophotocatalysis is the simultaneous use of US and photocatalysis on semiconductors. The simultaneous use of these two processes was reported to be more effective than their sequential application, though leading to additive effects, and to be more effective than sonolysis only when employing relatively low US intensity. Synergistic effects between two processes were observed when employing small particle size semiconductors or when operating at relatively low US frequency (30 kHz), known to be less effective than higher US frequency for the degradation of organic species (Mrowetz *et al.* 2003). Wang *et al.* (2006a;b) proposed that US irradiation can produce US-induced luminescence at a wavelength of below 375 nm, can excite the TiO_2 particle as a photocatalyst, and generate highly oxidative $HO^•$ radicals on the surface of the TiO_2 particles. The following explanations are suggested for US/TiO_2 and UV/US/TiO_2 systems: (i) US acts as a deaggregator by microstreaming causing microbubble collapse, which induced surface cleaning of the TiO_2 particles; (ii) the presence of an additional liquid-solid interface in the liquid bulk promotes cavitation; and (iii) US accelerates mass transfer between the bulk phase and the TiO_2 surface (Wu 2009).

1.4.2 Photochemical advanced oxidation processes

1.4.2.1 Photolysis of H_2O_2

UV irradiation has been the most known and widely used radiation method in initiating oxidative degradation processes. In direct photolytic reactions (UV photolysis) UV-C (λ = 200–280 nm) light irradiation is most commonly used, while UV-A (λ = 315–400 nm) is often employed for homogenous and heterogeneous photocatalytic degradation processes (Legrini *et al.* 1993). Some organic pollutants effectively absorb UV-C light directly, and absorption of this high-energy irradiation can cause destruction of chemical bonds and subsequent breakdown of the contaminant (Schwarzenbach *et al.* 2003). However, in most cases H_2O_2 addition is required to create AOP conditions that enhance contaminant degradation significantly. Baxendale and Wilson (1957) pioneered the UV-C photolysis of H_2O_2 in water that is the most direct method of $HO^•$ generation through the homolytic cleavage of H_2O_2:

$$H_2O_2 + h\nu \rightarrow 2HO^• \tag{1.47}$$

The quantum yield of the above reaction is 1.0 mol $HO^•$ per einstein absorbed by H_2O_2 (Prousek 1996). The molar absorption coefficient of H_2O_2 at 254 nm is only 18.6 M^{-1} cm^{-1} and hence the efficiency of the H_2O_2/UV process decreases drastically with increasing pollutant concentration and UV absorbance of the target chemical (CCOT 1995). For heavily polluted effluents, high UV doses and/or H_2O_2 concentrations are required, and this reduces the feasibility of this process for practical applications.

1.4.2.2 Photolysis of O_3 and the O_3/H_2O_2/UV system

The UV photolysis of aqueous O_3 is an indirect method for producing H_2O_2 that in turn reacts with O_3 and also absorbs UV-C light irradiation to initiate a free radical chain reaction leading to $HO^•$ formation (Glaze *et al.* 1987):

$$O_3 + H_2O + h\nu(< 310 \text{ nm}) \rightarrow H_2O_2 \tag{1.48}$$

$$2O_3 + H_2O_2 \rightarrow 2HO^{\bullet} + 3O_2 \qquad (1.49)$$

Eq. (1.49) represents the net reaction of $HO^{\bullet}$ production by the O_3/H_2O_2 process (Hoigné 1998). In addition, there are several other oxidative degradation mechanisms involved in the O_3/UV and O_3/H_2O_2/UV systems including direct UV-C photolysis, direct ozonation, direct oxidation with H_2O_2, and UV-C photolysis of H_2O_2 (Beltrán 2003). O_3 strongly absorbs in the UV region with a maximum molar absorption coefficient of 3300 M^{-1} cm^{-1} at 254 nm (Glaze *et al.* 1987; Legrini 1993). The addition of H_2O_2 to the O_3/UV process accelerates O_3 decomposition resulting in an increased rate of $HO^{\bullet}$ generation. The process combinations seem to be more efficient, however, the main drawback of the O_3/H_2O_2/UV treatment system is the low efficiency and high running costs associated with continuous O_3 production and UV irradiation throughout the process, as well as high capital costs associated with the implementation of O_3 generators and UV photoreactors (CCOT 1995; Mokrini *et al.* 1997).

1.4.2.3 Photo-Fenton processes

The combination of Fenton's reaction with UV light (180–400 nm), *i.e.* the so-called photo-Fenton process, enhances the reaction by producing additional $HO^{\bullet}$. The photo-reduction of Fe^{3+} (ferric hydroxo complexes; *i.e.* $Fe(OH)_n^{(3-n)+}$) to Fe^{2+} ion by UV light produces additional $HO^{\bullet}$ according to the following reaction (Pignatello and Huang 1993; Wadley and Waite 2004):

$$Fe^{3+} + H_2O \xrightarrow{hv} Fe^{2+} + HO^{\bullet} + H^{+} \qquad (1.50)$$

The formed Fe^{2+} reacts with H_2O_2 to produce more $HO^{\bullet}$ via Fenton's reaction. The continuous recycle of ferrous iron, as long as hydrogen peroxide is available, reduces the amount of iron salts required for the Fenton's reaction that makes the photo-Fenton treatment more attractive than the dark Fenton's oxidation process (Wadley and Waite 2004). Besides, if UV-C light is used for irradiation, H_2O_2 photolysis also contributes to $HO^{\bullet}$ production via the reaction shown in Eq. (1.50). At higher (near UV, even visible) wavelength ranges, ferric carboxylate complexes, $Fe(RCO_2)^{2+}$ may also produce additional Fe^{2+} and organic radicals ($R^{\bullet}$) by the following reaction (Pignatello and Huang 1993);

$$Fe(RCO_2)^{2+} \rightarrow Fe^{2+} + CO_2 + R^{\bullet} \quad (< \text{ca. } 500\text{ nm}) \qquad (1.51)$$

The quantum yield of Eqs. (1.50) and (1.51) are strongly pH- and wavelength-dependent (Safarzadeh-Amiri *et al.* 1996). As already mentioned for the dark-Fenton process, the main limitation of iron-based AOPs is the necessity to work at acidic pH (2–5) range, because at higher pH iron salts would begin to precipitate as ferric hydroxide (Wadley and Waite 2004). Furthermore, iron remaining in the solution has to be removed after the treatment.

1.4.2.4 Semiconductor-mediated heterogeneous photocatalysis

Semiconductor photocatalysts include titanium dioxide, zinc oxide, cadmium sulphide, bismuth oxide, ferric oxide, and tungsten oxide. Doped metal oxides have also been investigated such as silver or platinum over titanium dioxide (Ollis and Al-Ekabi 1993). Among the metal chalcogenides used in heterogeneous photocatalysis, titanium dioxide, which has suitable band gap energy (3.2 eV) is stable, relatively inexpensive and has a high photoactivity (Legrini *et al.* 1993). The decomposition of organic pollutants in titanium dioxide suspensions is due to photogeneration of valence band (VB) electrons and holes in the photocatalyst material under UV light irradiation mainly in the $\lambda = 300$–400 nm region (Ollis and Al-Ekabi 1993). Upon irradiation with photons < 384 nm wavelengths, electrons migrate from the valence band to the conduction band (CB) according to the following reaction (Wei and Wan 1992):

$$TiO_2 \rightarrow h_{VB}^{+} + e_{CB}^{-} \tag{1.52}$$

The holes react with electron donors in the electrolyte to produce $HO^{\bullet}$ (Wei and Wan, 1992):

$$h_{VB}^{+} + HO^{-} \rightarrow HO^{\bullet} \tag{1.53}$$

Peroxyl radicals ($HO_2^{\bullet}$) are also formed in this system via electron reduction when molecular oxygen is present, or alternatively, via hole oxidation (Wei and Wan 1992):

$$O_2 + e_{CB}^{-} + 2H^{+} \rightarrow H_2O_2 \tag{1.54}$$

$$2\,h_{VB}^{+} + 2H_2O \rightarrow 2H^{+} + H_2O_2 \tag{1.55}$$

$$H_2O_2 + HO^{\bullet} \rightarrow H_2O + HO_2^{\bullet} \tag{1.56}$$

The decomposition of pollutants can be accomplished either by holes or by free radicals produced in the bulk medium. Photogenerated holes and electrons must be continuously trapped by electron donors and acceptors (oxygen, hydrogen peroxide, transitions metals, *etc.*) to avoid charge recombination. Scavenging the conduction band electron prevents charge recombination and produces additional $HO^{\bullet}$ and superoxide ($O_2^{\bullet-}$) anions (Legrini *et al.* 1993):

$$H_2O_2 + e_{CB}^- \rightarrow HO^{\bullet} + HO^- \tag{1.57}$$

$$O_2 + e_{CB}^- \rightarrow O_2^{\bullet -} \tag{1.58}$$

Transition metals such as Fe^{3+} may also act as electron scavengers at suitable concentrations and reduce charge recombination (Balcioglu and Arslan 1997):

$$Fe^{3+} + e_{CB}^- \rightarrow Fe^{2+} \tag{1.59}$$

Subsequent free radical reactions result in the oxidation and eventual mineralisation of organic pollutants to CO_2, H_2O, and inorganic salts.

REFERENCES

Aieta, E. and Berg, J.D. (1986) A review of chlorine dioxide in drinking water treatment. *J. AWWA* **78**, 62–72.

Alvares, A.B.C., Diaper, C. and Parsons, S.A. (2001) Partial oxidation by ozone to remove recalcitrance from wastewaters-a review. *Environ. Technol.* **22**, 409–427.

Andreozzi, R., Caprio, V., Insola, A. and Marotta, R. (1999) Advanced oxidation processes (AOP) for water purification and recovery. *Catal. Today* **53**, 51–59.

Anipsitakis, G.P., Tufano, T.P. and Dionysiou, D.D. (2008). Chemical and microbial decontamination of pool water using activated potassium peroxymonosulfate. *Water Res.* **42**, 2899–2910.

Augugliaro, V., Litter, M., Palmisano, L. and Soria, J. (2006) The combination of heterogeneous photocatalysis with chemical and physical operations: A tool for improving the photoprocess performance. *J. Photoc. Photobio. C.* **7**, 127–144.

Bader, H. and Hoigné, J. (1981) Determination of ozone in water by the indigo method. *Water Res.* **15**, 449–456.

Balcioglu, I. and Arslan, I. (1997). Treatment of textile waste water by heterogeneous photocatalytic oxidation process. *Environ. Technol.* **18**, 1053–1059.

Baxendale, J. and Wilson, J. (1957). The photolysis of hydrogen peroxide at high light intensities. *T. Faraday Soc.* **53**, 344–356.

Beltrán, F. J. (2003) *Ozone Reaction Kinetics for Water and Wastewater Systems*, Lewis Publishers, Boca Raton, Florida, U.S.A.

Bunce, N.G., Merica, S.G. and Lipkowsky, J. (1997) Prospects for the use of electrochemical methods for the destruction of organochlorine wastes. *Chemosphore* **35**, 2719–2726.

Calgon Carbon Oxidation Technologies-CCOT (1995). The AOP Handbook. CCOT, Markham, Ontario.

Chen, R. and Pignatello, J. (1997) Role of quinone intermediates as electron shuttles in Fenton and photoassisted Fenton oxidations of aromatic-compounds. *Environ. Sci. Technol.* **31**, 2399–2406.

Darwent, B. deB. (1970) Bond dissociation energies in simple molecules. *Nat. Stand. Ref. Data Ser., Nat. Bur. Stand. (US)* **31**, 1–52.

Deng, Y. And Englehardt, J.D. (2007) Electrochemical oxidation for landfill leachate treatment. *Waste Manage.* **27**, 380–388.

DiCorcia, A., Constantino, A., Crescenzi, C., Marinoni, E. and Samperi, R. (1998) Characterization of recalcitrant intermediates from biotransformation of the branched akyl side chain of nonylphenol ethoxylate surfactants. *Environ. Sci. Technol.* **32**, 2401–2409.

Fenton, H.J.H. (1894) Oxidation of tartaric acid in presence of iron. *J. Chem. Soc.* **65**, 899–910.

Glaze, W. H., Kang, J. W., Chapin, D. H. (1987) The Chemistry of water treatment processes involving ozone, hydrogen peroxide and ultraviolet radiation. *Ozone-Sci. Eng.* **9**, 335–352.

Gogate, P. R. and Pandit, A. B. (2004) A review of imperative technologies for wastewater treatment II: hybrid methods. *Adv. Environ. Res.* **8**, 553–597.

Gracia, R., Aragües, J.L., Cortés, S. and Ovelleiro, J.L. (1995) In *Proceedings of the 12th World Congress of the International Ozone Association*, Lille, France, 15–18 May 1995, p. 75.

Haber, F. and Weiss, J. (1934) The catalytic decomposition of hydrogen peroxide by iron salts. *Proc. Roy. Soc. A.* **134**, 332–351.

Hendricks, D. (2006) *Water Treatment Unit Processes Physical and Chemical*, CRC Press, Taylor and Francis Group, Boca Raton, U.S.A.

Hoffmann, M.R., Hua, I., Hochemer, R. (1996) Application of ultrasonic irradiation for the degradation of chemical contaminants in water. *Ultrason. Sonochem.* **3**, 163–172.

Hoigné, J. (1998) Chemistry of aqueous ozone and transformation of pollutants by ozone and advanced oxidation processes, In *The Handbook of Environmental Chemistry, Volume 5, Part C. Quality and Treatment of Drinking Water. Part II* (ed. J. Hrubec), Springer, Berlin, Germany.

Hordern, B.K., Zió M. and Nawrocki, J. (2003) Catalytic ozonation and methods of enhancing molecular ozone reactions in water treatment. *Appl. Catal. B: Environ.* **46**, 639–669.

Ikehata, K. and Gamal El-Din, M. (2004) Degradation of recalcitrant surfactants in wastewater by ozonation and advanced oxidation processes: A review. *Ozone-Sci. Eng.* **26**, 327–343.

Ikehata, K. and Gamal El-Din, M. (2005) Aqueous pesticide degradation by ozonation and ozone-based advanced oxidation processes: A review (Part I). *Ozone-Sci. Eng.* **27**, 83–114.

Isse, A.A., DeGiusti, A., Gennaro, A., Falciola, L. and Mussini, P.R. (2006) Electrochemical reduction of benzyl halides at a silver electrode. *Electrochim. Acta* **51**, 4956–4964.

Kotronarou, A., Mills, G. and Hoffmann, M. R. (1991) Ultrasonic irradiation of para-nitrophenol in aqueous-solution. *J. Phys. Chem.* **95**, 3630–3638.

Legrini, O., Oliveros, E. and Braun, A.M. (1993) Photochemical processes for water treatment. *Chem. Rev.* **93**, 671–698.

Legube, B. and Karpel Vel Leitner, N. (1999) Catalytic ozonation: a promising advanced oxidation technology for water treatment. *Catal. Today* **53**, 61–72.

Malato, S., Blanco, J., Vidal, A. and Richter, C. (2002) Photocatalysis with solar energy at a pilot-plant scale: An overview. *Appl. Catal. B:Environ.* **37**, 1–15.

Mason, T. J. and Pétrier, C. (2004) Ultrasound Processes, In *Advanced Oxidation Processes for Water and Wastewater Treatment* (ed. S. Parsons), IWA Publishing, Padstow, Cornwall, UK, pp. 185–208.

Masschelein, W. J. (1992) *Unit Processes in Drinking Water Treatment.* Marcel Dekker, New York.

Mizuno, T., Goto, M., Kodoma, A. and Hirose, T. (2000) Supercritical water oxidation of a model municipal solid waste. *Ind. Eng. Chem. Res.* **39**, 2807-2810.

Mokrini, A., Ousse D. and Esplugas, S. (1997) Oxidation of aromatic compounds with UV radiation, ozone/hydrogen peroxide. *Water Sci. Technol.* **35**, 95–102.

Mrowetz, M., Pirola, C. and Sell, E. (2003) Degradation of organic water pollutants through sonophotocatalysis in the presence of TiO_2. *Ultrason. Sonochem.* **10**, 247–254.

Ollis, D.F. and Al-Ekabi, H. (1993) *Photocatalytic Purification and Treatment of Water and Air.* Elsevier, Amsterdam.

Oppenländer, T. (2003) *Photochemical Purification of Water and Air: Advanced Oxidation Processes (AOPs): Principles, Reaction Mechanisms, Reactor Concepts*, pp. 11–14 & 36–67, John Wiley and Sons, Inc., Chichester.

Park, T.J., Lim, J.S., Lee, Y.W. and Kim, S.H. (2003) Catalytic supercritical water oxidation of wastewater from terephthalic acid manufacturing process. *J. Supercrit. Fluids* **26**, 201–213.

Pignatello, J. (1992) Dark and photoassisted Fe^{3+} catalyzed degradation of chlorophenoxy herbicides by hydrogen peroxide. *Environ. Sci. Technol.* **26**, 944–951.

Pignatello, J. J. and Huang, L. Q. (1993) Degradation of polychlorinated dibenzo-p-dioxin and dibenzofuran contaminants in 2,4,5-T by photoassisted iron-catalyzed hydrogen peroxide. *Water Res.* **27**, 1731–1736.

Prousek, J. (1996) Advanced oxidation processes for water treatment. Chemical processes. *Chem. Listy.* **90**, 229–237.

Rice, R.G. and Wilkes, J.F. (1994) Ozone Chemistry Applied to Cooling Tower Water Treatment. In *Chemical Oxidation Technologies for the Nineties* (eds. W.W. Echenfelder, A.R. Bowers J.A. and Roth), Technomic Publishing Company Inc., Lancaster, United States.

Rodríguez, M. (2003) *Fenton and UV-Vis Based Advanced Oxidation Processes in Wastewater Treatment: Degradation, Mineralization and Biodegradability Enhancement.* Universitat De Barcelona Facultat De Química Departament D'enginyeria Química I Metal·Lúrgia, Spain.

Ruppert, G., Bauer, R. and Heisler, G. (1993) The photo-Fenton reaction - an effective photochemical wastewater treatment process. *J. Photochem. Photobiol. A.* **73**, 75–78.

Safarzadeh-Amiri, A., Bolton, J. R., Cater, S. R. (1996). The use of iron in Advanced Oxidation Processes. *J. Advan. Oxid. Technol.* **1**, 18–26.

Sagawe, G., Lehnard, A., Lubber, M. and Bahnemann, D. (2001) The insulated solar Fenton hybrid process: Fundamental investigations. *Helv. Chim. Acta.* **84**, 3742–3759.

Schwartz A.M., Perry J.W. and Berch J. (1977) *Surface Active Agents and Detergents Vol. II.* R. Krieger Pub. Co., New York, U.S.A.

Schwarzenbach, R., Gschwend, P. and Imboden, D. (2003). Photochemical transformation reactions. In *Environmental Organic Chemistry* (eds. R.P. Schwarzenbach, P.M. Gschwend, D.M. Imboden), pp. 437–484, Wiley, New York.

Sharma, V.K., Yngard, R.A., Cabelli, D.E. and Baum, J.C. (2008) Ferrate (VI) and ferrate (V) oxidation of cyanide, thiocyanate, and copper (I) cyanide. *Rad. Phys. Chem.* **77**, 761–767.

Shaw, R.W. and Dahmen, N. (2000) Destruction of toxic organic materials using supercritical water oxidation: Current state of the technology. In *Supercritical Fluids-Fundamentals and Applications* (eds. E. Kiran, P.G. Debenedetti and C.J. Peters), pp. 425–437, Kluwer Academic Publishers, Dordrecht.

Shemer, H. and Narkis, N. (2005) Trihalomethanes aqueous solutions sono-oxidation. *Water Res.* **39**, 2704–2710.

Staehelin, S. and Hoigné, J. (1983) Reaktions mechanismus und kinetik des ozonzerfalls in wasser in gegenwart organischer stoffe. *VomWasser.* **61**, 337–348.

Staples, C.A., Naylor, C.G., Williams, J.B. and Gledhill, W.E. (2001) Ultimate biodegradation of alkylphenol ethoxylate surfactants and their biodegradation intermediates. *Environ. Toxicol. Chem.* **20**, 2450–2455.

Stumm, W. and Morgan, J.J. (1996) *Aquatic Chemistry: Chemical Equilibria and Rates in Natural Waters*, 3rd edn, John Wiley&Sons, Inc., Canada.

Sun, Y. and Pignatello, J. J. (1993) Photochemical reactions involved in the total mineralization of 2,4-D by Fe^{3+}/H_2O_2/UV. *Environ. Sci. Technol.* **27**, 304–310.

Suslik, K. S. (1998) *Ultrasound: Its Chemistry, Physical and Biological Effects,* VCH Publishers, New York, U.S.A.

US EPA (1999) *Alternative Disinfectants and Oxidants Guidance Manual.* EPA 815-R-99-014, Office of Water, United States Environmental Protection Agency, Washington DC, U.S.A.

Veriansyah, B., Park, T.J., Limb, J.S. and Lee, Y.W. (2005) Supercritical water oxidation of wastewater from LCD manufacturing process: kinetic and formation of chromium oxide nanoparticles. *J. Supercrit. Fluids* **34**, 51–61.

Von Gunten, U. (2003) Ozonation of drinking water: Part I. Oxidation kinetics and product formation. *Water Res.* **37**, 1443–1467.

Wadley, S. and Waite T. D. (2004). Fenton Processes, In *Advanced Oxidation Processes for Water and Wastewater Treatment*, (S.A. Parsons, ed.), pp. 111–136, IWA Publishing, Cornwall.

Walling, C. (1975) Fenton's reagent revisited. *Acc. Chem. Res.* **8**, 125–131.

Wang, J., Pan, Z., Zhang, Z., Zhang, X., Wen, F., Ma, T., Jiang, Y., Wang, L., Xu, L., Kang, P. (2006a) Sonocatalytic degradation of methyl parathion in the presence of nanometer and ordinary anatase titanium dioxide catalysts and comparison of their sonocatalytic abilities. *Ultrason. Sonochem.* **13**, 493–500.

Wang, J., Ma, T., Zhang, Z., Zhang, X., Jiang, Y., Dong, D., Zhang, P. and Li, Y. (2006b) Investigation on the sonocatalytic degradation of parathion in the presence of nanometer rutile titanium dioxide (TiO_2) catalyst. *J. Hazard. Mater.* **137**, 972–980.

Weavers, L. K. and Hoffmann, M. R. (1998) Sonolytic decomposition of ozone in aqueous solution: mass transfer effects. *Environ. Sci. Technol.* **32,** 3941–3947.

Wei, T.Y. and Wan, C.C. (1992) Kinetics of photocatalytic oxidation of phenol on TiO_2 surface. *J. Photochem. Photobiol. A: Chem.* **69**, 241–247.

White, R., Jobling, S., Hoare, S.A., Sumpter, J.P. and Parker, M.G. (1994) Environmentally persistent alkylphenolic compounds are estrogenic. *Endocrinology* **135**, 175–182.

Wu, C.-H. (2009) Photodegradation of C.I. Reactive Red 2 in UV/TiO_2-based systems: Effects of ultrasound irradiation. *J. Hazard. Mater.* **167**,434–439.

Wu, C.-H., Kuo, C.-Y. and Chang, C.-L. (2008) Homogeneous catalytic ozonation of C.I. Reactive Red 2 by metallic ions in a bubble column reactor. *J. Hazard. Mater.* **154**, 748–755

Yang, C., Xu, Y. R., Teo, K. C., Goh, N. K., Chia, L. S. and Xie, R. J. (2005) Destruction of organic pollutants in reusable wastewater using advanced oxidation technology. *Chemosphere* **59**, 441–445.

Zhou, H. and Smith, D.W. (2001) Advanced technologies in water and wastewater treatment. *Can. J. Civ. Eng.* **28**, S49–S66.

Chapter 2

Textile industry

2.1 INTRODUCTION

The Chinese began cultivating silkworms and made fine silk garments around 2700 BCE; Egyptians were wrapping their mummies in fine linen cloth as early as 2500 BCE, whereas remains of woollen textiles dating far back as 6000 BCE have been found. All these are evidences of human textile activities reaching back into the ancient history. The modern textile industry manufactures upholstery, carpeting, draperies, linens, towels, and many other textile products besides clothing (Shahady 1998). Textile products are made of natural (cotton, linen, silk, wool, *etc.*) or synthetic (nylon, rayon, polyester, acetate, acrylic, *etc.*) fibres that are either used directly or spun into a yarn.

In the textile manufacturing process, the yarn or fibre is either knitted, interlocked or woven, thereafter dyed to obtain coloured products, rinsed and finally finished to provide special appearance and/or functions (production of anti-static, flame retardant, mildew resistant fabrics). Before a fabric is dyed, impurities are removed in a process called scouring (Shahady 1998). Bleaching is a chemical process employed to whiten the fabric, whereas finishes mentioned above are applied chemically or mechanically (Shahady 1998).

Each step in the textile manufacturing process generates some kind of solid, liquid or gaseous waste. The textile industries as well as the waste streams generated by textile processing activities are generally characterised by the different preparation (desizing, scouring, mercerising and bleaching), dyeing and finishing activities that are covered in the forthcoming sections in more detail. Figure 2.1 presents an overview of the major steps involved in textile manufacturing processes and operations taking place ***after fabric production***. These are generally called "textile finishing" processes and operations applied to produce knit or woven fabrics, stock or yarn.

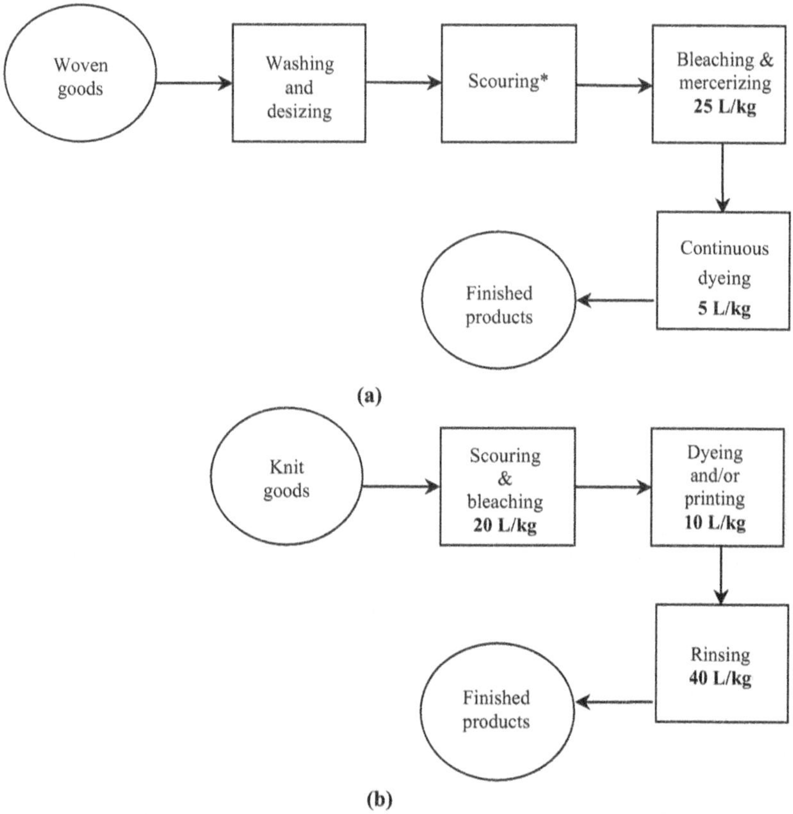

Figure 2.1 Main steps involved in the finishing of woven (a) and knitted (b) fabrics (**No separate data is available for this textile finishing step)* (generated by considering UNEP 1994; Shahady 1998).

Figure 2.1 also indicates the average water consumption rates per kg of finished product in each step involved. From Figure 2.1 it can be concluded that in the finishing of woven goods, 45 L water is consumed per kg of finished product, whereas for knit fabrics, 70 L waster is used in the average during the finishing operations and processes. More and more factories apply continuous finishing steps in their manufacturing operations and processes that on one hand decreases water consumption immensely, but on the other hand increases the pollution load per finished product.

2.2 POLLUTANTS EMERGING FROM TEXTILE PREPARATION, DYEING AND FINISHING ACTIVITIES

The receiving water discharge standards given in the Turkish Water Pollution Control Regulation (WPCR) consist of chemical and biochemical parameters such as BOD_5, COD, NH_4–N, TSS, Fe, Cr, S^{2-}, oil and grease for many industrial effluents. Only the fish toxicity test, based on the toxicity dilution factor (TDF), which indicates the toxicity of an effluent sample, was included in the Turkish Water Pollution and Control Regulation later in the year 1992 (WPCR 1988). The application of acute toxicity tests for the assessment of industrial effluents has not been initiated yet by either the Ministry of Environment and Forestry or other national authorities. Besides toxicity, no limit value exists yet in the Turkish WPCR for the colour parameter. COD, BOD_5 and TSS concentrations are the main textile industry wastewater characteristics basically regulated in the Turkish discharge consents. A large amount of refractory COD is caused by high molecular synthetic textile auxiliaries and dyes that are mainly coloured and/or toxic compounds. In some cases, textile dyes contain high adsorbable organically bound halogens (AOX) and heavy metal concentrations in their chemical structures (EC 1998). In particular, because particularly pigments are highly structured, complex polymers and toxic to microorganisms, biological treatment of these effluents that is the most common way of textile wastewater treatment, cannot remove all organic substances sufficiently.

In some countries like France, Germany, Austria, England and Italy, there are limit values for the effluent colouration. However, these countries use different parameter units, which make a comparison impossible. The oldest unit is the Hazen, in use since the beginning of the 20th century. In France, the current unit is mg/L of Pt-Co units. The colouration values are determined by a comparative analysis with model solutions prepared according to perfectly defined procedures. In France, a sample of the coloured effluent is diluted by a factor 30; if there is no visible colouration after dilution, the effluent is said to be complying with the norms (http://aida.ineris.fr/). The colour parameter is not considered in many countries like Morocco or India. Many federal countries, like Canada ((http://www.access.gpo.gov/cgi-bin/cfrassemble.cgi?title=200240; http://www.menv.gouv.qc.ca/indexA.htm) and Australia (http://www.epa.vic.gov.au/Water/EPA/#Other), have national environmental legislations, which, like in Europe, establishes the limit values to comply. Many countries, like Thailand, have adopted the U.S. regulations. Other countries have adopted the EU legislations, like Turkey or Morocco. In some countries, for example India, Pakistan, and Malaysia (www.ostc-was.org/environment/water.html#3.2.1), the emission limits are recommended, but not mandatory limit values. The six most important parameters for the textile industry are: BOD_5, COD, SS, salinity (conductivity, chloride or sulphate), colour and detergents or oil. It should be noted that sometimes, for some countries, information about a parameter is not given. This does not mean that this parameter is not taken into account in these countries. There may be a norm, which has not been made available to us. There is nothing in the table concerning the U. S., since the limit values for the six parameters are set by the state legislations and there are no average values.

Dyeing effluents generally contain two types of salt: sodium chloride (NaCl) and sodium sulphate (Na_2SO_4). Norms regarding NaCl are available in some country's legislation, but those regarding Na_2SO_4 and salinity are missing. However, in many countries there is no NaCl emission limit for discharge into a municipal wastewater treatment plants (Japan, Austria), into the receiving water bodies (Morocco) or whatever the destination of the effluent is (Austria, India, France, Mexico, and Argentina). Many countries consider salinity as a type of pollution. Countries that have normalized the discharge to sewage treatment plants have set their limits between 0.5 and 2000 mg/L for salinity. Discharge location seems to have little influence on limit values, which are of the same magnitude whether the effluents are discharged into municipal wastewater treatment plants or into the environment. In Italy, they are even similar. Chile accepts high salt concentrations for discharge into the environment (water body). These limit values are much lower than the salt concentrations in cotton exhaust reactive dyeing effluents even if the dyeing baths are mixed. These data show that salinity has not yet been given as much attention as other parameters such as BOD_5, COD, or TSS. In the nearest future this kind of emissions may lead to major ecological problems similar to those encountered for the production of drinking water from seawater. It is clear that a treatment and valourisation process, through the recycling of the mineral salts, may trigger the way to a standard.

The textile industry is one of the longest and most complicated industrial chains in manufacturing industry and still among the largest industries in the world regarding production rates and numbers of labour. Broadly speaking, the textile industry can be classified into cotton, woollen and synthetic fibre industries (UNEP 1998). The textile industry is not only by the huge amounts of water required, but also by the variety of chemicals used. Generally, there are several types of wet processing stages, fabric production profiles and hence fluctuations in effluent flow rates and compositions involved. Liquid wastes are the main contributors to waste emissions, whereas solid wastes mainly originate from fabric rejects, and air emissions are related to volatile organic compounds (VOCs). Waste emissions are typically hot, alkaline, with an apparent smell and colour due to the consumption of dyes and auxiliaries in the dyeing process. Textile effluent discharge into natural water bodies hence may lower the dissolved oxygen levels and deteriorate the aesthetic value of water sources owing to their strong colour and foaming nature. Although dyes are usually non-toxic, their metabolites may create serious ecotoxicological and health risks associated with the azo-, nitro-, sulphonate- and phenyl- functional groups in their structures. Some anaerobic metabolites (aromatic amines) have been declared as potentially toxic and carcinogenic (UNEP 1998).

The major environmental concern in the textile industry is about the amount of water being discharged as well as its chemical content and load. Other important issues are energy consumption, air emissions, solid wastes and odours. Because they have been controlled and regulated for a long time, there is a good database available for air emissions originating from specific textile processes and their control. On the other hand, for the variety of effluent streams originating from different process stages result in a complex combination of different chemicals and auxiliaries used that vary according to the finishing processes used (UNEP 1998). The type of dyestuffs and dye assisting chemicals used, for instance, depend on the type of fabric dyed that may even change during a day. Due to the fact that data available about effluents from specific processes is very poor, it has been preferred to compare overall mass streams between dye mills belonging to the same category. This approach makes it possible to assess specific water consumption and pollution emission levels of mills category-wise, starting from a brief overview of overall mass streams and ending in a more detailed analysis of single processes when data is available. Typical characteristics of process waste streams of textile mills are given in Table 2.1.

Table 2.1 Pollution profiles of different unit processes of the textile industry sector (arranged by considering UNEP 1994; 1998)

Unit process	Pollutant types found in the effluent	Characteristics of the effluent	Pollution load (kg COD/t fabric)
Desizing	Starch, glucose, carboxymethyl cellulose, polyvinyl alcohol, resins, oils, fats and waxes, biocides	High BOD, COD (30–50 % of total), SS	90
Scouring	Caustic soda, soda ash, waxes and greases, sodium silicate, sequestering agents	Strongly alkaline, dark colour, high BOD and COD (30 % of total)	40–80
Bleaching	Hypochlorite, caustic soda, hydrogen peroxide, acids, sequestering agents	High alkalinity, low BOD&COD (< 5 %)	10
Mercerizing	Caustic soda	High alkalinity, very low BOD&COD (< 1 %)	<1
Dyeing	Dyestuffs, reducing agents (sulphides, hydrosulphides), acetic acid, sequestering agents, surfactants, wetting agents	Strong colour, high alkalinity, high TDS, moderate to high BOD&COD	10–15 (Batch dyeing = 5 g COD/L; Continuous& semi-continuous dyeing = 2–200 g COD/L)
Printing	Starch, alcohols, ammonia, formaldehyde, aliphatic hydrocarbons, monomers *e.g.* acrylates, vinylacetate, styrene, colorants, gums, oil, mordants, metallic salt	Strong colour, high emission potential, oily appearance, moderate to high BOD&COD (10 % of total)	10–15
Finishing	Tallow, common salt, Glauber salt, antistatic agents, biocides, flame retardants, *etc.*	Slightly alkaline, low BOD&COD (< 1 %)	<1

As the environmental issues are becoming more and more complex every day, a comprehensive integrated approach is required to alleviate the problems associated with sustainable textile effluent management. In this respect, the main strategy is to minimise and if possible avoid waste generation thereby maximising the use of chemicals, water and heat. These precautions also require the implementation of ***best available practices and techniques*** (BAT-EC), good housekeeping procedures, and reuse applications. Some major directions can be listed as follows (EC 2003);

- counter-current rinsing to save water,
- conventional dyeing at lower liquor ratios,
- improved dyeing and after-rinsing techniques which require less water,
- recovery of sizes, caustic, dyes, lanoline, wax and grease via filtration, extraction and evaporation procedures,
- substitution of low biodegradability (bod) chemicals (sizes, sequestering agents and surfactants) with high biodegradability (bod) ones,
- modification of the product and process (development of new materials, methodology and technologies),
- heat and solvent recovery,
- purification of contaminated water for more general recycling.

The above listed major issues are also directly linked with process optimisation and the improvement of product quality. Changes in equipment and/or technology used for bleaching, dyeing, printing and rinsing has significantly modified waste characteristics and reduced effluent generation rates. Hence, treatment requirements have been limited due to the decreasing amounts of waste generated. However, at the same time, more concentrated effluent is expected requiring changes in the end-of-pipe treatment technologies applied.

On the other hand, nowadays textile products have particularly focused on health and environment related issues. This has led to the establishment of internationally recognized textile standards including MST and Eco-Tex criteria that serve as worldwide references for the sustainable production of textiles. Man-made fibres have thus been labelled according to their negative influences on public health and aquatic environment. Ecological fibres are now being produced by employing environmentally friendly processes and operations, saving time, space, chemicals *etc.* in order to minimise their negative effects on the environment and associated wastewater treatment and disposal costs. However, the textile industry is still confronted with serious environmental problems associated with the consumption of chemicals that have not been replaced yet for technical, technological and/or economic reasons including chemicals still being frequently used for textile preparation, dyeing and finishing activities (EC 2003);

- formaldehyde,
- some chlorinated solvents and (dye) carriers,
- biocides containing halogenated functional groups,
- heavy metal salts,
- metal complex (phythalocynanines, azo) dyes, azoic dyes,
- n-methylol resins,
- cationic surfactants, retarders and softeners,
- sequestering agents bearing polyphosphates,
- polyethylene glycol ethers,

The above mentioned chemicals are not biodegradable or difficult-to-biodegrade but still being used in significant quantities in textile finishing activities.

BAT-EC have been developed and proposed to optimise water and energy consumption as well as to minimise air, water and solid waste emissions with their associated treatment costs. In this chapter, it was preferred to focus on the BAT-EC tool "*selection and substitution of chemicals used*". Based on these tools, substitution of the harmful, inert (nonbiodegradable) substances is often an available option to reduce the environmental impact of a process (EC 2003);

Surfactants: These are used for many purposes in the textile industry (as detergents, lubricants, wetting agents, *etc.*). Some surfactants are considered as problematic because of their biodegradability and toxicity to aquatic species. Concerns currently focus on alkyl phenol ethoxylates (APEO) and in particular nonyl phenol ethoxylates (NPE) due to their endocrine disrupting behaviour. The main alternatives for APEO are fatty alcohol ethoxylates, but also for other surfactant types (anionic, cationic ones) substitutes are often available that are more biodegradable/bioamenable in the wastewater treatment plant and/or do not form toxic metabolites.

Sequestering agents: These are used in textile preparation and dyeing stages. They can often be avoided. However, chemicals are available that function as alternatives for conventional sequestering agents. These are readily biodegradable chemicals and also do not contain N or P in their structures (polycarbonates, gluconates, citrates and some sugar-acrylic acid copolymers). Costs and used quantities are comparable. Hot rinsing avoids the use of sequestering agents in the rinsing and neutralisation steps after dyeing. However, replacement of cold rinsing with hot rinsing leads to higher energy consumption, unless thermal energy after rinsing effluent is recovered.

Bleaching agents: Sodium hypochlorite bleaching gives rise of secondary reactions that form halogen compounds commonly measured as AOX, trihalomethanes (THMs) account to the bulk of the compounds formed. H_2O_2 is now the preferred bleaching agent for cotton and cotton blends as a substitute for the more selective sodium hypochlorite, although it is argued that sodium hypochlorite is still required to achieve high whiteness and for fabrics that are fragile and would suffer from de-polymerization. In this case, a two stage process first with hydrogen peroxide and thereafter with sodium hypochlorite can be applied to reduce AOX emissions. In this way, AOX emissions originating from the reaction between chlorine and fibre impurities/residues (pesticides) acting as haloform precursors can be reduced. AOX values up to 90-100 mg Cl/L have been typically observed from the second stage exhausted sodium hypochlorite-bleaching bath. Two stage bleaching processes with hydrogen peroxide are also possible; however, they have been reported to be 2-to-6 times more expensive than the conventional bleaching procedures. Chlorine dioxide is an excellent bleaching agent for synthetic fibres, flax, linen and other bast fibres that cannot be bleached using hydrogen peroxide alone. Technologies are available that produce ClO_2 without AOX generation (elemental chlorine-free bleach involving chlorine dioxide produced via hydrogen peroxide as the reducing agent of sodium chlorate).

Dye carriers: As noted earlier, the COD values of exhausted baths from the carrier dyeing of polyester are appreciably higher than those from all other batch dyeing processes. Many carrier formulations contain mixtures of active substances, together with auxiliary solvents and emulsifiers. In some cases, the ingredients other than the active (carrier) substance may contribute more to the dyebath effluent because most of the carrier is usually taken up efficiently by the fibre. Many carriers have offensive odours and chlorinated types exert significant toxicity to fish and biomass. For instance, the carriers butyl benzoate and benzyl phenol exert a COD of 1900–2000 mg per g of chemical, are biodegradable at a high degree (>85 %), but posses a significant bacterial and fish toxicity (LC_{50}) values of 200 and 6 mg/L, respectively. The carriers phenyl benzoate and o-phenyl phenol (1700–1750 mg COD/g) are also bioeliminated at a high degree (75–90 %); their bacterial (LC_{50}) toxicity is relatively low (> 300 mg/L). Di-and trichlorobenzenes are also still used in carrier formulations; these impart a lower COD (400–800 mg/g) to the discharged effluent, but are also less biodegradable (40–70 %).

Polyester (PES) dyeing carriers can be avoided (except for PES/WO and elastane/WO blends) by dyeing under high temperatures. Another option is the use of non-carrier dyeable PES fibres. However, due to differences in their mechanical and physical properties, these fibres cannot be regarded as substitutes for PET-based polyester fibres. When carriers are unavoidable, conventional active substances based on chlorinated aromatic compounds, or biphenyls and o-phenyl phenol, can be replaced with less harmful compounds including benzyl benzoate and N-alkylphthalimide.

Dispersing agents: These are typically present in disperse, vat and sulphur dye formulations. They have been improved by their partial substitution with optimised products base on fatty acid esters, or the use of mixtures of modified aromatic sulphonic acids. The first option is only applicable for liquid formulations of disperse dyes and hence finds limited applications. These dispersing agents are biodegradable compared to the conventional condensation products of naphthalene sulphonic acid with formaldehyde.

Dyestuffs: Poor dye fixation has been a chronic problem with reactive dyeing particularly in batch dyeing of cellulose fibres where a significant amount of salt is added to improve dye exhaustion. With sophisticated molecular engineering techniques it is possible to produce novel bifunctional and low salt reactive dyes that can attain fixation rates >95 % even for cellulose fibres and considerable reproducability and level dyeing (Blackburn and Burkinshaw 2002).

With pH-controllable acid and basic dyes it is advantageous to dye at isothermal conditions imposing a pH profile. It is preferable over temperature-controlled dyeing processes in that maximum exhaustion of dyes and insect resist agents can be achieved with a minimum use of organic levelling agents (Blackburn and Burkinshaw 2002). When dyeing wool with metal complex dyes, higher levels of exhaustion and fixation rate can be achieved by controlling the pH and by using auxiliaries with high affinity for the fibre and dyestuff (IPPC 2003). The higher exhaustion rate directly correlates with the reduced residual of chromium levels in the spent dyebaths; *e.g.* 10–20 mg/kg of treated wool, corresponding to 1–2 mg/L chromium in the spent dyebath at a liquor ratio of 1:10 (Blackburn and Burkinshaw 2002).

2.3 WATER CONSUMPTION

Water is usually obtained by textile mills from three principal sources, namely (i) surface water from reservoirs, rivers or canals; (ii) groundwater from well or bore-holes; and (iii) mains water from the public supply in urban areas. The ever increasing demand for water in the textile sector is deleterious to rivers, some of which have been heavily polluted, whilst water supply is threatened by further urbanization of the catchment areas (O'Neill 1999). The increasing legal and economic pressures have inspired many studies of water consumption and improved techniques for minimising losses in various sectors of the wet-processing industry. The three main purposes for which water is required are; (i) process water (60 %), (ii) cooling water (30 %) and boiler steam (10 %). Over the years, water consumption per kg of fibre decreased tremendously (by more than 60–70 %) by the introduction of more efficient textile finishing processes and partly eliminating water emissions from excessive rinsing activities. Detailed surveys conducted in the period 1990–2000 revealed that the water consumption and usage of textile mills vary markedly amongst the companies. Average consumption in the scouring (20–50 L/kg) and bleaching (70–100 L/kg) of cotton fabrics differed considerably according to the type of bleaching process employed. The same situation has also been observed for dyeing processes (10–150 L/kg) depending on the fabric dyed and dye class applied. Generally speaking, azoic, vat or sulphur dyeing need more water than acid, direct or disperse dyeing (Cooper 1998). Decision whether to introduce reuse and recycling techniques are made after considering factors such as the cost of fresh water, costs associated with the effluent recycling or disposal, limitations of water supplied and other local, socio-economic conditions. Not all water can be reused or recycled, since accumulation of soluble materials (total dissolved solids) in the system will negatively affect textile finishing processes and operations (Easton 1995). The use of standing dyebaths has long been known and investigated for dyes with relatively low substantivity (*e.g.* indigo and sulphur blacks onto cotton), but the first systematic dyebath reuse was carried out for solvent-assisted dyeing to minimise the consumption of expensive solvents. Repeated dyeing of fibres with spent dyebaths may offer considerable cost savings related with the use of dyes, auxiliaries and water. For dyebath reuse to be practicable, however, consistently high exhaustion is essential and widely contrasting colours must not be dyes in sequence. Dye auxiliaries present in the exhausted dyebath must not impair the performance of subsequent dyeings. The concentration of dyes and dye assisting agents must be determined each time to prevent accumulation and/or allow topping-up the bath for the next run. More difficulty is experienced if exhaustion is incomplete or when after treatment is necessary as for most cotton dyeing processes. The need for reservoirs to collect the spent dyebaths prior to replenishment and reuse will be another source of investment and any need for after-treatment may further complicate the whole procedure. Case studies have been reported where dyeing of nylon with acid dyes and polyester dyeing with disperse dyes up to 30 successive dyeing steps were possible for the same dyebath liquor that was controlled and adjusted for concentration. Reproducibility, levelness and fastness were not impaired in these trials and resulted in substantial cost savings in chemicals, water, energy as well as effluent charges (> 50 %). Research on possibilities for reuse of water in dyeing cotton and cotton blends with reactive dyes has been investigated in pilot and bulk-scale, but practical problems remain (Cooper 1998).

2.4 CONVENTIONAL END-OF-PIPE TREATMENT

Textile mills in Turkey are required to control their discharge and therefore have started installing treatment plants in the name of environmental protection. For the treatment of textile industry wastewater, biological treatment, chemical treatment and combinations of these are used. Textile wastewater treatment plants practice biological treatment rather than chemical processes claim that their preference is due to less excess sludge production, lower operational costs and better COD removal in biological treatment.

Textile wastewater treatment should start with effective balancing of flow and composition, together with cooling of hot (or heat recovery) from hot wastes. The usually highly alkaline effluent can be neutralised with sulphuric acid dosing or the carbon dioxide in the flue gases. Polyelectrolytes are attractive coagulants because they are highly effective and do not greatly contribute to the total weight of settled materials. The routine secondary (main) treatment of textile industry effluent is based on biological processes due to the biodegradable nature of most mixed effluent discharge ($BOD_5/COD \approx 0.25–0.40$). However, biotreatment does not remove sufficient colour, COD and dissolved solids to be satisfactory alone as a stand-alone treatment method (EC 2003).

Hardly biodegradable compounds can still be degraded in biological treatment plants under low food-to-microorganisms (F/M) ratio conditions, but non-biodegradable substances such as dyes are not degraded in biological treatment plants (Easton 1995). The removal mechanisms for hydrophobic dyes and cationic/nonionic surfactants are mainly sorption on biomass in activated sludge (Easton 1995; O'Neill 1999). Concentrated wastewater streams that contain non-or hardly biodegradable chemicals should be treated at source. Anaerobic/anoxic-aerobic biotreatment technologies are applied in countries where discharge consents require the effective removal of coloured compounds. Pretreatment with lime and ferrous sulphate or ferric chloride prior to mainstream activated sludge treatment is also commonly employed in particular for effluent originating from the printing activities (Cooper 1998). Tertiary (advanced) treatment stages are becoming more and more important for textile wastewater management due to the increasing water stress and legal requirements. As such, the need for more extensive treatment and greater sophistication and investment costs become apparent. Selective physicochemical methods for toxicity and colour abatement have been developed but hardly practiced due to technical and economic constraints. Promising physicochemical methods for advanced wastewater treatment are membrane processes, flocculation and adsorption onto activated carbon or cheap alternative materials (Vandevivere *et al.* 1998).

For the textile finishing industry, advanced oxidation processes (AOP) including Fenton-like reaction is proposed as a viable treatment technique to remove segregated, recalcitrant COD at higher loads ($\geq$ 5000 mg/L). Candidate wastewater streams are padding liquors, printing pastes, desizing, exhausted dyeing and finishing baths. Up to 70–85 % COD abatement can be achieved with a residual

COD being largely biodegradable due to the structural modification of the ingredients (Vandevivere *et al.* 1998). However, very strong residues (such as printing paste, padding liquors, carpet backing residues) can be more conveniently be kept out of the main effluent stream altogether and other alternative disposal routes may be used (precipitation/flocculation, wet air oxidation, dewatering followed by incineration of the resulting sludge). The incineration of the formed chemical sludge is a viable alternative to chemical oxidation such as wet air oxidation that is usually applied for high strength, biologically difficult-to-degrade wastewater. The following treatment methods are currently applied and recommended to achieve good performance for the treatment of mixed textile wastewater (Vandevivere *et al.* 1998);

- biological activated sludge treatment followed by tertiary treatment such as activated carbon adsorption with recycle of the activated carbon to the activated sludge system and destruction of the adsorbed non-biodegradable compounds by incineration or wet air oxidation of the exhausted carbon plus biomass.
- combined biological, physical and chemical treatments with the addition of powdered activated carbon and iron salt to the activated sludge system. reactivation of the excess (spent) sludge with wet oxidation or wet peroxidation if hydrogen peroxide is used during wet oxidation (thermal oxidation because of the high calorific value),
- ozonation of the biorecalcitrant compounds prior to biological treatment.

For wool scouring wastewater an evaporation plant to treat the effluent seems to be superior to a chemical coagulation-flocculation plant. However, the capital cost of the evaporation plant is much higher and payback takes 4–5 years for small mills including 3500 t/year. For medium side mill (15,000 t/year) evaporation is only slightly cheaper than chemical treatment with coagulation-flocculation over 10 years. If a dirt removal/grease recovery system is installed together with the evaporation plant the capital costs will be reduced significantly. The installation of a recovery loop also affects operating costs of the evaporation plant appreciably due to the grease market (EC 2003). Evaporation of the effluent, incineration of the sludge in combination with the full recovery of energy and water is the best option for the fine wool scouring mills, however, the complexity of this technique and the high initial costs make this treatment scheme more suitable for new installations, existing installations with no existing on-site effluent treatment plant and life-expired effluent treatment plants that have to be replaced by new installations (EC 2003). Although no precise information is being submitted from the running treatment plants, it is known and reports that most treatment plants dealing with dyehouse effluent practice biological activated sludge treatment alone or in combination with a preliminary anaerobic treatment system (Vandevivere *et al.* 1998).

2.5 ADVANCED AND EMERGING TREATMENT PROCESSES

In this section, a global overview of more recently developed and employed chemical treatment processes for the abatement of colour, toxicity, inert COD and other parameters that cannot be efficiently treated via conventional physicochemical or biological treatment methods are evaluated. Special emphasis is given to the application of electrochemical (including electrocoagulation) and advanced oxidation processes that have proven to be very effective in the degradation of dyehouse effluent. A brief overview of available advanced/emerging treatment options is followed by a detailed economic analysis were mainly operating costs are taken into consideration.

2.5.1 Electrocoagulation

Besides wastewater characteristics such as conductivity and pH, or process variables such as current density, operating time and electrode material, another important process design parameter that is crucial for the complete techno-economic evaluation of the electrocoagulation system is the electrode connection mode. In a study conducted by Kobya *et al.* (2007), a wastewater (initial COD = 2031 mg/L; TSS = 102 mg/L; conductivity = 2310 μS/m; turbidity = 671 NTU; pH = 8.88) that was a mixture of exhaust dyeing solutions obtained from a textile factory located in Turkey (production capacity = 1000 m^3/d) was subjected to electrocoagulation (wastewater volume = 740 mL) using aluminium (Al) and iron (Fe) electrodes. COD and turbidity removals were considered as the performance criteria. Electrodes were connected in three modes; namely monopolar-parallel (MP-P), monopolar-serial (MP-S) and bipolar-serial (BP-S). In the MP-P electrode connection, the current is divided between all the electrodes in relation to the resistance of the individual cells. Hence, a lower potential difference is required as compared with serial connections. In the MP-S mode, each electrode pair is internally connected with each other. Because the cell voltages are summed up, a higher potential difference is required for a given current. In the BP-S connection, there is no electrical connection between the inner electrodes (bipolar), only the outer electrodes (monopolar) are connected to the power supply. Effects of pH (5, 6, 7, 9), current density (30, 40, 50, 60 A/m^2) and operating time (5, 10, 15, 20, 25 min) were evaluated for the different electrode connection modes in terms of COD and turbidity removal efficiencies, electrical energy requirements and sludge production for the electrocoagulation of textile wastewater using Al and Fe electrodes. Energy (electrode)-chemical coagulant consumption, metal hydroxide sludge production, operating time, initial-final pH values, treatment performance and associated operating costs were compared with those obtained for chemical coagulation with $FeCl_3 \cdot 6H_2O$, $Fe_2(SO_4)_3 \cdot 7H_2O$, $AlCl_3 \cdot 6H_2O$, and $Al_2(SO_4)_3 \cdot 18H_2O$. All treatment systems exhibited similar trends with respect to energy consumption (operating costs) and sludge production. Experimental results have indicated that an acidic to neutral pH medium results in higher COD removals and that the treatment performance for iron electrodes is more dependent on the connection mode. For a high turbidity removal, Al electrodes connected in BP-S mode performed best under acidic pH (= 5), whereas for the iron electrode connection pH = 7 resulted in more filterable flocs and thus higher turbidity abatement in the MP-S mode. The MP-P mode performed better at low current densities, whereas at high current densities, the three modes perform nearly equal. Generally speaking the connection mode had by far more influence on COD removal efficiencies, but almost negligible effect on turbidity removals. In the case of electrocoagulation with Al electrodes, removal rates reached a steady-state in 5 min, whereas the process required 10 min to achieve steady removal efficiencies for treatment with Fe electrodes. The final acidic pH values and high chloride concentrations are the two major drawbacks of chemical coagulation that resulted in almost 10 % higher COD removal performance than electrocoagulation. Considering these two issue ferrous sulphate should be preferred among the studied coagulants. In addition, the operating cost of chemical coagulation (€ 0.50–0.72 /m^3) is 3.2 times higher than that of electrocoagulation (€ 0.19/m^3 for Fe and € 0.30/m^3 for Al electrodes). Other important data that are essential to evaluate the feasibility of electrocoagulation (MP-P mode; current density 30 A/m^2; treatment time 15 min; initial pH = 5 for Al and pH = 7 for Fe electrodes, respectively) for textile wastewater treatment are (Bayramoglu *et al.* 2007);

- Electrode consumption: 0.126 kg/kg COD (Fe), 0.096 kg/kg COD (Al)
- Metal hydroxide sludge production: 0.309 kg/m^3 (Fe), 0.168 kg/m^3 (Al)
- COD removal: 65 % (Fe), 63 % (Al)
- Turbidity removal: 83 % (Fe), 80 % (Al)

Aluminium and iron are suitable electrode materials for the treatment of textile wastewater. These two electrode materials (mild steel with a carbon content of 0.3 % and aluminium alloy with a copper content at 4 %) have been employed to treat two textile wastewater samples (Zongo *et al.* 2009); one was obtained from a textile industry plant located in France, the other sample was collected from a water treatment local station where a mixture of different dyehouse effluents originating from different small factories was received.

The plant wastewater had the following characteristics; pH: 7.06; turbidity: 115 NTU; COD: 1787 mg/L; conductivity: 2.8 mS/cm; carbonate: 2191 mg/L; nitrate: 3.91 mg/L; phosphate: 1.9 mg/L; sulphate: 144 mg/L; chloride: 84.3 mg/L. The station wastewater's characteristics were; pH: 9.04; turbidity: 110 NTU; COD: 1237 mg/L; conductivity 2.4 mS/cm; carbonate 1475 mg/L; nitrate: 11.1 mg/L; phosphate: 2.6 mg/L; sulphate: 56 mg/L; chloride: 27.9 mg/L. The electrocoagulator was a parallel plate electrochemical cell (2 litres) made of poly methyl acrylate with two electrodes made of iron or aluminium. Batch experiments were conducted with liquid recirculation (2.5 cm^3/s) in the circuit, consisting of a peristaltic pump, the electrolytic cell and a double walled tank for temperature control at around 20°C. Gentle agitation was also provided in the electrocoagulator and the treatment time was selected as 1 h. The current was fixed in the range of 50–200 A/m^2 using a DC power supply where the cell voltage was continuously recorded. When Al electrodes were used the pH was adjusted to 6.5 to ensure the highest amount of $Al(OH)_3$ formation and precipitation. During the experiments the absorbance (colour) at 436 nm and the metal content of the treated samples were also measured. In the treatability experiments, the variation of voltage with the current density was determined for several treatment runs and the current efficiency of metal dissolution was calculated from the amount of metal ion in the liquid sample using Faraday's Law and plotted versus electrical charge. For Al, the current efficiency was found to be near 2, whereas iron dissolution occurred quantitatively at the anode surface, as revealed by the current efficiency being larger than 80 %. The COD abatement was found as 88 % for the wastewater collected in the station and ranged from 74 to 78 % for that from the plant. Absorbance values reached values close to zero (< 0.1). Turbidity removals were rapid and as the other parameters increased with increasing applied current density. Considering the treatment of a wastewater with a conductivity at 2 mS/cm at 100 A/m^2; the calculated cell voltage attained 2.5 V. Taking into account the stoichiometry of the metal dissolution, the molecular weight of these metals and the current efficiency of metal dissolution, the electrical energy consumption for the treatment of 1000 mg/L COD was estimated to be 1.75 kWh/m^3 with Fe and 1.53 kWh/m^3 with Al. These values could be reduced by 30 % by a two-fold increase in the electrical conductivity upon addition of 1 g/L NaCl. Taking into account the molecular weights of these metals (Fe = 56 g/mol; Al = 27 g/mol); the number of electrons exchanged in the anode process together with the current efficiency , treatment via Al or Fe requires a similar amount of electrical energy. However, the use of iron electrodes usually results in the formation of fine brown particles that are more difficult to settle than the gel-like (gelatinous) aluminium hydroxide flocs.

The influence of the key process variables such as current density, electrolysis time and initial pH on the efficiency of treatment of industry textile wastewater containing reactive dyes was investigated in another study (Alinsafi *et al.* 2005). Electrocoagulation experiments were carried out according to a 2^3 full factorial experimental design using Al electrodes (process independent variables: electrolysis time, initial reaction pH, applied current density). Electrocoagulation was optimised according to a 2^2 factorial experimental design (Doehlert Matrix) to improve the efficiency of decolourisation and organic matter (DOC) removal. Uniform Doehlert Networks were generated from a Simplex. The effect of electrocoagulation performed under optimised reaction conditions on the biodegradability of the effluents was also investigated. In the experimental runs, rectangular aluminium flat electrodes (height = 100 mm; width = 50mm and distance between electrodes = 20 mm) were used in a 100 mL electrocoagulator made out of Pyrex glass. NaCl was added to the reaction solution in all experimental runs to acquire a conductivity of 4.7 mS/cm. Cell voltage and current were measured by a digital voltmeter and ampermeter, respectively. The reaction solution (total volume = 500 mL) was continuously circulated in the system with the help of a peristaltic pump at a flow rate of 370 mL/min. The current efficiency (φ) for the production of dissolved Al^{3+} was calculated by comparison of the weight loss of the Al electrodes during the experiment with the theoretical amount of Al consumed ($m_{theo} = I \times t \times M_{Al}(3F)^{-1}$) according to Faraday's Law. The specific energy consumption (SEEC) was also calculated as a function of the applied cell voltage. 50 mg/L aqueous solutions of reactive textile dyes, synthetic textile wastewater (containing commercial reactive dyes hydrolysed at pH 12 and at 80°C for 1.5 h, starch, $(NH_4)_2SO_4$ and Na_2HPO_4) and two real textile wastewater samples (with a COD of 450 and 620 mg/L) obtained from a knit cotton factory located in Morocco and using the same type of dyes were also tested. The pH of the final synthetic textile wastewater samples were adjusted to 10 and the conductivity to 4.7 mS/cm. In order to evaluate the effect of electrocoagulation on the biodegradability of the wastewater, short-term respirometry (microbial oxygen uptake rate) experiments were conducted. For that purpose, activated sludge obtained from a local domestic wastewater treatment plant was used as the heterotrophic biomass and spiked twice (before and after sample introduction) with a sodium acetate solution, which acted as the reference carbon source and the wastewater sample to be tested. Changes in the oxygen uptake rates of the reaction samples were determined by the establishment of respirograms that were characterised by the maximum oxygen uptake rates and the total volume of oxygen consumed for 20 min after the injection of the wastewater sample. For the aqueous reactive dye solutions, colour removal was in the range of 72–98 %, and DOC abatement occurred in the range of 46–54 %. Experimental design and process optimisation results have indicated that reaction time and current density are more effective process variables, whereas pH (selected range 7–10; that was later on fixed to 10, *e.g.* the original pH of real textile effluent) has no pronounced impact on colour and organic carbon (COD, DOC) removals. Highest treatment efficiencies were obtained when the electrolysis time was between 90 and 120 min and for a current density of about 12 mA/cm^2, corresponding to COD and colour removals of 35 % and 92 %, respectively. For the real textile wastewater samples the following removal efficiencies were obtained;

- Synthetic Wastewater Sample SWS (COD = 70 mg/L; 51.2 % DOC, 75 % COD, 83.9 % colour removals)
- Sample I (COD = 450 mg/L; 22.8 % DOC, 65.2 % COD, 83.3 % colour removals)
- Sample II (COD = 620 mg/L; 34.2 % DOC, 74.8 % COD, 98.6 % colour removals)

Colour refers to the average absorbance value at the three wavelengths of 620, 525 and 436 nm. In all studied samples, an increase of the biodegradability (oxygen uptake rates) was observed after electrocoagulation. However, as was observed for COD and colour removal efficiency, it strongly depended upon the sample composition.

In a study conducted by Arslan-Alaton *et al.* (2007), treatment of real reactive dyebath effluent (average values: COD = 510 mg/L; filtered COD = 430 mg/L; absorbance at 436 nm = 2.95 cm^{-1}; 525 nm = 5.7 cm^{-1}; 620 nm = 0.70 cm^{-1}; pH = 11.22; conductivity = 27.6 mS/cm; Cl^- = 7800 mg/L; TSS = 235 mg/L) comprising of an exhausted reactive dyebath and its sequential rinses with electrocoagulation using Al and stainless steel (SS) electrodes was investigated. The electrocoagulation unit consisted of a 2000 mL capacity polyethylene reactor (length = 34.3 cm, width = 12.5 cm, height = 28.3 cm) equipped with six 304 SS or Al electrodes used as both anode and cathode materials. The distance between the electrodes was 2 mm and the effective area of the electrodes was 38.13 cm^2. EC experiments were run for up to 60 min with Al electrodes and for 90 min for SS electrodes depending on the time whenever the temperature in the EC unit started to increase. The experimental study focused on the effect of applied current density (22–87 mA/cm^2; at an initial, optimum pH of 5.5) on decolourisation and COD removal rates using Al and SS as electrode materials. Results have indicated that the treatment efficiency was enhanced appreciably by increasing the applied current density when Al electrodes were used for EC, whereas no clear correlation existed between current density and removal rates for EC with SS electrodes the treatment efficiency could only be improved when the applied current density was in the range of 33–65 mA/cm^2. For instance, at a current density of 33 mA/cm^2, 50 min were needed to achieve 97 % reduction in colour (absorbance), whereas 95 % decolourisation was obtained in only 15 min when the current density was elevated to 87 mA/cm^2 (rate coefficient = 0.2 min^{-1} at 525 nm). From the obtained findings it was also evident that the kinetic regime for colour removal shifted from mass-transfer to kinetically-limited at current densities above 55 mA/cm^2 owing to the fact that colour abatement rates did not change further when the current density was elevated from 55 to 65 mA/cm^2. On the other hand, 72 % COD removal was obtained after 25 min EC at 87 mA/cm^2 (rate coefficient = 0.1 min^{-1}) similar overall COD removal efficiencies were observed for current densities in the range 33–65mA/cm^2. Decolourisation was much faster during EC with SS electrodes; 95 % colour was removed after 30 min even at 22 mA/cm^2 and 99 % in only 10 min EC at a current density of 87 mA/cm^2). This can be explained by the evidence for direct oxidation at the anode and azo dye reduction at the cathode in case of steel electrodes, provided that the dissolved O_2 level drops to less than 1 mg/L in the reaction medium as a consequence of oxygen depletion being favourable for negative reduction potentials. The main difference observed for EC employing SS electrodes was that decolourisation profiles did not change as a function of the applied current density in the range of 33–65 mA/cm^2 and hence practically the same decolourisation efficiencies were obtained at these current densities. COD abatement rates on the other hand reached asymptotic values after 15–30 min (resulting in 65 % COD removal after 30 min). An average overall COD removal efficiency was obtained as 70 % after 90 min EC. Colour and COD removals rate coefficients were highest in the range of 43–65 mA/cm^2 (rate coefficients: 0.24 min^{-1} for colour at 525 nm and 0.06 min^{-1} for COD abatements). Most appropriate working conditions were established as an initial pH of 5.5, a current density of 43 mA/cm^2 and a treatment time of 20 and 30 min for Al-EC and SS-EC, respectively. In conclusion, it could be established that EC with SS electrodes was superior in terms of decolourisation kinetics (99–100 % colour removal after 10–15 min EC at all studied current densities), whereas EC with Al electrodes was more beneficial for COD removal in terms of electrical energy consumption, *i.e.* 5 kWh/m^3 wastewater for EC with Al electrodes instead of 9 kWh/m^3 wastewater for EC with SS electrodes.

2.5.2 Electrochemical treatment (anodic oxidation)

In a study of Radha and Kalaivani (2009), the organic content of real textile industry wastewater (original values: COD = 20,840 mg/L; TOC = 8430 mg/L; TS = 14,960 mg/L; TDS = 11,380 mg/L; Cl^- = 4816 mg/L; pH = 3.56) was treated electrochemically by using a graphite rod as the anode and stainless steel as the cathode material. The effect of pH (1.3, 1.7, 2.9, 3.5, and 5.2) as well as current density (5, 9, 14, 19, 23, and 28 mA/cm^2) was evaluated in terms of the collective environmental parameters colour, COD, TS and TDS (% removal efficiencies). Experiments were carried out in a square packed bed electrochemical reactor equipped with the DC power supply at a fixed potential difference of 8 V. The gap between the electrodes was kept constant at 6 mm. The authors observed that at the highest studied current density, 68 % COD together with nearly 49 % TS and 51 % TDS were removed. Current density had no significant effect on TOC removals that were obtained as 97 %. COD removals were lower most probably due to the chloride interference with the applied COD test procedure. Due to the fact that throughout the experimental runs, active free residual chlorine remained nearly constant, it was concluded that oxidation in the reaction solution was not accelerated by the action of active chlorine species. COD abatement rates could be fitted to pseudo-first order kinetics and the highest coefficient was obtained at 28 mA/cm^2 as 0.559 min^{-1} (variation coefficient was 4.95 %). Experiments were also conducted at different electrolysis times (30, 60, 90, 120, 150, 180, and 210 min) to study colour removal efficiencies. A maximum colour removal efficiency of 96 % was obtained for varying current densities and the time required to achieve the ultimate colour removal rate decreased with increasing current density from 210 min at the lowest one (5 mA/cm^2) to 60 min at the highest tried current density. The pH of the reaction solution also exhibited a significant effect on COD and colour removal efficiencies with an increasing treatment performance with decreasing initial pH's (68 % COD removal at an initial electrolysis pH of 1.3). This is attributable to the instability of formed $HO^\bullet$ radicals at basic pH;

$$CO_x(\text{adsorbed } HO^\bullet) \Rightarrow CO_x + \tfrac{1}{2}\,O_2 + H^+ + e^- \qquad (2.1)$$

The above equation shows the anodic reaction on the graphite anode where adsorbed free radicals ($HO^\bullet$) are leading to higher oxide formation.

Electrooxidation over anodes made of graphite, Pt, TiO_2, IrO_2, PbO_2, several Ti-based alloys, and more recently boron doped diamond electrode in the presence of a supporting electrolytes (typically NaCl) has been employed for the decontamination of various industrial effluents including dyes, synthetic dyehouse effluent and real textile wastewater. The study carried out by Chatzisymeon *et al.* (2006) is one of the few reports dealing with the electrochemical treatment of real and synthetic dyehouse effluent. The aim of their study was to investigate of these textile effluents over a Ti-Ta-Pt-Ir anode regarding the effect of varying operating conditions such as current, type and initial electrolyte and solution pH on decolourisation, reduction of COD and energy consumption. The effect of electrochemical treatment on the effluent acute toxicity using the marine photobacteria *Vibrio fischeri* was also studied.

The synthetically prepared effluent was a mixture of 16 dyes with a total concentration of 361 mg/L. The effluent was near-neutral, strongly coloured and highly toxic with a COD content of 281 mg/L. The real textile effluent also contained Na_2CO_3 (440 mg/L), Na_2SO_4 (5500 mg/L) and NaOH (110 mg/L) besides several dyestuffs and H_2O_2 as the bleaching agent. The COD and total solids content of the real wastewater were 404 mg/L and 75 mg/L. The pH was alkaline (9.5) with a dark green blue colour. Its EC_{50} value was higher (75 % instead of 4.1 %, in v/v). The real wastewater also contained organic dye auxiliaries. The experiments were conducted in an electrolytic cell comprising a stainless steel 316 L cathode, a titanium anode covered by a thin film of tantalum, platinum and iridium alloy and a power supply unit. I a typical run, the effluent was mixed with an appropriate amount of NaCl (or another electrolyte) batch loaded in a vessel and continuously pumped into the electrolytic cell by a pump at a flow rate of 0.81 L/s. The flow rate could be adjusted by valves. In some experimental runs, the initial pH was adjusted to acidic values with sulphuric acid solution. Experimental results have indicated that increasing the salinity from 0.5 to 1–4 % increased decolourisation from 39 to 50–70 %, respectively after 5 min reaction of synthetic effluent. Stepwise addition of 0.05 % NaCl to the originally 0 % salinity effluent in 6 steps had a marked effect on colour removal efficiency. In all cases of salinity addition (minimum 0.5 %) more than 95 % colour removal was obtained within 10–15 min. In the case of real wastewater that had a natural salinity due to its sodium salts content, only 60 % colour removal occurred after 180 min without electrolyte addition. Increasing the electrical current from5 to 10–14 A improved colour removal from 39 to 68–87 %, respectively, after 5 min with the synthetic effluent. However, over 98 % colour removal was obtained for the real effluent after 10–15 min regardless of the applied current. On the other hand, in terms of the COD parameter increasing the applied current had a significant impact on COD removals. For instance, the COD reduction for the synthetic effluent was 55, 65 and 86 % at 5, 10 and 14 A, respectively; for the actual effluent 29, 34 and 39 % COD removals were reached at 5, 10 and 20 A, respectively. The observation implied that dye intermediates that were more resistant to electrooxidation were formed in the reaction medium. The higher resistance of the real wastewater to electrochemical oxidation was attributable to the presence of organic dye assisting chemicals making the effluent less prone to electrooxidation. Acidic conditions (pH 3 instead of pH 6 or pH 9.5) were favoured in terms of COD removal for both samples. However, colour removal was not affected by varying solution pH values. In further experiments the effect of iron containing electrolytes (ferric chloride and ferrous sulphate) on treatment efficiency was studied considering that they also may serve as homogenous catalysts and reduce hydrogen peroxide to hydroxyl radicals (Fenton type reactions). In these experiments 0.7 % iron salt was used providing 0.5 % chloride in the case of ferric chloride. Both COD (10 %) and colour (85 %) removal occurred slowly in the presence of ferrous sulphate emphasizing the beneficial effect of chloride as the oxidising agent. In the presence of ferric chloride complete colour accompanied with 50–60 % COD removal was obtained after 180 min; changing the flow rate only slightly affected COD removal efficiencies. Specific energy consumption (SEC values) were computed for colour and COD abatements obtained after 15 and 180 min of reaction, respectively. Since dye degradation was rapid, speaking for a low electrical energy requirement in terms of decolourisation. SEC values were higher for the synthetic effluent than for the actual effluent that contained a higher electrolyte concentration and extra electrolytes in addition to NaCl. Increasing the current density at constant salinity increased the SEC values for both colour and COD removal efficiencies and both effluent types. Although, NaCl mediated electrochemical treatment was capable of complete colour and partial COD removal, effluent toxicity values were always found to be high implying that toxic organochlorine compounds and resistant oxidation products that were formed as well as secondary residual oxidants that remained in the reaction solution even after prolonged oxidation periods. Even under conditions leading to 90 % COD removal, effluent toxicity was higher than that of the original effluents.

Dimensionally stable anode (DSA) materials are interesting for the use in treatment systems as they present relatively long lifetimes and do not necessarily undergo the same poisoning phenomena associated with traditional electrode materials such as platinum. The most commonly used DSA material is $Ti/Ru_{0.3}Ti_{0.7}O_2$ anode which has been extensively used in chlor-alkali industry for several years. DSA are classified as active (RuO_2, IrO_2) and non-active (SnO_2, PbO_2) leading to selective or non-selective (free radical initiated) oxidation. In a work conducted by Malpass *et al.* (2007), the effluent obtained directly from a dyehouse effluent (*e.g.* a balance tank collecting discharges from all process stages) having a relatively high organic carbon concentration (225 mg/L) was treated electrochemically by employing DSA plate anode and a stainless steel cathode. The reactor was mounted by positioning the electrodes between Teflon and Viton spacers of varying thickness. The electrolyte was pumped vertically through the cell provided by an electric pump at a flow velocity of 425 cm^3/min unless otherwise stated. The cell also contained a membrane that provided an in-situ probe connected to the reference electrode. All potentials were measured against a reversible hydrogen electrode. Colour (absorbance at 490, 620 and 665 nm, *i.e.* max. absorption bands) and TOC parameters were followed during electrolysis. For each galvanostatic electrolysis, a volume of 250 mL wastewater (pH_o = 6.9; original colour = blue-green; electrical conductivity = 2.1 mS/cm) was transferred to the electrolyte reservoir. Electrolyte (NaCl or Na_2SO_4) was added at a concentration of 0.0–0.10 mol/L (keeping a constant ionic strength for both electrolyte types) to achieve and ultimate NaCl concentration of 5.85 g/L. Cyclic voltametry was also used to provide a valuable insight into the reactions occurring at the electrode surface and in-situ characterisation of oxide electrodes. The 10-fold difference in the voltametric profiles (current per areas) between pure solutions with added electrolyte and the effluent solution bearing the same electrolyte concentration indicates that there are components in the textile effluent that probably interact at the electrode surface and inhibit oxide transitions. Without electrolyte addition no significant TOC removal (around 4–9 %) and limited colour abatement could be achieved at 40 as well as 60 mA/cm^2. In the presence of Na_2SO_4, colour removal did not differ appreciably from that obtained for the as-received effluent (7–9 % and 21–22 % at 40 and 60 mA/cm^2, respectively). In the presence of NaCl, colour removal followed first order and TOC abatement zero order kinetics. Colour removal increased with increasing electrolyte concentration up to 1.45 g/L NaCl, TOC removal continued to increase from 9 % to 34 % by increasing the electrolyte concentration from 0 to 5.84 g/L (NaCl). Increasing the current density (electron input) in the range of 20–160 mA/cm^2, TOC removal increased from 20 to 58 % at a NaCl concentration of 0.10 mol/L. The removal of colour was complete at higher current densities and the most efficient one was observed to be 60 mA/cm^2, corresponding to an energy consumption of 154 kWh/kg TOC. The flow rate is extremely important to evaluate whether reactions occur with a mass-transfer-controlled or electron-transfer-controlled mechanism. Increasing the flow rate from 32 to 425 L/min did not affect colour but TOC abatement rates revealing that TOC removal is controlled by mass transfer. In the present work the involvement of active chlorine (Cl_2, HOCl, OCl^-) in the degradation process is emphasized and the risk of producing chlorinated, toxic disinfection by-products (chlorinate organics) is emphasized.

2.5.3 Photoelectrocatalysis

Paschoal *et al.* (2009) investigated the photoelectrocatalytic oxidative treatment of textile wastewater bearing disperse dyes. The process utilised titania thin film electrodes prepared by the sol-gel method and used as a photoanode (applied voltage = + 1.0 eV). Treatment performance was evaluated in terms of colour, TOC removals as well as high performance liquid chromatography (HPLC) results. The photoelectrocatalytic oxidation experiments were performed for 480 min in a 250 mL reactor equipped with water refrigeration and an ultra-thermostatic bath. The photoactive anode (12 cm^2) was illuminated with UV-C light of a 125 W medium pressure lamp that was placed in the centre of a quartz sleeve. Compressed air was also continuously bubbled into the reactor. A gauze electrode was used as the counter electrode. Two different types of textile wastewater were collected and assigned as the sample before (COD = 3900 mg/L; TOC = 910 mg/L; conductivity = 11.31 mS/cm) and after conventional biological treatment (COD = 3264 mg/L; TOC = 687 mg/L; conductivity = 15.56 mS/cm) by the industry. The effluent contained disperse dyes (77 ppm Dispersil Orange and 72 ppm Dispersil Ruby and 92 ppm Dispersil Red; all patented azo dyes) and dispersing agents (a commercial anionic derivative of a naphthalene compound) and had a pH of 9.3. Prior to photoelectrocatalytic treatment, sodium sulphate (0.1 M) was added as a supporting conductive electrolyte. Dye oxidation products and intermediates were identified on a HPLC equipped with a diode array detector (DAD). According to the experimental results, the dispersing agent present in the dye's content was removed immediately, whereas it took longer treatment periods to degrade the disperse dyes (> 300 min for dispersil orange and ruby to 1.5–2 pp, 250 min for Dispersil Red for complete removal). The chromatographic results indicated that no measurable intermediate compounds were identified at 200–800 nm using the DAD. Highest TOC removal was obtained were in the range of 45–63 %. For the real wastewater samples containing disperse dyes and dispersing agents, 80 % (before conventional treatment) and 89 % colour removal after conventional treatment (peak absorbance at 440 nm) was obtained after 360 min accompanied with 47–50 % TOC and 100 % dispersing agent abatements (peak absorbance at 230–232 nm). HPLC analysis of the real textile wastewater samples indicated that the disperse dyes slowly and gradually disappeared from the textile effluent and no signal for possible organic intermediates was detectable. Electrical energy consumptions per unit mass of COD removed were calculated as 0.0044 and 0.0040 kWh/kg for untreated and biotreated textile effluents, respectively.

2.5.4 TiO_2-mediated heterogeneous photocatalysis

Solar photocatalysis (TiO_2/sunlight) in the presence of H_2O_2 was investigated for the treatment of real textile effluent by Garcia *et al.* (2009). Treatment performance was evaluated in terms of the parameters COD, inorganic ions (ammonium, sulphate, nitrate) and analysis of characteristic UV-visible light absorption bands in the 200–800 nm range. Degussa P-25 grade titania (TiO_2) were used as the photocatalyst (BET = 50 m^2/g; 30 nm particle size; 75–80 % anatase and 20–25 % rutile crystal phase). Five effluent samples were collected from a clothing factory located in Brazil (average values: pH at dyeing machine outlet = 13; pH in the balancing tank = 7.5; NH_4–N = 5.1 mg N/L; N_{org} = 13.4 mg/L; COD = 368.4 mg/L). 0.25 g/L suspensions of titania and 10 mM H_2O_2 was used in the experiments. The pH was adjusted 3.50 by 0.1 M HCl prior to photocatalysis to achieve the best adsorption level. The reactor consisted of an open cylindrical borosilicate glass with suspensions containing 500 mL textile effluent. The average temperature and solar radiation flux were 30°C and 2.85 MJ/m^2, respectively. UV-Vis absorption spectra results implied that in the colour region absorbance values completely disappeared after 150 min, whereas for the UV region longer treatment periods (6 h) were necessary and still absorption bands were observed at 228, 254 and 280 nm. H_2O_2 consumption was in the range of 88–100 % for all textile effluent samples. COD removals in the range of 62–100 % could also be achieved after 6 h solar photocatalytic treatment. Not the TOC parameter, but inorganic ions were measured to follow the extent of mineralisation. In all textile effluent samples, variable final concentrations of ammonia, nitrate and sulphate was detected after 6 hours of photocatalytic treatment. The formation of gaseous nitrogen cannot be ruled out. Both nitrogen and sulphate species are expected as oxidation end products of textile dyes present in the dyehouse effluent. Negatively charged ions such as nitrate and sulphate are expected to be adsorbed on the positively charged titania surface at acidic pH's.

2.5.5 Ozonation

Ozone is known to effectively decolourise textile dyes in aqueous solution and in real/simulated dyehouse effluent at mass-transfer limited rates. In a paper authored by Chu *et al.* (2008) micro bubble technology was employed to increase the mass transfer rate of ozone and enhance the ozone oxidation of real textile industry wastewater (COD = 530–600 mg/L; BOD_5 = 80–116 mg/L; pH = 6.9–8.7; TSS = 60–80 mg/L). Enhancement of the ozone utilisation would decrease the amount of ozone supply and lower the off-gas (unreacted) ozone concentration. Experiments were performed using a micro bubble generator (consisting of a recycling pump, an accelerator and an injector) and an ordinary bubble contactor featuring a conventional air diffuser, which is commonly used in ozonation systems, for comparative purposes (wastewater volume = 20 L). Ozone was produced from pure oxygen by an ozone generator. The micro bubble generator produced a milky and high intensity micro bubble solution, which could reach a higher oxygen transfer rate at a lower input gas flow rate (investigated range = 0.02–1.50 L/min). A volumetric oxygen transfer rate of 0.086–0.413 kg/(m^3×h) and a total mass transfer coefficient of 0.1072–0.4859 min^{-1} were obtained at the studied airflow rate range. The decolourisation capacity was determined by scanning the optical spectra of a sample from 200 to 800 nm and integrating the area below the absorbance curve. During the ozonation of real textile wastewater by using the micro bubble oxidation system, the input ozone could be almost completely utilised, and the rate of decolourisation and organic reduction were much faster than those of the bubble contactor. The time required for 80 % colour removal from real textile wastewater was about 140 and 280 min by ozone micro bubble and conventional bubbles, respectively. The COD removal efficiency in the micro bubble system was higher by 20 %; it reached 70 % (at 200 min) in the micro bubble system and was only 50 % at 360 min for the bubble contactor. pH decreased gradually during ozonation to from the original neutral to acidic values as a consequence of acidic intermediate production. The results revealed that in addition to the enhancement of the mass transfer of ozone, micro bubbles could improve the oxidation of actual textile wastewater.

In another study conducted by Somensi *et al.* (2010), ozonation of real, untreated textile wastewater (COD = 1505 mg/L; BOD = 91.2 mg/L; surfactants = 1.18 mg/L; colour as A_{455} = 0.754; pH = 9.1; phenolic compounds = 0.090 mg/L; total iron = 0.77 mg/L; nitrate = 2 mg N/L; phosphate = 12 mg P/L; sulphate = 345.3 mg/L) was conducted in a pilot-scale plant (an air dryer, compressor, ozone generator and a 18 L cylindrical PVC/PE made semi-batch contactor) and the efficiency of this treatment was evaluated based on the parameters colour and soluble organic matter removals the latter being measured as COD at two different pH values 3.0 and 9.1). Ozone gas was produced at a concentration of 20 g/m^3 and sparged counter-current in relation to the textile wastewater circulation for a treatment period of 4 h. Identification of intermediate and final degradation products of ozonation were also carried out via gas chromatography-mass spectrometry (GC-MS). Besides, the final ecotoxicity (Lumistox® test conducted with marine photobacteria to obtain EC_{50} values) of the pretreated wastewater was measured. After 4 h of ozone treatment with wastewater recirculation at a flow rate of 0.45 m^3/h, the average efficiencies for colour removal were 68 % (at pH 9.1) and 41 % (at pH 3.0), while COD removal was 26 % (at pH 9.1) and 19 % (at pH 3.0) for an ozone feed rate of 20 g/h. Ozonation had a positive impact on the BOD_5/COD (a rough estimate of ultimate biodegradability) ratios obtained for the textile wastewater by a factor of 7. A GC-MS analysis of the ozonated textile wastewater indicated that some oxidation products were present at the end of the pretreatment time. Among the intermediates, phenols, quinones, hydroquinones, phthalic, muconic, fumaric, maleic were identified, but they disappeared completely after 60 min of ozone treatment. The presence of some aliphatic alcohols originated from the partial hydrolysis of nonylphenol ethoxylated (NPE) compounds, which are widely used in textile mills as dispersants, humectants and emulsifiers are observed in the raw textile wastewater. It should also be noted here that GC-MS methods are not appropriate for the detection of polar or large molecular weight compounds. Thus, it is very probable that some polar degradation products were generated which could not be detected with the analytical method used. In spite of this fact, the bacterial luminescence inhibition test showed a significant toxicity reduction (decrease in the EC_{50} values from 3.4 % to 28.6 % based on volumetric dilution ratios) upon comparing the untreated and ozonated textile wastewater's toxicity. In conclusion, ozonation pretreatment of textile industry wastewater is an important step in terms of improving wastewater biodegradability as well as reducing acute toxicity, which should be removed completely via sequential biological treatment.

2.5.6 Homogenous and heterogeneous Fenton's processes

Until now, numerous papers have been published dealing with the advanced oxidation of textile dyes, simulated dyehouse effluent and textile wastewater. Most of these papers report effective colour removal accompanied with partial COD and TOC abatements. In some cases, an appreciable biodegradability improvement was observed, that was measured as the BOD_5/COD or BOD_5/DOC ratio. Among them, Fenton's reagent has been the focus of several scientific studies where textile effluent is being treated under varying process conditions. In the study authored by Papadapoulos *et al.* (2007), textile wastewater samples taken from a factory located within the vicinities of Athens, were characterised for its colour (measured spectrophotometrically as absorbance), pH (7.9), BOD (1200 mg/L), COD (8100 mg/L) and TOC (3010 mg/L) contents. Then, Fenton batch oxidation experiments were conducted with 2-L samples using a Jar Test Apparatus to study the effect of Fe^{2+} and H_2O_2 concentrations (molar ratios = 0.492–4.927) at an optimum pH of 3.0–3.5. The overall reaction time was 24 h, but sample aliquots were taken at regular time intervals (0.5, 1, 2, 3, 4, 5 and 6 h). Four different H_2O_2 and Fe^{2+} concentrations were selected and combined. The optimum oxidant concentration was determined as 0.882 mM in the presence of 40–50 mg/L Fe^{2+} where around 49–50 % COD removal was achieved. An increase in the H_2O_2 to 1.764 mM did not improve oxidation efficiency any further; indicating that excessive amounts of either oxidant or catalyst inhibited the process or at least did not enhance oxidation rates. The maximum colour reduction that could be achieved was 72 %. Most of the COD removal occurred within the first 2–3 h of Fenton's oxidation process. In parallel to this observation, 40 % and 60 % TOC and BOD removals were obtained after 6 h, respectively. The BOD/COD ratio was not improved during treatment (remained in the range of 0.0119–0.0970after 0.5–6.0 h oxidation, respectively), and the average oxidation state increased from –0.04 to +0.21 within 6 h Fenton's reagent. The COD/TOC ratio first slightly increased at t = 0.5 h to 3.310 and gradually decreased thereafter to 2.528 after 6 h.

A Fenton's oxidation study was conducted with three samples obtained from a carpet mill (Gulkaya *et al.* 2006) that had the following dyeing lines; acidic (300 m^3/d; COD = 2770 mg/L; BOD_5 = 890 mg/L; pH = 4.83), basic (200 m^3/d; COD =

2576 mg/L; BOD_5 = 580 mg/L; pH = 4.49), and disperse dyeing (300 m^3/d; COD = 2150 mg/L; BOD_5 = 1100 mg/L; pH = 4.76). The wastewater samples taken as 24 h composite from these three individual wastewater lines were mixed by a flow rate balancing approach. This composite sample had a COD about 2400 mg/L, a BOD_5 of 890 mg/L, a TOC of 2000 mg/L and a pH of 4.8. The Fenton's oxidation treatment of the composite effluent was optimised in terms of temperature (25–70°C); pH (2–5.5), H_2O_2 (19.3 and 577.5 g/L) and $FeSO_4$ (1.1–10.9 g/L) ranges. The experiments were conducted with 100 mL samples. Oxidation was continued for 30 min, followed by 30 min coagulation at pH = 7. Samples were analyzed after 2 h precipitation-decantation for colour (UV-Vis absorbance), COD and TOC. COD was only measured after pH adjustment and coagulation, due to the well-known positive interference of residual (unreacted) H_2O_2 with COD analysis. Experimental results have indicated that temperature had no effect on removal efficiencies that slightly increased above 40°C from around 90 % to 94 %, and up to 95 % at a temperature range of 50–70°C. COD and TOC removals were not sensitive towards pH in the 2–5 range, with slightly higher efficiencies obtained at pH 3, known as the optimum pH for Fenton's reaction. COD and TOC removal efficiencies levelled off beyond concentrations above around 385 and 200 g/L H_2O_2, respectively. Highest COD and TOC removals were reached at a $FeSO_4$ concentration of 5.5 g/L, where 95 % treatment efficiency was observed. All parameters including UV absorbance were more effectively removed after Fenton's coagulation than Fenton's oxidation. Different sets of experiments were conducted at varying oxidant to catalyst doses by keeping a specific mass ratio constant. An optimum H_2O_2:Fe molar ratio (140–427:1; in molar concentration units) was reported in the range of 95–290 (at pH 3 and 50°C) independent from the applied concentrations. This observation implied that provided the ratio is optimal, considerably lower reagent doses can be used for the achievement of highest treatment performance. Experiments performed at three different H_2O_2:Fe mass ratios (96, 192 and 287 g/g) indicated that H_2O_2 concentration should be kept around 96–116 g/L and $FeSO_4$ at 1.1 g/L (Fe = 0.405 g/L) to obtain the highest COD removal rates (90–95 %). The biodegradability expressed as the BOD_5/COD ratio was improved by 30–40 % after Fenton's treatment at a H_2O_2:Fe mass ratio of 192 g/g (316 mol/mol).

Catalytic wet oxidation is a reaction involving an organic compound in water and oxygen over a catalyst. A heterogeneous reaction takes place over a solid catalyst that is usually limited by mass transfer but does not require separation after the treatment is completed. Since hydrogen peroxide is a stronger oxidant than oxygen, catalytic wet peroxidation usually works at milder reaction conditions (temperature and pressure). However, catalyst leaching and deactivation still remain challenges for developing catalysts for oxidation of wastewaters. In the study conducted by Dantas *et al.* (2006), simultaneous adsorption and catalytic wet peroxidation using two different composite Fe_2O_3/carbon catalysts (one carbon and one carbon with iron oxide deposited on it) was explored. The adsorbent carbons were characterised through the determination of their specific BET area and iron oxide content. Additionally, **Scanning electron microscope** (SEM) analysis and the zero point of charge determination were carried out. The effects of pH (3–9), H_2O_2 concentration (100–300 g/L) and catalyst dosage (500–1000 mg/L) on treatment efficiency (changes in COD, BOD, colour and aromaticity-UV_{254}) were evaluated. The textile wastewater (COD = 1000 mg/L; BOD_5 = 300 mg/L; TS = 2545 mg/L; volatile solids = 880 mg/L; NH_3–N = 57 mg N/L; nitrate-N = 1.2 mg N/L; pH = 9) was obtained from a plant located in Brazil. In a typical run, 400 mL textile wastewater was subjected to catalytic wet peroxidation for 90 min-25 h. Separate blank experiments were also carried out in the absence of hydrogen peroxide to determine the adsorptive capacity of the solid catalysts. Leaching tests were also performed to establish whether iron is dissolved in the aqueous reaction medium. Results have indicated that preferably aromatic compounds are removed by adsorption on the iron oxide-loaded catalyst whose adsorption capacity is twice that of the carbon catalyst in terms of COD abatement (around 60 %). In addition, the difference of adsorption capacity at varying pH (3–9) is negligible revealing that adsorption does not relate to the electrostatic phenomenon. No iron leached out of the solution proving that the homogenous Fenton process did not play a role in the oxidation of the textile wastewater. It was observed that catalytic oxidation slowed down within the first hour of oxidation followed by a slow reaction period. For the iron oxide loaded catalyst more than 70 % COD removal occurred with a parallel conversion of H_2O_2 up to 100 %. Due to the increasing amount of active sites for hydrogen peroxide decomposition, COD removal increased with catalyst dosage. Removal of colour decreases with increasing pH, but the heterogeneous Fenton reaction is less sensitive to pH than the homogenous Fenton's process. Increasing the hydrogen peroxide concentration from 500 to 1000 mg/L improved aromaticity removal from 58 to 69 %, but had no effect on COD abatement; the lowest concentration was sufficient for 71 % COD removal with the iron oxide loaded carbon and did not increase further (300 g/L catalyst; pH = 3; T = 25°C).

2.5.7 AOP Combinations

Several can be found in the scientific literature dealing with the comparison of AOP for the treatment of dyes or textile wastewater. In the work of Garcia *et al.* (2007), real textile wastewater samples (500 mL) originating from the cotton dyeing process with reactive dyes (3 types; average pH in the dyeing outlet after pH adjustment = 7.5; NH_4^+ = 4.9 mg/L; N_{org} = 12 mg/L; COD = 772 mg/L) were subjected to homogenous and heterogeneous AOP (H_2O_2/UV, Fe^{2+}/H_2O_2/UV, TiO_2/UV and H_2O_2/TiO_2/UV) using 0.25 g/L titania (Degussa P25); 20 mM H_2O_2 and 10 mg/L Fe^{2+} at pH 3.0–3.5 using three 250 W high pressure lamps that were cooled by four fans. Sulphate was detected in all treated effluents, indicating that the naphthalene sulphonates present in the effluent originally having no sulphate content was oxidised to sulphate during application of AOP. Organic nitrogen originating from the azo and amino groups was oxidised to ammonium and nitrate, whereas molecular nitrogen gas formation is also possible. The highest efficiency was obtained by the H_2O_2/TiO_2/UV process reaching up to 90 % COD removals. It was also established that degradation of the sulphonate groups to sulphate occurred quickly, during the initial stages of oxidation. The increasing order of COD and UV absorbance removal efficiencies was obtained as follows; H_2O_2/UV < TiO_2/UV < Fe^{2+}/H_2O_2/UV < H_2O_2/TiO_2/UV.

2.5.8 Wet air oxidation

Wet air oxidation is very useful for treating a variety of refractory organic pollutants in wastewater, but the high pressure (20–210 bar) and temperature (200–325°C) required for its efficient operation limits its practical applications (Pintar and Levec 1992). Catalytic wet air oxidation has been proposed to work under relatively mild oxidation conditions (P = 20–50 bar; T = 130–250°C). In the last decades, various heterogeneous catalysts including noble metals deposited on the supports and transition metal oxides have been developed (Belkacemi *et al.* 2000). In these studies, superior catalytic activity in the wet air oxidation of organic pollutants and real wastewaters was demonstrated (Gomes *et al.* 2002). The catalytic performance of 3 % w/w copper supported on carbon nanofibres (CNFs) in the wet air oxidation of textile wastewater (TOC = 5.9 g/L; COD = 17.9 g/L; BOD_5 = 5.5 g/L; TSS = 23.9 g/L; TDS = 12 g/L; pH = 8.7; EC_{50} = 2.8 % v/v) has been tested by Rodríguez *et al.* (2008) using a batch stirred tank microreactor made of Hastelloy C-276 (a nickel-molybdenum-chromium-tungsten alloy with excellent general corrosion resistance and good fabricability) in order to determine the decolourisation and TOC removal efficiency and kinetics. CNFs are pure, mechanically strong and mesoporous support materials and hence ideal for liquid phase reactions. Their diameter was in the range of 20–50 nm and their lengths were 50–100 nm. The catalyst was prepared by excess wet impregnation of the CNF support with copper or iron metal precursors (nitrate salts), activated under nitrogen and hydrogen gas in the furnace to reduce the metal salt to its corresponding ground state. Previously, the surface chemistry of the carbon nanofibres was modified by acidic treatment by introducing carboxylic acid

(CH_3COO–) groupings. A BET survey indicated that a specific surface area of 244 and 188 m^2/g for CNF support and Cu-CNF catalyst, respectively. Thermal decomposition started above 450°C and reached a maximum value at 680°C with almost complete weight loss. The reactor had a capacity of 100 mL and 75 mL sample was oxidised in each run. Toxicity was measured via Microtox® protocol and GC-MS analysis was carried out to evaluate the formation of oxidation intermediates. Thermal decomposition (TOC abatement) was found to be negligible in the absence of catalyst at 140°C and 8.3 bar oxygen partial pressure. TOC removal reached 21.5 % when oxygen was fed to the reactor in the presence of CNF only. Stirring speed improved TOC removals when the agitation speed was increased from 500 to 1000 rpm (studied range = 500–1300 rpm). In order to study the effect of temperature, experiments were conducted in a temperature range of 120–160°C and two oxygen partial pressures of 6.3 and 8.7 bar at 1000 rpm and 3 % w/w catalyst load. Colour, TOC removal and toxicity reductions were found to be 97 %, 74 % and 43 %, respectively at 140°C and 8.7 bar, after 180 min catalytic wet air oxidation. Increasing the catalyst load from 6.6 g/L to 13.3 g/L and to 20 g/L fairly increased TOC removals from 74 % to over 80 % for 20 g/L Cu-CNF, respectively. The main degradation intermediates detected in the wastewater were decanoic acid, methyl ester and 1,2-benzene dicarboxylic acid, that were further degraded by means of Cu/CNF catalyst. Treatment of the textile effluent at 160°C and 8.7 bar of oxygen partial pressure indicated that the use of a Cu-CNF catalyst significantly improved TOC and colour removal efficiencies and it could be considered as an option for a pretreatment step in the treatment of dyehouse effluents.

2.5.9 Supercritical water oxidation

Supercritical water oxidation (SCWO) shows substantial promise for clean and efficient decontamination of many aqueous organic wastes (Shaw and Dahmen 2000; Mizuno and Goto 2000; Anikeev *et al.* 2004; Veriansyah *et al.* 2005). SCWO can rapidly and efficiently destroy organic substances in wastewater into H_2O and CO_2 with very high conversion rates of 99–99.99 %, at 500–650°C in 1–100 s, which are significantly short residence times (Portela *et al.* 2001). Unlike wet air oxidation and other AOP that usually result in partial treatment (incomplete oxidation of accumulation of partial oxidation intermediates), this method enables the reuse of treated water, thus cutting off the water consumption costs and discharge fees. However, it should be considered that real textile wastewater contains a variety of organic as well as inorganic chemicals that may hinder the practical application due to serious technical limitations and high capital investments (Shaw and Dahmen 2000). Studies devoted to SCWO are focusing on catalytic and non-catalytic SCWO (Park *et al.* 2003; Koech and Hatakeda 2002).

Sogut and Akgun (2007) investigated the supercritical oxidation of aqueous Disperse Orange 25 in a 316 SS continuous flow coiled tubular reactor at 400 – 600°C and a fixed pressure of 25 MPa. H_2O_2 was used as the oxygen source and the initial oxidant concentration was kept between 36.75–183.75 mM. The selected process parameters were the temperature, initial dye concentration (24.25–121.25 mM in terms of COD), oxidant concentration and flow rate. Two high pressure feed streams (1–5 mL/min) that were separately pre-heated in the first section of the reactor inlet and thereafter mixed using a mixing tee, were used to load the reactor with the aqueous dye sample and the H_2O_2 solution. Besides colour and COD, product gas samples were analyzed by a GC-MS. Type of aqueous nitrogen products were identified using an ion chromatograph. The results demonstrated that supercritical water oxidation decreases the COD up to 99 % in very short reaction times (reactor residence time = 4–12 s). Under all studied experimental conditions COD removals were higher than 87 % and typically in the range of 93–98 %. The global pseudo-first-order rate constant was determined with an activation energy of 27.8 kJ/mol and a pre-exponential factor of 34.3s^{-1} at a 95 % confidence level. Gas phase products were detected as CO, CO_2, NO, NH_3, CH_4, C_2H_6 and C_4H_{10} by qualitative GC-MS analysis. In the liquid phase nitrate (4.5 mg/L) and nitrite (< 0.3 mg/L) were detected indicating that the disperse dye was completely degraded and converted to oxidation intermediate and end products.

2.5.10 Electron beam irradiation

Kim *et al.* (2007) investigated the treatment of real textile industry wastewater via electron beam irradiation. Electron beam irradiation technology was applied to enhance the biodegradability of textile wastewater (COD = 1805 mg/L; BOD_5 = 1184 mg/L; TN = 38 mg/L; TP = 2.5 mg/L; pH = 11.0; T = 42°C) prior to an activated sludge process. The activated sludge process using both irradiated and non-irradiated textile wastewater was applied in order to investigate the effect of electron beam radiation on a combined biological treatment process. The mixed and return sludge textile industry wastewater treatment plant was used as the seed source in the biotreatment experiments. The MLSS was maintained at 2000–2700 mg/L. Dehydrogenase activity (DHA) was also determined for the evaluation of microbial activity in untreated and treated effluent samples. Besides, low molecular weight organic acids which were produced as by-products of electron beam irradiation were qualitatively identified via HPLC. The biodegradability that was also measured as the BOD_5/COD ratio increased from 0.68 to 0.79 at a dose of 1.0 kGy. The number of organic acids increased upon increasing the irradiation dose from 0.4 to 1.0 kGy. The main organic acid was identified as oxalic acid. In the biological treatability experiments, the BOD_5 and COD removal efficiencies were higher for the irradiated wastewater than for the non-irradiated wastewater. Regardless of the HRT variation, about 80 % BOD_5 and COD removal efficiencies were obtained when the irradiated wastewater was treated by an activated sludge process. However, about 75 % BOD_5 removal efficiencies were obtained when the non-irradiated wastewater was treated biologically. When the HRT was reduced to 18 h, the BOD_5 removal efficiency of the effluents after activated sludge treatment of the non-irradiated wastewater was less than 70 %. The DHA values at the same operating time for the irradiated wastewater were higher than those of the non-irradiated wastewater. It can be inferred that the substrates in the wastewater were previously converted into lower molecular weight compounds or higher biodegradability compounds by electron beam radiation.

2.5.11 AOP involving power ultrasound

Power ultrasound and combinations with hydrogen peroxide, ozone, UV light, heterogeneous and homogenous catalysts have been applied to treat dyes and dyehouse effluent. In a paper authored by Mahamuni and Adewuyi (2009) numerous papers devoted to the treatment of industrial wastewater by ultrasound were reviewed. The paper focuses on estimation of costs for the application of various AOP. In order to develop a cost estimation methodology and calculate operating-capital costs, operating conditions have to be exactly known. From the available results or directly from the calculations of the investigators, reaction kinetics (kinetic rate coefficients) can be found, since one of the main parameters affecting operating costs is treatment time (or, more specifically reactor residence time) required to achieve a target, specific removal efficiency. The electrical energy consumption is usually expressed per unit volume (m^3) of wastewater to be treated. Labour, lamp replacement and chemicals (oxidant, catalyst, chemicals used for pH adjustment) costs also have to be considered when estimating operating costs. In this paper capital cost calculations were also included where AOP reactors, piping, valves, contractor, contingency and engineering costs were estimated. From the several review papers it is evident that (i) total costs for ultrasonic reactive azo dye degradation are too high to be practically implementable; (ii) general speaking, the costs associated with sonolysis of industrial pollutant (even phenol) are at least one or two orders of magnitude higher than those calculated for treatment via other AOP; and (iii) only treatment combinations with other, more conventional AOP may reduce total costs to more realistic values. More specifically, the combination of power ultrasound with UV, ozone, photocatalysis or UV/H_2O_2 enhances reaction kinetics significantly thus reducing the associated treatment costs to more acceptable levels as compared with sonolysis only. Still, hybrid ultrasonic systems are one or two orders of magnitude more costly than the more conventional (currently well established) AOPs such as H_2O_2/UV-C or O_3/H_2O_2 processes. Among the hybrid systems involving power ultrasound, sonolysis combined with UV irradiation and hydrogen peroxide appeared to be the most economical solution for wastewater containing reactive azo dyes. Main reasons for the high costs related with ultrasound applications are long treatment periods, very high capital costs and in particular the extremely low efficiency of converting electrical energy to cavitation energy (< 34 %). Alternative ways of cavitational effects production could be the use of hydrodynamic cavitation, the use of additives or the involvement of hybrid methods. The low reaction volumes treated in the applications combined with sonolysis are another factor that results in extreme treatment costs when available data was extrapolated to real-scale dimensions. Since most textile dyes are very water soluble, have complicated structures and remain in the bulk solution. Consequently, textile dyes are degraded outside the cavitation bubbles such that $HO^{\bullet}$ transport to the contaminant becomes the rate controlling step in the oxidation mechanism. Since $HO^{\bullet}$ are not stable and immediately recombine to less powerful oxidants, the whole process becomes very ineffective for the degradation of hydrophilic pollutants found in textile wastewater.

Since no study reporting the sonolytic treatment of real textile effluent is available in the scientific literature, application of ultrasound (US) to treat textile dyestuffs has been exemplified to evaluate the merits of colour removal from dyehouse effluent by using power ultrasound. Tezcanli-Güyer and İnce (2004) comparatively investigated the degradation of the model azo dye Acid Orange 7, frequently used as a probe chemical in advanced oxidation studies, by employing ultrasonic irradiation and its combinations with ozone and ultraviolet light. The 1200 mL capacity hexagonal reactor used in the experiments was equipped with three piezoelectric transducers (600 W; optimised power = 360 W); 6 low pressure UV-C lamps (18 W each; total incident light flux $I_0 = 2.1 \times 10^{-4}$ einstein/s), and an ozone gas inlet, from which ozone produced from pure air was provided at a flow rate of 0.25 L/min (optimum ozone feed rate = 40 g/m^3). The initial azo dye (molecular weight = 350 g/mol) concentration was selected as 20 mg/L. TOC and BOD_5 parameters were also followed in this work. First-order decay rate coefficients were determined for colour and UV absorbance removal rates. In order to demonstrate the extremely high electrical energy requirements of a treatment process involving ultrasound irradiation, electrical consumption rates were calculated by considering t_{90} values (time required to achieve 90 % decolourisation, calculated from the pseudo- first order colour abatement rate constants), power inputs of the UV-C lamps and transducers, as well as energy equivalents of ozone generators (around 15 kWh per kg ozone production). Based on the treatment systems run under optimised reaction conditions; the electrical energy requirements were obtained in the following increasing order; O_3 (4.3 kWh/m^3) < O_3 + UV-C (48.8 kWh/m^3) < O_3+US (147.6 kWh/m^3) < O_3+UV+US (190.8 kWh/m^3) < US only (540 kWh/m^3) revealing that the application of mere ozonation might be more beneficial from the technical/economical point-of-view and although colour abatement is enhanced in the presence of US and UV, it does not outweigh the operating costs associated by the introduction of US and UV systems requiring an extra energy input. It should be noted here that capital investments were not considered at all in these calculations.

2.6 CONCLUDING REMARKS

From the above reviewed and evaluated work it is clear that the extreme complexity and variety of dyehouse effluent samples makes the overall evaluation and global unification of treatability studies devoted to textile industry wastewater rather difficult. For instance, operating costs of electrochemical, photochemical, photocatalytic systems as well as treatment processes where oxidant (ozone, electron beam irradiation, sonolysis, *etc.*) production is mainly based on the use of electrical energy are assessed on the basis of electrical energy consumption calculations. On the other hand, in solar irradiation applications the costs of the collector material per unit area may play a decisive role in determination of investment costs. In other applications (wet air oxidation, wet peroxidation, supercritical water oxidation, or catalytic wet air oxidation) capital investments are relatively high; in that case, reactor material and associated equipment (pressure-tight valves, high pressure pumps, heating devices, *etc.*) are mainly contributing to the overall costs, but energy self-sustainable systems can be operated below a certain calorific value provided by high-strength waste streams. Some general conclusions could be drawn and are summarised below;

The colour parameter that is one of the major concerns of dyehouse effluent due to ecotoxicological risks imposed by metabolites in the discharged environment and aesthetic reasons. All the above exemplified chemical oxidation processes are quite effective in the removal of colour from dyehouse effluent. In other words, if the major target is to remove colour (dyestuffs) from textile wastewater, application of chemical oxidation is technically and economically relatively feasible as compared to the removal of other collective parameters.

On the other hand, COD and TOC (organic carbon) abatements are in most cases rather incomplete and require longer treatment periods and/or extended oxidant/irradiation doses. In most cases complete oxidation cannot be achieved under technically/economically feasible treatment conditions. From the above mentioned case studies it can be concluded and in particular the more energy (cost)-intensive advanced treatment process should be applied in combination with conventional ones for pre- or post- treatment purposes.

Depending on the chemical oxidation process employed, more recalcitrant and/or toxic oxidation intermediates or end products can be formed during treatment of textile wastewater. For instance, investigators of electrochemical treatment processes have reported the formation of active chlorine species as well as organically-bound organic halogens (AOX) that increase the toxicity of the treated wastewater to levels being more detrimental than that of the original dyehouse effluent not being subjected to chemical oxidation. Considering the fact that the main purpose of chemical oxidation is to reduce the toxicity and recalcitrance (inertness) of textile industry wastewater, it is important to verify in preliminary treatability experiments whether and under what chemical reaction conditions potentially toxic oxidation intermediates build up the reaction solution.

Most of the emerging/advanced chemical oxidation processes mentioned in this chapter are still not affordable in large scale for textile industry wastewater treatment; in particular no application exists for treatment of real dyehouse effluent via supercritical oxidation or sonolysis. Studies devoted to these advanced oxidation processes are limited to aqueous dye solutions. More efficient means of cavitational energy production from different sources and the development of more efficient heterogeneous catalysts that promote the application under milder, *e.g.* sub-critical conditions could be encouraged in future studies.

REFERENCES

Alinsafi, A., Khemis, M., Pons, M.N., Leclerc, J. P., Yaacoubi, A., Benhammou, A. and Nejmeddine, A. (2005) Electrocoagulation of reactive textile dyes and textile wastewater. *Chem. Eng. Process.* **44**, 461–470.

Anikeev, V., Yermakova, A. and Goto, M. (2004) Decomposition and oxidation of aliphatic nitro compounds in supercritical water. *Ind. Eng. Chem. Res.* **43**, 8141–8147.

Arslan-Alaton, I., Kabdaşlı, I., Hanbaba, D. and Kuybu, E. (2008) Electrocoagulation of a real reactive dyebath effluent using aluminium and stainless steel electrodes. *J. Haz. Mater.* **150**, 166–173.

Bayramoglu, M., Eyvaz, M. and Kobya, M. (2007) Treatment of the textile wastewater by electrocoagulation: economical evaluation. *Chem. Engin. J.* **128**, 155–161.

Belkacemi, K. Larachi, F., Hamoudi, S. and Safari, A. (2000) Catalytic wet oxidation of high-strength alcohol-distillery liquors. *Appl. Catal. A* **199**, 199–209.

Blackburn, R. S. and Burkinshaw, S. M. (2002) A greener approach to cotton dyeings with excellent wash fastness. *Green Chem.* **4**, 47–52.

Chatzisymeon, E., Xekoukoulotakis, N. P., Alberto, C., Kalogerakis, N., Mantzavinos, D. (2006) Electrochemical treatment of textile dyes and dyehouse effluents. *J. Hazard. Mater.* **137**, 998–1007.

Chu, L.-B., Xing, X.-H., Yu, A.-F., Sun, X.-L. and Jurcik, B. (2008) Enhanced treatment of practical textile wastewater by microbubble ozonation. *Proc. Safety Environ. Protection* **86**, 389–393.

Cooper, P. (1998) Removing colour from dyehouse waste waters-a critical review of technology available. *J. Soc. Dyers Colour*. **109**, 97–100.

Dantas, T. L. P., Mendonça, V. P., José, H. J., Rodrigues, A.E. and Morei, R.F.P.M. (2006) Treatment of textile wastewater by heterogeneous Fenton process using a new composite Fe_2O_3/carbon. *Chem. Eng. J.* **118**, 77–82.

Easton, J. R. (1995) The dye maker's view. In *Colour in Dyehouse Effluent* (ed. P. Cooper), The Society of Dyers and Colourists, Alden Press, Oxford, pp. 6–21.

EC-IPPC (2003) (Integrated Pollution Prevention and Control) Reference document on Best Available Techniques in the Textile Industry, European Commission, Joint Research Centre, Institute for Prospective Technological Studies, Seville.

Garcia, J. C., Simionato, J. I., da Silva, A. E. C., Nozaki, J. and de Souza. N. E. (2009) Solar photocatalytic degradation of real textile effluents by associated titanium dioxide and hydrogen peroxide. *Solar Energy* **83**, 316–322.

Garcia, J. C., Oliveira, J. L., Silva, A. E. C., Oliveira, C. C., Nozaki, J. and de Souza., N. E. (2007) Comparative study of the degradation of real textile effluents by photocatalytic reactions involving UV/TiO_2/H_2O_2 and UV/Fe^{2+}/H_2O_2 systems. *J. Hazard. Mater*. **147**, 105–110.

Gomes, H. T., Figueiredo, J. L., Faria, J. L., Serp, P. and P. Kalck, P. (2002) Carbon-supported iridium catalysts in the catalytic wet air oxidation of carboxylic acids: kinetics and mechanistic interpretation. *J. Mol. Catal. A* **182–183**, 47–60.

Gülkaya, İ., Sürücü, G. A. and Dilek, F. B. (2006) Importance of H_2O_2/Fe^{2+} ratio in Fenton's treatment of a carpet dyeing wastewater. *J. Hazard. Mater.* **136**, 763–769.

Kim, T.-H., Lee, J.-K. and Lee, M.-J. (2007) Biodegradability enhancement of textile wastewater by electron beam irradiation. *Radi. Phys. Chem*. **76**, 1037–1041.

Koech, G. C. and Hatakeda, K. (2002) Degradation of reactive dyes by supercritical water oxidation in a batch reactor. *Color. Tech.* **118**, 112–114.

Kobya, M., Bayramoglu, M. and Eyvaz, M. (2007) Techno-economical evaluation of electrocoagulation for the textile wastewater using different electrode connections. *J. Hazard. Mater*. **148**, 311–318.

Mahamuni, N.N. and Adewuyi, Y.G. (2010) Advanced oxidation processes (AOPs) involving ultrasound for waste water treatment: a review with emphasis on cost estimation. *Ultrason. Sonochem.* **17**(6), 990-1003.

Malpass, G.R.P., Miwa, D.W., Machado, S.A.S. and Motheo, A.J. (2008) Decolourisation of real textile waste using electrochemical techniques: effect of electrode composition. *J. Hazard. Mater.* **156**, 170–177.

Mizuno, T., Goto, M., Kodoma A. and Hirose, T. (2000) Supercritical water oxidation of a model municipal solid waste. *Ind. Eng. Chem. Res.* **39**, 2807–2810.

O'Neill, C., Hawkes, F. R., Hawkes, D. L., Lourenço, N. D., Pinheiro, H. M. and Delée. W. (1999) Colour in textile effluents - sources, measurement, discharge consents and simulation: a review. *J. Chem. Technol. Biotechnol.* **74,** 1009–1018.

Papadopoulos, A. E., Fatta, D. and Loizidou, M. (2007) Development and optimization of dark Fenton oxidation for the treatment of textile wastewaters with high organic load. *J. Hazard. Mater.* **146**, 558–563.

Park, T .J., Lim, J. S., Lee, Y. W. and Kim, S.H. (2003) Catalytic supercritical water oxidation of wastewater from terephthalic acid manufacturing process. *J. Supercrit. Fluids* **26,** 201–213.

Paschoal, F. M. M., Anderson, M. A. and Zanoni, M. V. B. (2009) The photoelectrocatalytic oxidative treatment of textile wastewater containing disperse dyes. *Desalination* **249**, 1350–1355.

Pintar, A. and Levec, J. (1992) Catalytic oxidation of organics in aqueous solutions. I. Kinetics of phenol oxidation, *J. Catal.* **135**, 345–357.

Portela, J. R., Nebot, E. and Martinez de la Ossa, E. (2001) Generalized kinetic models for supercritical water oxidation of cutting oil wastes. *J. Supercrit. Fluids* **21**, 135–145.

Radha, K. V., Sridevi, V. and Kalaivani, K. (2009) Electrochemical oxidation for the treatment of textile industry wastewater. *Biores Technol.* **100**, 987–990.

Rodríguez, A., Ovejero, G., Romero, M. D., Díaz, C., Barreiro, M. and García, J. (2008) Catalytic wet air oxidation of textile industrial wastewater using metal supported on carbon nanofibers. *J. Supercrit. Fluids* **46**, 163–172.

Shahady, T. D. (1998). The Textile Industries. In *Preserving the Legacy. Industrial Process and Waste Stream Management* (ed. H.H. Guyer), pp. 403–416, John Wiley and Sons, Inc., New York.

Shaw, R.W. and Dahmen, N. (2000) Destruction of toxic organic materials using supercritical water oxidation: current state of the technology. In *Supercritical Fluids-Fundamentals and Applications* (eds. E. Kiran, P.G. Debenedetti and C.J. Peters), Kluwer Academic Publishers, Dordrecht, pp. 425–437.

Sogut, O. O. and Akgun, M. (2007) Treatment of textile wastewater by SCWO in a tube reactor. *J. Supercrit. Fluids* **43**, 106–111.

Somensi, C.A., Simionatto, E.L., Bertoli, S.L., Wisniewski, A. Jr. and Radetski, C.M. (2010) Use of ozone in a pilot-scale plant for textile wastewater pre-treatment: physico-chemical efficiency, degradation by-products identification and environmental toxicity of treated wastewater. *J. Hazard. Mater.* **175**(1-3), 235-240.

Tezcanli-Güyer, G. and İnce, N. H. (2004) Individual and combined effects of ultrasound, ozone and UV irradiation: a case study with textile dyes. *Dyes Pigments* **42**, 603–609.

Turkish Water Pollution Control Regulation (WPCR) (1988) Official Gazette Numbered 19919 and dated 4 September 1988, Ankara.

UNEP (United Nations Environmental Programme) (1994) *The Textile Industry and the Environment*. UN Sales No.: E93-III-D5, ISBN 92-807-1367-1, Paris.

UNEP (United Nations Environmental Programme) (1998) *A Workbook for Trainers: Cleaner Production in Textile Wet Processing, Part 3: Introduction to Current Environmental Issues and Aspects in Textile Wet Processing*, UNEP, Paris.

Vandevivere, P. C., Bianchi, R. and Verstraete, W. (1998) Treatment and reuse of wastewater from the textile wet-processing industry: review of emerging technologies. *J. Chem. Technol. Biotechnol.* **72,** 289–302.

Veriansyah, B., Park, T. J., Limb, J. S. and Lee, Y. W. (2005) Supercritical water oxidation of wastewater from LCD manufacturing process: kinetic and formation of chromium oxide nanoparticles. *J. Supercrit. Fluids* **34**, 51–61.

Zongo, I., Maiga, A. H., Wéthé, J., Valentin, G., Leclerc, J.-P., Paternotte, G. and Lapicque, F. (2009) Electrocoagulation for the treatment of textile wastewaters with Al or Fe electrodes: Compared variations of COD levels, turbidity and absorbance. *J. Hazard. Mater.* **169**, 70–76.

http://aida.ineris.fr/ http://aida.ineris.fr/
http://www.access.gpo.gov/cgi-bin/cfrassemble.cgi?title=200240
http://www.access.gpo.gov/cgi-bin/cfrassemble.cgi?title=200240
http://www.menv.gouv.qc.ca/indexA.htm http://www.menv.gouv.qc.ca/indexA.htm
http://www.epa.vic.gov.au/Water/EPA/#Other
http://www.epa.vic.gov.au/Water/EPA/#Other
http://www.ostc-was.org/environment/water.html#3.2.1

Chapter 3

Leather tanning industry

3.1 LEATHER TANNING INDUSTRY

3.1.1 Overview

Leather tanning is conversion of animal skins or hides into leather and leather products. The industry is classified by U.S.A. Department of Commerce Standard Industrial Classification as SIC 3100 "*Leather and Leather Products*" (US EPA 1982). In the EU document IPPC Directive 2008/I/EC, the industry is defined as "*Plants for Tanning Hides and Skins*" and technical definition of "*Tanning as Stabilising Hides and Skins and All Associated Activities*" and referred to "*Classification of Economic Activities in the European Community*" (NACE) classification of 441.1, 441.2, 441.3 (EC 2003). The industry applies numerous processes to raw skins and hides to produce leather for shoes, garments, upholstery, sporting goods, and a variety of other uses. Majority of skin and hides treated in tanneries is cattle hides, sheepskins, and pigskins. Other skins and hides used in the industry are mainly horses, goats, deer, and other animals. Globally about 60 million tonnes of raw hides on a wet salted basis were processed. For goat and sheep's worldwide 646,800 tonnes raw skins on a dry basis were converted into leather products. Europe produced about 71,700 tonnes of heavy leather and 2473 million square feet of light leather (2001–2003 figures, EC 2009). Developing countries hold over 78 % of the world bovine herd and about 65 % of the global number of sheepskins. The leather industry is a common industrial branch worldwide (EC 2009). The capacity of the leather tanning plants is variable. In the U.S.A., 40 % of the leather tanning plants had fewer than 50 employees (US EPA 1982). In the Europe, tanneries are also small and medium enterprises.

3.1.2 Production processes

There are a number of processes in the leather tanning and finishing industry. These processes vary depending on several factors of which following are of importance:

- sort of skin or hide *e.g.* cattlehide, sheepskin;
- application of unhairing operation;
- the extent of pre-processing;
- type of tanning agent employed.

Generally cattle hides do not require degreasing operation and can be tanned using either chrome or purely vegetable tanning agents. Unhairing operation is not applied to sheepskins if double side leather *i.e.* fur-suede is produced. Leather processing may begin with raw hide, pickled skins, wet blue, or crust. Two common tanning methods are chrome and vegetable tanning. In general, a leather tanning is realised in three main steps: (i) beamhouse; (ii) tanning; and (iii) wet finishing. There are four processes which do not involve water use or their wastewater production is negligible. The first group of processes is applied to raw skins or hides which are commonly salt cured (green-salted). These processes are ahead of beamhouse and include operations sorting, trimming of heads, long shanks, and cutting the hides in halves, and other mechanical preparation operations. Fleshing is a mechanical operation to remove excess flesh, fat, and muscle from the interior of the skins. Fleshing is carried out in the beamhouse, although, some applications may be realised through the stage of degreasing. Removal of flesh prior to liming is referred to as green fleshing, and after liming is known as lime fleshing. The third stage operations follow the tanning process where leathers are, drained, rinsed, summed, shaved and split. The fourth dry operations follow retanning or wet finishing processes, and are called as mechanical finishing operations. These operations include conditioning, staking, buffing, dry milling, polishing, plating, grading. The processes realised in beamhouse are briefly introduced as follows.

Soaking is made to clean and re-hydrate the skin which is lost its moisture during the curing process. A pre-soaking step is usually applied to remove salt, blood, urine, and dirt, and facilitates the water penetration. During soaking, a salt balance is established between the environment and skins, moisture balance is restored, and clean and softened leather is made ready for subsequent operations. During soaking, interfibrillary material (non-collagenous proteins) is also partly removed. The process is carried out in drums, paddles, pits, or raceways, and may last from several hours to a few days. The chemicals used in the process are: surfactants, bactericides/fungicides, alkalis, and particularly for short soaking protease based enzymes. The main goal of unhairing process is to eliminate the hair on the skin. The process is conducted using sodium sulphide and lime. The sulphide pulps the hair and epidermis, and high pH (~13) causes the hides to swell the collagen fibre network to open facilitating removal of non-structural proteins and flesh. Hair linkages to other keratin proteins are weakened by the chemical action, and can be mechanically removed. For sheepskins, the unhairing involves smearing the flesh side with sulphide followed by the mechanical removal of hair using unhairing machines. During liming process, interfibrillary components and epidermis are removed, and fibre structure is opened up to make the leather ready for subsequent operations, mainly for tanning process. Liming process makes use of in addition to sulphide and lime surface active agents and amino-based liming aids. It is carried out in drums, paddles, mixers, or pits, and lasts from 18 hours to days depending on mixing conditions. The aim of deliming is to remove the alkali and to adjust the pH to reduce swelling, further loosening of collagen fibres, softening the leather and render it porous to be prepared for bating. Deliming is commonly carried out using weak acids. Ammonium salts are cheap, and able to make the pH adjustment easier and penetrate fast, therefore, they are the most common chemicals used in deliming process. Bating is the process involving the removal of interfibrillary proteins in the grain side of the leather together with unwanted hair roots and scuds by enzymatic degradation. For this purpose, proteases are used. Deliming and bating are subsequent processes and are carried out usually in drums. Degreasing is mostly used for sheepskin and pigskin. The purpose of degreasing is to remove natural fat to avoid its further movement to the surface of the finished leather and causing fat spew and also to prevent the formation of insoluble chrome soaps in the tanning step. Degreasing follows the bating or in the case of pickled skin as the starting material after depickling process. Three different methods are commonly used for degreasing: (i) degreasing in an aqueous medium with an organic solvent and non-ionic surfactant; (ii) degreasing in an aqueous medium with a non-ionic surfactant; and (iii) degreasing in a solvent. The most commonly used solvent is kerosene, however sometimes white spirit or halogenated paraffin can also be used. Sodium sulphate or sodium chloride may also be used to break the emulsion. Pickling is a process in which the leather is converted into a cationic matrix by acid environment. In this process, the pH is adjusted to 2.5–2.8. Sulphuric acid or formic acid is used for acidification. Sodium chloride is added to prevent swelling in the collagens. Depending on purpose, biocides, pre-tanning substances, fatliquoring chemicals may also be used. Mixers, drums or paddles are used for pickling. In the tanning process, collagen fibres are stabilised by cross-linking action of tanning agents and leather acquires stability against heat, enzymatic biodegradation, and thermo-mechanical stress. Various agents can be used for tanning; mineral tanning agents, vegetable tanning agents, and alternative tanning agents such as syntans, aldehydes, and oil tanning. Among the tanning agents, most commonly used is chrome. Chrome tanning is realised using trivalent chromium salts. Tanning is completed with basification by increasing the pH from 2.5–3.0 to 3.8–4.0 using sodium formate, sodium carbonate, or magnesium oxide. Basification leads to irreversible binding of trivalent chromium to the protein. Tanning is carried out in the pickling liquor, and no additional water is used. Tanned hides are brought to a pH suitable for subsequent processes in the neutralisation process with which wet finishing operations begin. Basic salts and neutralising tannins are used in the neutralisation process. Bleaching is used to remove stains or to reduce colouring before dyeing operation for vegetable tanned skins or double side leather using sodium bicarbonate and sulphuric acid. Retanning is made to fill the looser and softer parts of the leather to obtain uniform physical characteristics and to impart other characteristics required for the finished leather such as improving resistance to alkali and perspiration, assisting in the production of corrected grain leathers, and improving dyeing characteristics. Mineral tanning agents mostly chromium, syntans, aldehydes, vegetable tanning agents and resins are used in the process. The dyeing process is carried out to produce a homogeneous colouring over the whole surface of leather. Water-based acid dyes are commonly used for dyeing. Leathers are lubricated in the fatliquoring process to regain the fats lost in previous processes, and to have the characteristics targeted for final product. Animal or vegetable origin and synthetic mineral-based oils and chemical emulsifiers are used in the process (US EPA 1982; EC 2009).

3.1.3 Wastewater generation and characterisation

Assessment of the amount of wastewater and their characterisation, as well as pollution control strategies in the leather tanning industry are made utilising classification of the industry. The main criteria on which the subcategorisation is based have been given in the previous subsection as four factors: sort of skin; application of unhairing; the starting raw material; and type of tanning agent. Tünay *et al.* (1995) discussed subcategorisation of the industry and proposed a systematic approach on the basis of structure of the leather tanning industries. US EPA (1982) defined seven subcategories for the leather tanning industry. These are shown in Table 3.1. Main conventional and non-conventional pollution parameters of the leather tanning industry are: BOD_5; COD; oil and grease; TSS; sulphide; TKN; pH; pH-alkalinity; TDS; chloride; volatile solids; nitrate; and nitrites (US EPA 1982).

Table 3.1 Wastewater amount and characteristics by subcategory (winter and summer averages) (US EPA 1982)

Subcategory	Wastewater m^3/ton	BOD_5 kg/ton	TSS kg/ton	Oil & grease kg/ton	Total Cr kg/ton	Sulphide kg/ton
Hair pulp/chrome tan/ retan-wet finish	37–37	63–56.7	137.5–102.8	28.2–269	2.88–3.71	1.89–2.86
Hair save/chrome tan/ retan-wet finish	45–60	56.8–47.4	127.2–94.2	12–23	1.8–2.4	2.76–2.43
Hair save/non-chrome tan/ retan-wet finish	18–30	36.7–44.6	64.8–60.1	0.02–19.7	0.87–0.44	1.13–3.0
Retan-wet finish	16–20	21.9–57.4	71.9–120	8.3–71.1	6.27	7.6
No beamhouse	15–12	8.4–11.7	12.3–9.8	6.56–4.04	1.45–0.9	0.005
Through the blue	29–26	135.1–65.3	284.7–341.1	65.9–183.9	3.38–1.66	0.03
Shearling	93–105	99.9–58.4	141.0–91.3	81.1–30.6	0.72–4.0	0.05

IPPC document for Tanning of Hides and Skins of EU states that "*the Best Available Technology (BAT-EC)*" based permit conditions, and emission limit values are determined taking into account of "*local side-specific-factors such as the technical characteristics of the installation concerned, its geographical location and the local environmental conditions*". The classification is of importance to determine the source characteristics for the assessment of information about amount and characteristics of wastewaters, and pollution control application (EC 2009).

All processes and their washings in beamhouse, tanyard, and finishing stages contribute to wastewater generation except for undumped liquors like pickling. Amount of wastewaters generated and wastewater characteristics in terms of main pollution parameters are given on the basis of subcategories by US EPA (1982) as presented in Table 3.1. Figure 3.1 shows input and output materials on a conventional chrome tanning bovine process basis in terms of all waste categories (EC 2009). European Commission IPPC document provides also detailed information about chemicals used and waste generation and characteristics on process basis for bovine and ovine leather tanning. Table 3.2 shows summary information of wastewater amount and characteristics (EC 2009). There are comprehensive studies investigating water use and wastewater characteristics in the literature. Water use and waste loads were evaluated in a United Nations report (UN 1997). A review of leather making processes and their waste characteristics was given by Thanikaivelan *et al.* (2005). Floqi *et al.* (2007) evaluated water pollution from Albanian tanneries. Joseph and Nithya (2009) made an assessment of material flows in the life cycle of leather. Nazer *et al.* (2006) studied the impact of unhairing-liming process. Elsheikh (2009) evaluated tannery wastewater pretreatment conducting experimental quality assessment of wastewaters. Tünay *et al.* (1999) studied the use and minimisation of water in leather tanning processes. Comprehensive studies on wastewater characterisation of leather tanning processes were conducted in Turkey (Kabdaşlı *et al.* 1993; 1999; 2002; Tünay *et al.* 1994; 1995; Ateş *et al.* 1997; Zengin *et al.* 2002).

Micropollutants in the wastewater are originated from raw skins and hides, and the chemicals used in the processes. The most important toxic matter is chromium. The major sources of micropollutants are: syntans particularly based on naphthalene and phenol, biocides (important agents are pentachlorophenol, 2,4,6-trichlorophenol, cresol based biocides and 4-nitrophenol), and heavy metals used in metal-complex dyes (US EPA 1982). US EPA reported a comprehensive list of micropollutans from the above sources, however the following are the toxic pollutants found in treated effluents at more than two plants above the limits of detection: 2,4,6-trichlorophenol, chloroform, 1,2-dichlorobenzene, naphthalene, 1,4-dichlorobenzene, methylene chloride, pentacholorophenol, phenol. The IPPC approach to minimise the micropollutant in wastewaters is BAT-EC substitution application.

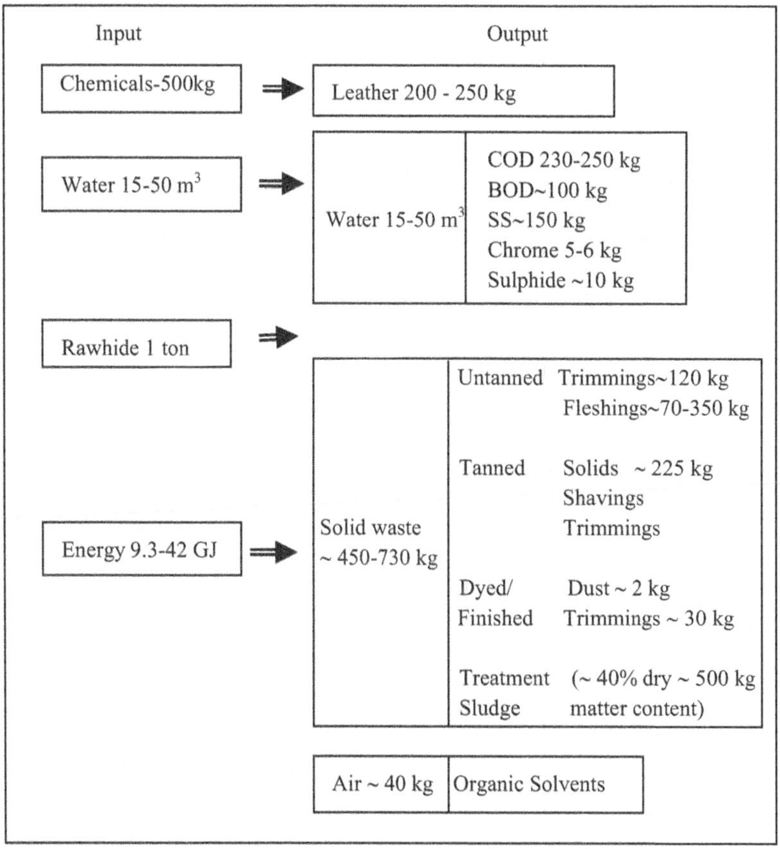

Figure 3.1 Input/output overview for a conventional (bovine chrome-tanning) process (EC 2009).

Table 3.2 Summary of pollution loads discharged in effluents from individual processing operations (EC 2009)

Process	Technology	Pollution load kg/t raw hide								
		SS	COD	BOD_5	Cr	S^{2-}	NH_3-N	TKN	Cl^-	SO_4^{2-}
Soaking	C	11–17	22–33	7–11	–	–	0.1–0.2	1–2	85–113	1–2
	A	11–17	20–25	7–9	–	–	0.1–0.2	1–2	5–10	1–2
Liming	C	53–97	79–122	28–45	–	3.9–8.7	0.4–0.5	6–8	5–15	1–2
	A	14–26	46–65	16–24	–	0.4–0.7	0.1–0.2	3–4	1–2	1–2
Delim.-	C	8–12	13–20	5–9	–	0.1–0.3	2.6–3.9	3–5	2–4	10–26
bating	A	8–12	13–20	5–9	–	0–0.1	0.2–0.4	0.6–1.5	1–2	1–2
Tanning	C	5–10	7–11	2–4	2–5	–	0.6–0.9	0.6–0.9	40–60	30–55
	A	1–2	7–11	2–4	0.05–0.1	–	0.1–0.2	0.1–0.2	20–35	10–22
Post-tanning	C	6–11	24–40	8–15	1–2	–	0.3–0.5	1–2	5–10	10–25
	A	1–2	10–12	3–5	0.1–0.4	–	0.1–0.2	0.2–0.5	3–6	4–9
Finishing	C	0–2	0–5	0–2	–	–	–	–	–	–
	A	0–2	0	0	–	–	–	–	–	–
Total	C	83–149	145–231	50–86	3–7	4–9	4–6	12–18	137–202	52–110
	A	35–61	96–133	35–51	0.15–0.5	0.4–0.8	0.6–1.2	5–8	30–55	17–37

C: conventional technique, A: advanced technique

3.1.4 In-plant control and clean technologies

Best management applications and good housekeeping practices are important and common issues to all in-plant control efforts and programs (US EPA 1982; EC 2009). Water conservation is among other common applications. General approaches for water conservation are: using batch instead of running water washes and modifying existing equipment to use short floats. Among the reuse applications, use of wash water for solution make-up, wash/soak water reuse are important yet easy practices. Another in-plant control tool is substitution of material to increase treatability of wastewater and to reduce the environmental impact. There are numerous applications in this respect in the leather tanning and finishing industry such as substitution of alkylphenol ethoxylates with alcohol ethoxylates and substitution of halogenated organic compounds which have been used in many process steps. Other practices including specific substitution of substances and application of new and emerging technologies are summarised on the process basis as follows.

Curing of leather and leather receiving operations in a tannery are important to protect leather quality and facilitate treatment. Substitution of biocides and salt, and mechanical removal of salt are the important control applications to be accounted for.

Beamhouse: Use of soaking liquor in pre-soaking and green fleshing before liming are advised as general applicable practices. However, the important control alternatives in the beamhouse are about the unhairing process. Use of hair-save process, particularly for bovine leather unhairing, is important and advantageous, since a significant organic-matter reduction can be realised through separation of undestroyed hair which may also be recovered. There are several processes developed for hair-save, the most common one is use of only alkali without the use of sulphide. Another application is low-sulphide unhairing which is based on using organic sulphur compounds including thioglycolate, mercaptoethanol, and thiourea derivatives. Enzymes and amines can be added to facilitate unhairing and reducing sulphide consumption. Recycling of sulphide liquor by filtration is another method which requires close inspection of the composition. Enzymes may be added at the end of the process to facilitate clarification. Use of solely enzymes is another application tested for a long time. Lime-free fibre opening methods using strong alkalis and enzymes are also considered. Lime splitting for heavier hides may considerably reduce the chemicals used. In-situ sulphide oxidation is another application which also facilitates the subsequent wastewater treatment operations (US EPA 1982; UN 1997; Thanikaivelan *et al.* 2005; EC 2009). Nazer *et al.* (2006) conducted an experimental study to recycle unhairing effluent four times without affecting the quality of final product, by using several reactors. Thanikaivelan *et al.* (2002) proposed a new enzyme unhairing method assisted with very low amount of sodium sulphide. Li *et al.* (2010) tested a cleaner beamhouse process involving the use of thioglycolate and urea as well as swelling agents such as sodium silicate, sodium hydroxide, guanidine, and protease enzymes. Jian *et al.* (2010) used ultrasound to accelerate the diffusion of enzymes through the skin enhancing the efficiency of enzyme unhairing. There are also methods for recovery of sulphide as elemental sulphur by chemical, microbiological electrochemical, and photocatalytic methods (Sabate *et al.* 1990; Steudel 1996; Waterston *et al.* 2007).

Tanyard: There have been several methods developed to replace the ammonia salts at the deliming step. The most common replacement method is carbon dioxide. Carbon dioxide is used with a low amount of ammonia, its reaction is slower, auxiliary agents such as organic acids are used for particularly thick leathers. Uses of boric acid and magnesium lactate in deliming have long been considered. Deliming with magnesium sulphate is another method. Use of hydrochloric acid in conjunction with sodium hydrogen carbonate is a subject of interest. Sodium hydrogen carbonate can be used as the main chemical for deliming. Lactic acid was the first organic acid used for deliming. Formic acid, acetic acid, and esters of organic acids have been used as deliming agents (US EPA 1982; UN 1997; Thanikaivelan *et al.* 2005; EC 2009). Recently developed ammonium-free bating methods utilizing new enzyme preparations are also available (UN 1997).

Salt-free/reduced pickling methods using non-swelling polymeric sulphonic acids have been developed. Pickling liquor recycling with filtration and fungicide addition was also practiced (EC 2009). Bes-Piá *et al.* (2008) studied reclamation of pickle liquor using nanofiltration. Reuse of chrome tanning liquors in the pickling baths is another reuse method. Waste control methods for degreasing include: (i) substitution of nonyl phenyl ethoxylate-based surfactants; (ii) substitution of chlorinated solvents; and (iii) use of closed-loop machines for solvent degreasing (UN 1997; EC 2009). Tanning process is a key to both quality of leather and the impact of its waste on the environment. There are several methods employed as in-plant measures to increase the process efficiency and to minimise chromium discharges. The first measure is increasing the efficiency of chrome tanning which can be realised by the optimisation of process parameters such as pH, temperature, time, and using short floats. High-exhaustion chrome tanning is an effective method for chromium control. The method can be applied either by increasing the collagen activity using aromatic dicarbon acids such as glyoxylic acid or by incorporating reactive groups into the chrome-tannin complexes (EC 2009). Another high-exhaustion method uses an aluminium based syntan together with chrome and recycles the spent solution for pickle float (closed pickle-tan loop (Sreeram and Ramasami 2003)). There are other external chromium exhaust aids such as polyamides, polycarboxylates, long chain alkanolamines, polyelectrolytes, and ion-exchange resins (Sreeram and Ramasami 2003). Combination of chrome with other tanning agents like aluminium, zirconium, titanium, or glutaraldehyde was also proved to be effective in decreasing chromium content in spent tanning solutions. Use of monoethanolamine as catalyst to increase the exhaustion of chromium was also proposed. Recycling of tanning liquor can be applied directly or after chromium separation. Direct recycle has been considered a technically viable process (Sreeram and Ramasani 2003; Thanikaivelan *et al.* 2005). Recycling of tanning liquor to pickling is another alternative, however, it may cause surface fixation of chromium. Chrome recovery and reuse is an advantageous process in terms of both economy and reducing the environmental impact. The process is well defined. It is based on precipitation of chromic hydroxide at alkali pH, separation of the chromic hydroxide solid phase, and redissolving by acidification. A three-step process involving precipitation, extraction and electro-deposition process as well as a solvent extraction technique has also been developed (UN 1997; Sreeram and Ramasami 2003; Thanikaivelan *et al.* 2005; EC 2009). Electrochemical methods have also been used for the removal and recovery of chromium from tanning liquors and leather tanning wastewaters. Sirajuddin *et al.* (2007) recovered chromium from tannery wastewater using electrolysis process with lead anode and copper cathode, and obtained efficiencies over 99 %. Ouejhani *et al.* (2008) studied electrooxidation to convert Cr (III) to Cr (VI) over a titanium-platinum anode in an electrolysis unit using tanning liquor as electrolyte for further selective recovery of Cr (VI) in tributylphosphate phase. Pretanning with aluminium salts sometimes combined with polyacrilates, glutaraldehyde derivatives, syntans, titanium salts, or colloidal silica is considered an alternative to improve chrome uptake (EC 2009). Use of organic tanning agents to produce chrome-free leathers has been considered for specific applications. Among these applications, vegetable tanning, mostly applied as Liritan process (Sreeram and Ramasami 2003), syntans, and aldehyde based tanning agents are promising. In addition, mineral tanning agents such as Al (III), Si (VI), Fe (III), Ti (IV), Zr (IV), and Ce (III) or (IV) have been studied as alternatives to chrome tanning. Some process modifications such as two-stage chrome tanning and pickless tanning were also introduced to improve chrome-tanning process (Sreeram and Ramasami 2003; Thanikaivelan *et al.* 2005; EC 2009). On the other hand, avoiding formation of hexavalent chromium during and after tanning is an important issue (EC 2009). In-plant controls applicable in the finishing stage are mostly based on process selection to increase the fixation of chemicals. Reduction in metal content can be realised with the use of high-exhaustion chrome tanning in the retanning process. Replacement of metal-complex dyes with acid dyes also helps reduction of metal content of wastewaters. The organic matter discharges can be reduced by providing high-exhaustion of chemicals in post-tanning operation. High-exhaustion of dyes using dyeing auxiliaries, use of high affinity retanning chemicals, use of low phenol and low formaldehyde syntans, and high-exhaustion of fatliquors by the use of amphoteric polymers are among the applicable alternatives (EC 2009).

In addition to the aforementioned in-plant controls and process modifications, there are recent developments and new process approaches. Biocatalytic fibre opening, three-step tanning technique (Thanikaivelan *et al.* 2005), use of ultrasound technology to facilitate the leather processing (Sivakumar *et al.* 2009), reversal of leather processing sequence (Saravanabhavan *et al.* 2006) and new approaches to cleaner production are among these developments.

3.1.5 Wastewater treatment

The first step in proper and secure treatment of leather tanning wastewaters is stream segregation. At least two waste streams, sulphide containing wastewater and chromium containing wastewater are segregated, pretreated and then combined with other wastewaters. Combination of flows without stream segregation is possible but generally not preferred due to difficulties of sulphide and chromium removal as well as the risk of H_2S evolution. Sulphide containing wastewater includes unhairing and its washing(s). Chromium containing wastewaters generally include only tanning bath, and retanning liquor is discharged with other wastewaters. Depending on treatment/recovery applications, segregation of other streams is realised. A general wastewater treatment scheme is given in Figure 3.2 (US EPA 1982). Unhairing waste stream is first undergone sulphide removal. Sulphide can be removed conventionally either by iron sulphide precipitation (mostly in combined wastewaters) or sulphide oxidation. Sulphide oxidation is the method of choice since no sulphide containing sludge is produced by the application of this method. Sulphide oxidation methods will be explained in detail in Subsection 3.2.1. Sulphide recovery methods have also been briefly described in in-plant control and sulphide oxidation subsections. Carbonation of unhairing wastewaters using carbon dioxide, usually from stack gas, although an additional step, may be useful to remove excess alkalinity and suspended solids. The process, particularly with the aid of coagulants removes organic matter. US EPA (1982) reported 84 % BOD_5 removal with flue gas carbonation. Tünay *et al.* (1994) found that carbonation of segregated and sulphide-oxidised unhairing liquor with carbon dioxide and using $FeCl_3$ as coagulant provided 20 % COD removal. Protein removal by acidification may follow carbonation. US EPA reported

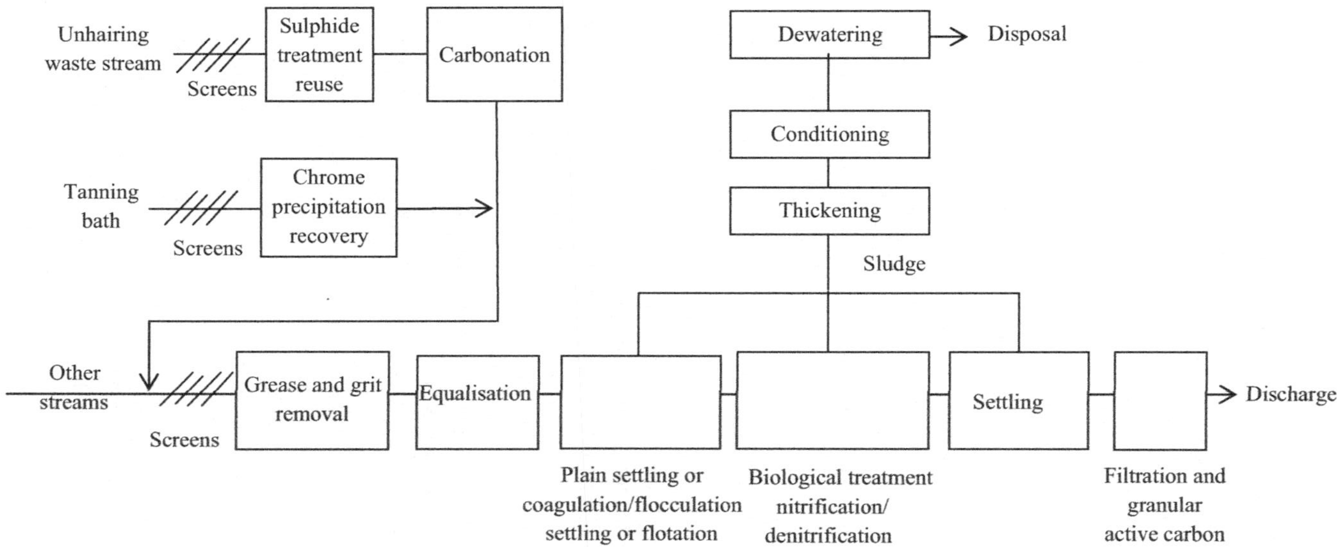

Figure 3.2 A general wastewater treatment scheme for leather tanning industry wastewaters (US EPA 1982).

COD and TKN removals as high as 84 and 86 %, respectively (US EPA 1982). Tünay *et al.* (1994) applied protein removal by acidification and using $FeCl_3$ and anionic polyelectrolyte for coagulation, and obtained around 20 % removal for both COD and TKN. Following combination of pretreated chromium and sulphide bearing wastewaters with other waste streams, a preliminary treatment involving grease and grit removal and an aerated equalisation, a plain settling or a settling after coagulation flocculation is employed. The efficiency for plain settling with the use of small amount of anionic polymer and/or pH adjustment is high and between 84–90 % for suspended solids and 60–67 % for BOD_5 (US EPA 1982). Chemical coagulation followed by sedimentation is a common application, and alum, lime, ferric chloride, and poly aluminium chloride can be used in the process. Alum provided high BOD_5 and suspended solid removals of over 90 % (US EPA 1982). Tünay *et al.* (1994) found around 50 % COD removals using $FeCl_3$ as coagulant in an application to tannery wastewater. Haydar and Aziz (2009) used alum for chemical coagulation of leather tanning effluent, and obtained 53 and 96 % removals for COD and suspended solids, respectively.

Biological treatment follows plain settling or coagulation settling, and is the main unit for organic matter removal. Aerobic biological processes are preferred and although biological filters, rotating biological contactors, and aerated lagoons can be employed, the activated sludge process is the most common biological process. The efficiency of BOD removal with extended aeration activated sludge system is high and over 90 %. However, due to the composition of tannery effluent COD removal never approaches to completion. Evaluation of COD fractions of tannery effluent indicated that soluble inert and particulate inert COD fractions are 9 and 13 % of the total COD, respectively (Doğruel *et al.* 2006). Therefore, depending on the strength of tannery wastewater biological treatment effluent may not satisfy COD limits of 200–300 mg/L defined for direct discharge. That is why several stages of polishing treatment were accounted for as additional COD removal steps.

Another important function of biological treatment is the removal of nitrogen by nitrification-denitrification process. Biological nitrogen removal of tannery wastewater is applicable but it is costly and nitrogen removal steps are sensitive to both environmental conditions such as temperature and common components of the tanning wastewater such as heavy metals. Zengin *et al.* (2002) proposed the use of magnesium ammonium phosphate (MAP) precipitation for nitrogen removal and recovery in leather tanning wastewaters. They have made experimental assessment of the applicability of MAP precipitation as both source-based nitrogen removal and recovery, and nitrogen removal from combined wastewaters. Both approaches proved promising. In the application to combined wastewaters nitrogen levels were reduced to nutrient requirement of aerobic biological treatment (Kabdaşlı *et al.* 2002).

Polishing treatment for additional COD removal includes upgrading biological treatment with activated carbon, multimedia filtration of biological treatment effluent and use of granular activated carbon columns after filtration. Application of membrane filtration may provide another alternative (US EPA 1982; Purkait *et al.* 2009). However, recently chemical oxidation methods have been successfully applied to leather tanning wastewater not only as a polishing treatment but also as a means of organic matter control and enhancement in the biodegradability. These applications will be reviewed and discussed in the next subsection.

3.2 OXIDATION PROCESSES

3.2.1 Sulphide oxidation

3.2.1.1 Oxidation of sulphide by air oxygen

Catalytic oxidation of sulphide, or more precisely reduced sulphur species, by dissolved oxygen is a common process for industrial wastewaters such as tanneries, textile, and pulp and paper. The theory of the process, in fact, is sophisticated due to a number of sulphur species involved in the process and their complex interactions through the course of reactions. Therefore, the application is rather on an empirical basis. However, both theoretical and experimental studies conducted in the field so far provided an insight into the process and defined a framework on which some of the design and operation criteria are based. Aqueous hydrogen sulphide (H_2S) is a diprotic weak acid. First stage dissociation constant of the acid (pK_1) is 7.0 at 25°C (O'Brien and Birkner 1977), however there are different values given for the second stage dissociation constant (pK_2) and pK_2 ranges from 11.96 to 17–19 at 25°C (Fischer *et al.* 1997).

Then, H_2S is dominant only below pH 6, while sulphide (S^{2-}) ion is only accounted for above pH 12 or higher. Between these pH values (6–12) only HS^-, bisulphide or sometimes named as monosulphide ion, is of importance, and it is the main reacting species with dissolved oxygen through the sulphide oxidation process. The main reactions with oxygen can be given as follows (Nielsen *et al.* 2003):

$$2HS^- + O_2 \rightarrow 2S^0 + 2OH^- \tag{3.1}$$

$$2HS^- + 2O_2 \rightarrow S_2O_3^{2-} + H_2O \tag{3.2}$$

$$2HS^- + 3O_2 \rightarrow 2SO_3^{2-} + 2H^+ \tag{3.3}$$

$$2HS^- + 4O_2 \rightarrow 2SO_4^{2-} + 2H^+ \tag{3.4}$$

The mechanism of the process has not been fully understood, although, there have been several approaches to explain the reaction pathways. Abel (1956) proposed a stepwise reaction sequence involving the formation of species HSO_2^- as an intermediate. Chen and Morris (1970) considered the sulphide oxidation reactions as a chain reaction. O'Brien and Birkner (1977) proposed an overall model consisting of a series of parallel reactions to form the major reaction products; sulphite, thiosulphate and sulphate. Hoffmann and Lim (1979) in their analysis of catalytic sulphide oxidation postulated reaction steps yielding sulphate, thiosulphate, and polysulphides beginning with HSO_2^- species. The authors considered a free radical pathway in the overall reaction unlikely based on the fact that no dithionate and tetrathionate were found as products. Boyce *et al.* (1983) also supported the heterolytic mechanism and did not consider free radical intermediates based on the reasoning that no dithionate was formed. Kotronarou and Hoffman (1991a) suggested free radical formation reactions initiated by $HS^\bullet$ for catalytic sulphide oxidation. Steudel (1996) proposed also a $HS^\bullet$ initiated reaction mechanism in which $HS^\bullet$ or $S^{\bullet -}$ triggered the formation of polysulphur compounds for aqueous sulphide oxidation catalysed by transition-metal ions. Fischer *et al.* (1997) suggested a reaction mechanism for the oxidation of aqueous sulphide solution by oxygen based on disulphide as a primary intermediate and conducted experimental studies at pH values 9 and 14 to verify the reaction scheme. They proposed that the oxidation reactions were realised through the formation of polysulphides.

Polysulphide species are formed when neutral sulphur atoms combine with monosulphide species as presented by the general reaction:

$$HS^- + (x-1)S^0 \rightleftharpoons S_x^{2-} + H^+ \tag{3.5}$$

General formula for polysulphides are H_2S_x, HS_x^- and S_x^{2-} where $x = 2–5$. In neutral and slightly alkaline solutions tetrasulphide and pentasulphide are the predominant forms (O'Brien and Birkner 1977). Kleinjan *et al.* (2005) found that in the presence of excess amount of sulphur, the average polysulphide chain length x, was 4.91 (21°C).

Comprehensive studies were conducted on polysulphide decomposition, equilibrium and oxidation reactions which helped the understanding of the role of polysulphide reactions through sulphide oxidation (Licht and Davis 1997; Kamyshny *et al.* 2004; 2007; Kleinjan *et al.* 2005; Filpponen *et al.* 2006). Kleinjan *et al.* (2005) studied the rate of polysulphide oxidation, and found the reaction as first order with respect to polysulphide concentration between pH values 7 and 12. Oxidation products were determined to be sulphide and thiosulphate between pH values 7 and 9. While sulphide oxidation was interfered with polysulphide oxidation at pH values less than 7, more than stoichiometric amount of sulphide and thiosulphate were formed at pH values greater than or equal to 10 as a result of hydrolysis of elemental sulphur with hydroxide ions. The developments in the field summarised above indicated that polysulphide formation and oxidation played a key role in the sulphide oxidation by oxygen and brought explanation several important phenomena observed in the process such as induction period, effect of pH, and the changes in the rate of oxidation. Sulphide oxidation by dissolved oxygen is highly susceptible to catalysis as well as to inhibition. Chen and Morris (1970) found catalytic activity of metals in the order of $Ni^{2+} > Co^{2+} > Mn^{2+} > Cu^{2+} > Fe^{2+} > Ca^{2+}$ or Mg^{2+}, and stated that 0.1 mM Ni^{2+} caused a decrease in the time for oxidation of 0.01 M sulphide at pH 8.65 from several days to a few minutes. Although, sometimes as high as several hundred mg/L metal catalyst doses arc required. Martin and Rubin (1972) used iron, copper, manganese, and nickel salts all being equivalent to 1 mg/L as manganese in solution at pH 11.2 for air oxidation of sulphide both in distilled water and in an industrial wastewater, and 70 to 96 % sulphide oxidation levels were achieved in 37 to 130 minutes. Theoretical analysis of metal catalysts were made using transition metal–4,4′,4′′,4′′′ tetrasulphophthalocyanine complexes for Fe^{2+}, Mn^{2+}, Co^{2+}, Ni^{2+}, Cu^{2+}, and V^{4+} for sulphide oxidation. The oxidation was found to proceed via an activated complex that bound O_2 and S_3^{2-} and the rate of oxidation could be analyzed using Michaelis-Menten kinetic model (Hoffmann and Lim 1979; Boyce *et al.* 1983; Kotronarou and Hoffmann 1991a). Dalai *et al.* (1999) used activated carbon catalyst for sulphide oxidation. In addition to metals, some organic compounds such as phenolics accelerated the oxidation reaction while complex formers such as ethylene diamine tetraacetic acid (EDTA) inhibited the oxidation (Chen and Morris 1970). Use of catalysts in wastewater needs to be carefully evaluated considering the wastewater composition and suspended solids content and their interactions with catalysts as well as with the oxidation process.

The pH is an important variable in sulphide oxidation. Generally, the rate of oxidation increases with increasing pH up to 8. Above this pH the reactions and pH variation are more complex and depend on many factors such as HS^- ionisation, sulphide to oxygen ratio, and reaction intermediates and products. Chen and Morris (1970) in their experiments found that after pH 8–8.5 the rate of sulphide oxidation decreased sharply until a minimum at pH 9 and then, increased to a second maximum about equal in magnitude to the one at pH 8–8.5 near pH 11 and then decreased slowly in more alkaline solutions. However, Millero *et al.* (1987) found the rate independent of pH over the pH value of 8.0. The initial $[S(-II)]_T/[O_2]$ ratio affects the reaction rate and products. O'Brien and Birkner (1977) suggested that initial production of sulphite increased as oxygen became the limiting reactant. Chen and Morris (1970) concluded that sulphur precipitation was favoured by a high $[S(-II)]_T/[O_2]$ ratio. In acid to neutral solutions sulphur and sulphate are major products. In alkaline solutions between pH 7.5–11 the main products are sulphite, thiosulphate and sulphate. Among them, thiosulphate is the major product, particularly, in higher pH values and it is not oxidised further for a long period of time (O'Brien and Birkner 1977).

The rate of sulphide oxidation increases with increasing temperature. Millero *et al.* (1987) reported activation energy of 57 kJ/mol which corresponded to 2.2 times increase in reaction rate as the temperature rose from 10°C to 20°C. They provided equations for calculation of rate constant with varying temperatures for both water and seawater. The rate of sulphide oxidation also increases significantly with increasing ionic strength (O'Brien and Birkner 1977; Millero *et al.* 1987; Tünay *et al.* 1998). Millero *et al.* (1987) provided following equation that is valid over a pH range from 4 to 8, a temperature from 5–65°C and an ionic strength from 0 to 6 M:

$$\log \mathrm{k} = 10.50 + 0.16\,\mathrm{pH} - (3.0 \times 10^3)/\mathrm{T} + 0.44\,\mathrm{I}^{1/2} \tag{3.6}$$

where k is oxidation rate constant, T is temperature (K), and I is ionic strength of solution (M).

The kinetics of sulphide oxidation by oxygen has been given in the form:

$$-\mathrm{d}\left(\mathrm{S}(-\mathrm{II})_\mathrm{T}\right)/\mathrm{dt} = \mathrm{k}\left[\mathrm{S}(-\mathrm{II})_\mathrm{T}\right]^\alpha [\mathrm{DO}]^\beta \tag{3.7}$$

where $S(-II)_T$ is total sulphide (total of H_2S species) concentration, (mg/L); DO is dissolved oxygen concentration, (mg/L); k is the reaction rate constant, ($[mg/L]^{1-\alpha}[mgO_2/L]^{-\beta}\,h^{-1}$); α and β are reaction orders with respect to sulphide and oxygen concentration, respectively (Chen and Morris 1970; Almgren and Hagström 1974; O'Brien and Birkner 1977; Millero *et al.* 1987; Nielsen *et al.* 2003). An alternative approach to the sulphide oxidation is the use of bisubstrate Michaelis-Menten rate law (Hoffmann and Lim 1979; Wilmot *et al.* 1988). However, the use of Eq. (3.7) is more common and there have been several studies to determine the order of reaction. Chen and Morris (1970) found the overall reaction order as 1.9, with the orders of 1.34 and 0.56 with respect to total sulphide and oxygen, respectively. O'Brien and Birkner (1977) found 1.02 for the order with respect to reduced sulphur and 0.8 with respect to oxygen. Buismann *et al.* (1990) reported values 0.597 and 0.642 with respect to total sulphide and oxygen, respectively. Millero *et al.* (1987) assumed reaction order to one for both total sulphide and oxygen. The variation in the given values is due to varying experimental conditions.

Experimental studies for catalytic sulphide oxidation that have been conducted on industrial wastewaters are rather limited. Pascke *et al.* (1977) reported 90 % sulphide removal by catalytic air oxidation from a cellulose sponge-making wastewater within 1–2 hours at pH 8.5–9.0 with 9.9 m^3/minute air feeding and using 100 mg/L $MnSO_4$ dose. Ölmez *et al.* (2000) applied catalytic air oxidation to indigo dyeing spent bath containing 8660 mg/L sulphide and 11,560 mg/L sulphite. $MnSO_4$ was used as catalyst at 200 mg/L dose, and pH was maintained at 11.09 using phosphate buffer and NaOH addition. 24 hour aeration provided almost total sulphide removal and sulphite concentration reduced to 1875 mg/L.

Nacheva (1989) conducted catalytic sulphide oxidation using $MnCl_2$ as catalyst on Na_2S solution which was simulated unhairing wastewaters with initial sulphide concentrations 680–1900 mg/L (by H_2S) and pH range 12-12.6. Aeration was provided at the rate of 64 m^3/m^2.h. As the catalyst dose as Mn to S ratio varied between 0.06–0.131 within 3 hours sulphide concentration decreased down to around 50 mg/L which could not be further reduced as the aeration was extended to 7 hours. Residual sulphide level could be reduced to half the 50 mg/L level by increasing the Mn to S ratio to 0.31 and over. In all experiments, pH was reduced to levels 10.5–11 after one hour and by the end of tests reached about neutral conditions. In all cases polysulphide appearance was observed.

Germann (1995) summarized the experimental studies conducted on sulphide containing wastewater of tanneries. In all trials, membrane aeration providing 22–44 m^3/h air and with 50 mg/L $MnSO_4$ as catalyst, initial sulphide levels of 231–462 mg/L were reduced to 2–6 mg/L in 7 hours, however reduction of sulphide to 2–3 mg/L required 24 hours. In another experiment, 800 mg/L initial sulphide was reduced to 10–40 mg/L in one hour, further reduction to 5–10 mg/L required additional 7 hours.

Tünay and co-workers (1998) conducted several studies on actual tannery wastewaters for the assessment of catalytic air oxidation of sulphide. An experimental study was conducted on liming liquor with initial sulphide concentration of 1950 mg/L. The pH of the sample was 12. The experiment conducted without catalyst yielded 265 mg/L sulphide at the end of 9 hours. Another experiment was conducted with the addition of 100 mg/L $MnSO_4$ catalyst and 100 mg/L sulphide concentration remained at the end of 9 hours. In the experiment with catalyst initial reaction rate was faster than that of the system without catalyst, up to 6 hours. However, following 6 hours the oxidation rate diminished and reaction rates of the two systems became close to one another. The authors pointed out that a single rate coefficient could not adequately describe the whole sequence of the reactions (Tünay *et al.* 1998). Another study was realised on total wastewaters of a group of tanneries. A wastewater sample having 100 mg/L initial sulphide was aerated with the addition of 100 mg/L $MnSO_4$ for 6 hours and final sulphide concentration was found 25 mg/L (Kabdaşlı *et al.* 1999). Another sample containing 56 mg/l sulphide was again aerated for 6 hours with the addition of 200 mg/L $MnSO_4$ and 24 mg/L sulphide concentration was achieved at the end of the experiment indicating that lower reaction rate was compensated by the increase in the catalyst concentration (Kabdaşlı *et al.* 1999). In a study segregated beamhouse wastewaters containing 2670 mg/L sulphide was air oxidised using 200 mg/L $MnSO_4$ as catalyst. In 6 hours, almost complete oxidation was obtained (Tünay *et al.* 1994).

A high oxidation rate which enabled to eliminate 2500 mg/L initial sulphide from limehouse using 0.05 to 0.1 Mn/S ratios was reported by US EPA (1982). In a study unhairing liquor of a tannery containing 230 mg/L sulphide was totally air-oxidised using $MnSO_4$ as catalyst at Mn/S ratio of 0.72 (US EPA 1982).

Sulphide oxidation of tannery wastewater, although, there are some applications for the treatment of combined wastewaters, is applied to segregated wastewater which includes unhairing operation and its subsequent washing discharges. Four important design parameters are; pH, oxidation duration, air and mixing requirements, and catalyst addition.

pH of the segregated sulphide containing flow of tanneries is usually over 11 and this pH is proper for the process as far as the rate of process is concerned. However, pH reduction and loss of alkalinity are frequent observations (US EPA 1982; Tünay *et al.* 1998; Nacheva 1989). This is particularly important for wastewaters having relatively low pH of below 10. In such conditions, another operating problem of H_2S stripping may also be faced (Tünay *et al.* 1998; Nacheva 1989). Therefore, pH monitoring and a means of alkalinity addition should be considered as a part of design of sulphide oxidation systems using air oxygen.

As related literature and design experience indicated, reaction time of air oxidation of sulphide is variable depending on wastewater character and system design parameters. There are oxidation times as low as 2–4 hours as well as a long duration of 24 hours as reported by US EPA (1982). However, majority of the design values as reported by US EPA (1982) is 7–8 hours of oxidation times. These oxidation durations are also consistent with literature data, particularly, those conducted on tannery wastewaters (US EPA 1982; Tünay *et al.* 1994; 1998). Therefore, as a general guidance a minimum of 8 hours oxidation time can be accepted to ensure complete sulphide oxidation.

Air and mixing requirements are key parameters to a successful operation of sulphide oxidation systems. As the literature information and process kinetics indicated, both the rate and products of reaction is a function of dissolved oxygen level in the oxidation reactor. On the other hand, mixing is quite important for the transport and functioning of both oxygen and catalyst. Mechanical mixing is also advantageous to cope with foaming which is a frequent and acute operation problem. Thiosulphate is considered as the primary end product of sulphide oxidation. Oxygen requirement for conversion of sulphide to thiosulphate is 1 kg of oxygen per kilogram of sulphide oxidised. Thiosulphate is a stable end product and it is not further oxidised during the course of reaction. However, as reported in the literature, other products of oxidation mainly sulphite and sulphate may be formed in alkaline conditions. Conversion of 1 kg sulphide to sulphite and sulphate requires 1.5 and 2.0 kg oxygen, respectively. There is no basis of prediction for the ratios of these products in the literature. Therefore, on the basis of oxygen requirement the figure selected may be close to 1.5 kg oxygen per kilogram of sulphide to ensure complete oxidation. Mixing of sulphide oxidation tanks is usually provided with the aeration system. For diffused aeration a design value 18.3 $m^3/m^2.h$ is given in terms of air feed, but the values given as power requirement vary in a great range of 4–137 kW/1000 m^3. For mechanical aerators which are assumed to be more efficient for mixing the design values are between 34–47 kW/1000 m^3 aeration volumes (US EPA 1982).

Sulphide oxidation by oxygen requires the presence of catalyst to achieve applicable rates of reaction. The catalysts being used are mainly transition metal ions of which cobalt and nickel have been shown to be quite effective. However, in the practice, these ions are avoided due to their inhibitory effect on subsequent biological treatment and adverse effects on the environment as well. Manganese salts are both effective and cheap, therefore, they are of the most common use. Potassium permanganate ($KMnO_4$), although, claimed to be more effective than Mn(II) salts in lower doses, in some cases, its dose is needed to be increased up to 750 mg/L and it is also a very expensive material with respect to Mn(II) salts. Manganese chloride ($MnCl_2$) and manganese sulphate ($MnSO_4$) are presently applied commonly, in varying doses. In the practice, doses vary between 30–300 mg/L as $MnSO_4$ or 0.05–0.72 as Mn/S ratio (US EPA 1982; Tünay *et al.* 1998; Nacheva 1989). Since the effectiveness of the catalyst depends on wastewater characteristics and suspended matter content, the right dose needs to be assessed on a case by case basis.

Catalytic sulphide oxidation is usually carried out in batch systems. Therefore, use of two aeration tanks instead of one tank provides a great flexibility but increases the capital cost. Capital cost of the treatment can be evaluated considering the above-discussed design criteria. The main items of the system are; concrete tank for the reactor(s), a storage and pumping facility, floating mechanical aerator(s), effluent pumps, water and electricity supply and electrical control equipment. Sometimes a neutralisation facility may be required.

Operation cost involves the main elements of power supply for aerator(s) and pumping, and use of catalyst. Based on the figures 39.5 kW/1000 m^3 aerator, 8 hours detention time, two effluent pumps (one spare), 0.4 kg Mn/kg S^{2-} catalyst dose as $MnSO_4$ and neutralisation with H_2SO_4 (if required), US EPA (1982) has given graphs showing cost versus daily flow (operation cost was based on operation 260 days per year). Capital cost calculated by the use of the graph is €75, 000 for wastewater flow of 100 m^3/day.

3.2.1.2 Sulphide oxidation by hydrogen peroxide

Hydrogen peroxide is an effective oxidant for sulphide. It reacts with sulphide to yield elemental sulphur or sulphate depending on pH (Cadena and Peters 1988):

$$H_2S + H_2O_2 \rightarrow 2H_2O + S^0 \qquad pH < 7.5 \qquad (3.8)$$

$$H_2S + 4H_2O_2 \rightarrow SO_4^{2-} + 4H_2O + 2H^+ \qquad pH > 7.5 \qquad (3.9)$$

Cadena and Peters (1988) reported that in wastewaters hydrogen peroxide first reacted with bacterial catalase producing oxygen and water, and sulphide was oxidised by oxygen. Hoffmann (1977) proposed that hydrogen peroxide and sulphide reaction proceeded via a nucleophilic displacement by sulphide on hydrogen peroxide with the formation of polysulphides, and reaction order was first order with respect to both hydrogen peroxide and sulphide. Millero *et al.* (1989) reported that at concentrations of $H_2O_2 > 10^{-5}$ M the oxidation of H_2O_2 with H_2S became competitive with oxygen. They proposed that the reaction involved several elementary steps involving one-electron transfers. They also stated that the reaction rate was independent of ionic strength (0–6 M) at pH 8, the effect of temperature was slight in that between 10°C and 20°C the rate constant changed only about 1.6 %. The reaction rate was first order with respect to both H_2S and H_2O_2. The reaction rate increased with pH 2 to 7.5 and decreased between 7.5 and 13. The increase up to pH 7.5 was due to ionisation of H_2S because HS^- species was more reactive than H_2S. The decrease above pH 7.5 was caused by formation of HO_2^- from H_2O_2 ionisation which reacted by HS^- slowly. Ahmad *et al.* (2009) used ferric oxide catalyst for hydrogen peroxide oxidation of aqueous sulphide at pH 11.2, and found the reaction rate was proportional to the catalyst and H_2O_2 concentrations. The rate was found to be higher at lower sulphide concentrations. Since oxidation to sulphate requires four times more hydrogen peroxide than oxidation to elemental sulphur (S^0), reducing the pH below 7.5 is preferred in practice (US EPA 1982). 20–30 minutes of reaction times are required for a completed reaction (Cadena and Peters 1988).

The advantages of the process are small reactor volumes and feeding equipment. However, hydrogen peroxide is expensive. Therefore, hydrogen peroxide oxidation of sulphide is mostly used as a polishing step for low sulphide concentrations (US EPA 1982). Germann (1995) reported that in a tannery wastewater treatment after a rapid reduction of sulphide by catalytic air oxidation down to 10–40 mg/L in one hour, hydrogen peroxide was used to comply with the limit of 2 mg/L sulphide. Since hydrogen peroxide reacts with other substances present in wastewater, the required amount may be greater than stoichiometric dose.

In the study of Germann (1995), 0.5 L hydrogen peroxide per cubic meter wastewater was used to remove 10 mg/L sulphide. US EPA (1982) reported that in a tannery wastewater sulphide concentration of 60 mg/L was reduced to zero by treatment with sulphuric acid and 100 mg/L hydrogen peroxide. Chemical costs, including acid, were approximately € 7.5 per m^3 of wastewater treated.

3.2.1.3 Electrochemical oxidation

Electrochemical methods have long been known, though, recently they have been employed for the treatment of a great variety of wastewaters, particularly industrial effluents (Murugananthan *et al.* 2004; Khudaish and Al-Hinai 2006; Dutta *et al.* 2008; Särkka *et al.* 2009).

Särkka *et al.* (2009) applied electrochemical oxidation to sulphide removal from a paper mill wastewater. In this study, 4–32 mg/L sulphide concentration in the wastewater was removed in an electrolysis unit using mixed-metal electrodes. They explained the mechanism as oxidation by the electrochemically generated oxygen in the system. Murugananthan *et al.* (2004) studied removal of sulphide, sulphate and sulphite ions by electrocoagulation process. Use of the titanium electrodes provided oxidation of sulphide to elemental sulphur and sulphate, while aluminium and iron electrodes produced metal sulphides which have been removed from the water by coagulation-flotation. Khudaish and Al-Hinai (2006) applied vanadium pentaoxide as a deposited film at the glassy carbon electrode to promote the electrochemical oxidation of aqueous sulphide at pH 10.28. The product of oxidation was elemental sulphur which deposited on the electrode. Dutta *et al.* (2008) used a fuel cell made up of graphite anode and cathode to electrochemically oxidise sulphide while producing energy. Elemental sulphur was identified as the predominant final oxidation product that was deposited on the anode. The sulphide oxidation rate was not affected by the presence of an organic electron donor. Şengil *et al.* (2009) experimentally investigated the removal of COD, sulphide and oil and grease from tannery liming drum wastewater by electrocoagulation using mild steel and aluminium electrodes. Elemental sulphur was formed at anodic sites as the steel electrodes were used. Aluminium electrodes were not as effective as steel electrodes. They determined

90 % sulphide removal at pH 3 with mean energy consumption of 0.5–24 kWh/m^3. Waterston *et al.* (2007) used boron-doped diamond anode and graphite cathode to electrochemically treat sulphide in a synthetically prepared sample. Up to 60 mM sulphide was totally oxidised within 5.5 hours. The product was sulphate. Reaction kinetics was first order with respect to current density and zero order with respect to sulphide concentration.

3.2.1.4 Other sulphide oxidation methods

There are other methods for sulphide oxidation. Most of these methods have been tested at lab-scale. Some methods seem to have application potential. These methods are summarized below.

Ozone is a powerful oxidant and has a great variety of application in both water and wastewater treatment. There are several applications of ozone in leather tanning industry wastewaters such as inert COD removal, oxidation prior to biological treatment to increase the biodegradability of wastewaters. There is a study carried out at lab-scale to test the applicability of ozonation for sulphide oxidation in the leather tanning industry. Bouzid *et al.* (2008) applied ozone to a liming float wastewater stream. Initial 1317 mg/L sulphide concentration of the wastewater was eliminated by ozonation in two hours with an ozone dose of 3.5 gram ozone per gram of sulphide. Thiosulphate and elemental sulphur were intermediate products which were finally converted to sulphate. In the application COD removal was 16 % and BOD_5/COD ratio increased from 0.22 to 0.36. Cadena and Peters (1988) studied chlorine and $KMnO_4$ to oxidise sulphide in domestic wastewater having sulphide concentration of 1.4 mg/L. Complete sulphide oxidation was realised in 5 minutes for both oxidants. The doses required for total oxidation were 7.1 and 12.5 gram oxidant per gram of sulphide for chlorine and $KMnO_4$, respectively. Betterton and Hoffmann (1990) used peroxymonosulphate (Oxone®) for the oxidation of sulphide in synthetically prepared sample at 10^{-5} M concentration. Oxidation completed in 120 seconds and produced sulphate at pH 8 and at high $[HSO_5^-]/[S(-II)]$ ratios, and a mixture of sulphate and elemental sulphur at low $[HSO_5^-]/[S(-II)]$ ratios and at pH 4. Kotronarou and Hoffmann (1991b) found that in wastewater and at neutral pH sulphide could be oxidised with peroxymonosulphate at $[HSO_5^-]/[S(-II)]$ dose of 2.1 in 2 minutes of reaction time. The reaction was found to be faster than hydrogen peroxide oxidation of sulphide. Valeika *et al.* (2006) used manganese (IV) oxide for the oxidation of sulphide in tannery liming liquor. Sulphide oxidation was realised in 80–120 minutes at pH 9 and 13, respectively. The dose of MnO_2 was 0.2–0.5% of solution mass. Optimum temperature was 30°C. The authors proposed a reaction mechanism. Sharma *et al.* (1997) applied ferrate (VI) oxidation for sulphide removal. The dose was found to be $[H_2S]_T$

/ [Fe (VI)] = 1.51 at pH 7 and 2.5 at pH 9 and 11.3. At pH 7, the main product was thiosulphate at higher pH values a mixture of thiosulphate, sulphite, and sulphate were formed. Kotronarou *et al.* (1992) applied ultrasonic irradiation to oxidise sulphide in synthetic solution. Initial sulphide concentration of 200 μM was rapidly oxidised by sonolysis. At pH 10, reaction of $HS^{\bullet}$ and $HO^{\bullet}$ was the mean pathway while at pH $\leq$ 8.5 thermal decomposition of H_2S became important. Linkous *et al.* 2004 realised UV photochemical oxidation of aqueous sodium sulphide. The products of reaction were hydrogen and sulphur. The reaction rate was independent of pH within the range 8.5–13.3.

3.2.2 Oxidation applications to raw wastewater

3.2.2.1 Electrochemical oxidation and electrocoagulation

Electrochemical processes, as explained in Subsection 1.3, are quite effective and versatile. Due to several mechanisms acting simultaneously in these processes several pollutant parameters having different nature and demanding different means for their removal can be handled properly. Therefore, electrochemical processes have, particularly in the last decade, become quite common, at least as a subject of research at laboratory scale and as small scale applications. Leather tanning wastewaters with their strong character and bearing different kind of pollutants are good candidates for the application of electrochemical techniques. One of the first studies in this field was carried out by Vlyssides and Israilides (1997) using an electrolysis reactor made up of stainless steel cathode and platinium alloy covered titanium alloy anode. The wastewater being treated in the electrolysis was a typical strong tannery wastewater containing high level organics, suspended solids, chromium, and sulphide. Chloride content of the wastewater was also high and 18,500 mg/L. Working pH and temperature of the system was kept constant at 9.0 and 45°C, respectively, during the experiments. Electrolysis was applied for 3 hour duration, however within the first 30 minutes complete removal of sulphide and chromium together with 95 % reduction in phenolics, 65 % reduction in ammonia, and 52 % COD removal were realised. BOD_5/COD ratio was increased from 0.2 to 0.27. At the end of three hours, COD removal reached 84 % and BOD_5/COD ratio increased to 0.44. The energy consumption was presented as 4.8 to 11 kWh per kg COD removed for first 10 minutes then increased first 58 then a high number of 200 kWh per kg COD removed. Szpyrkowic *et al.* (1995) applied electrochemical process to medium strength raw tannery wastewater with COD values between 1774–3200 mg/L and chloride concentrations between 2560–4500 mg/L. There was no sulphide in the wastewater. They used Ti/Pt and Ti/Pt/Ir electrodes. Current densities were 3–4 A/dm^2. In 45 to 60 minutes, with Ti/Pt electrodes and working pH values between 6.5–7.5, COD and TKN removals were 46–70 and 47–85 %, respectively. Only 10 % COD removal was realised with Ti/Pt/Ir electrodes in 30 minutes. Energy consumptions were given as 39.8–46.9 kWh/m^3 for Ti/Pt electrodes and 13.4 kWh/m^3 for Ti/Pt/Ir electrodes. Costa *et al.* (2008) studied treatment of tannery wastewater sampled from the equalisation tank of a leather plant processing wet-blue leathers in an electrolysis unit running at 20 mA/cm^2 current density with dimensionally stable anodes containing tin, iridium, ruthenium and titanium. Chloride content of wastewater was 787 mg/L and pH was 4.0. Although, total phenolic compounds and UV-vis absorbance were greatly removed, TOC removal was only 40.5 % after 5 hours of electrolysis time at 100 m A/cm^2. However, trivalent chromium was oxidised to hexavalent chromium through the process which was overcome by the addition of sodium sulphite. Espinoza-Quinones *et al.* (2009) electrochemically treated raw tannery effluent having a COD concentration over 17 g/L but no sulphide, using aluminium electrodes and the results of their experiments were optimised using ANOVA technique. The results were also compared to a previous study carried out with iron electrode. Under optimum operating conditions of 45 minutes electrolysis time, 68 mA/cm^2 current density, and initial pH range 6.5–8.5, COD removal was at the level of 80 % together with 96 % chromium and high level of turbidity removals. Operation cost was found accounting for energy and electrode material expenses depending on the applied current density. For 68 mA/cm^2 current density the cost was €1.42/m^3 which was just half of the that calculated value of conventional alum coagulation. Kurt *et al.* (2007) applied electro-Fenton process to raw tannery wastewater having 2810 mg/L COD, 62 mg/L chromium and 89 mg/L sulphide, at pH 3.0, 5.0 and 7.2 (original wastewater) with varying doses of H_2O_2. The best result was obtained at pH 3.0 with 1670 mg/L H_2O_2 dose in 12 minutes as 73 % COD removal. The corresponding energy consumption was 3.1 kWh/kg COD removed. Sulphide removal was almost complete in 10 minutes.

3.2.2.2 Ozone oxidation

Raw wastewaters from beamhouse and tanyard were subjected to ozonation to improve biological degradation characteristics. The optimal specific ozone consumption was found as 2 g O_3/g DOC for both batch and continuous experiments. The ozonated effluents proved to be biologically degradable with full biological nitrification (Jochimsen *et al.* 1997). Preethi *et al.* (2009) studied ozonation of raw tannery wastewater and found that up to 70 % COD could be removed at pH 11 in 90 minute ozonation with 6×10^{-3} m^3/min ozone dose. COD removals reduced at lower pH values. The BOD_5/COD ratio increased from 0.18 to 0.4 by ozonation. Ozonation reaction was found to obey first order kinetics.

3.2.3 Oxidation applications to pre-treated wastewater

3.2.3.1 Ozone oxidation

Doğruel *et al.* (2004) applied ozone oxidation to pretreated tanning wastewater having 1785 mg/L COD with ozone feed rate of 42.8 mg O_3/min. At pH 8 and in 30 minutes feeding time only 20 % COD was removed. In a further study by the authors, another pretreated tanning wastewater was ozone oxidised in a similar manner at the same ozone feeding rate and reaction time and 15 % COD was removed (Doğruel *et al.* 2006). In this study COD fractions were also determined. The authors concluded that pre-ozonation was not effective either in terms of total COD or inert COD removal efficiency. Another important conclusion was that by pre-ozonation the target fraction of inert COD remained unaffected, while biodegradable part was removed.

3.2.3.2 Other chemical oxidation methods

Schrank *et al.* (2004) applied advanced oxidation methods UV, TiO_2/UV, O_3, and O_3/UV to tannery wastewater pre-treated by alum coagulation. COD of the wastewater was 2365 mg/L. UV light applied without a catalyst did not change the parameters COD, TOC, and toxicity. Ozone application was effective only at pH 7 and 11 providing COD removals 11 and 17 %, respectively, but even at pH 11 original BOD_5/COD ratio of 0.42 did not increased but decreased to 0.34. TiO_2/UV application at 1 g/L TiO_2 and 120 minutes of reaction time, yielded a maximum COD removal of 7 % at pH 3. The highest efficiency of 21 % COD removal was obtained with the application of O_3/UV (ozone dosage 2.6 g O_3/h for 30 minute reaction time), but BOD_5/COD ratio again decreased to 0.34. Toxicity test conducted using *Daphnia magna* and expressed as Lowest Ineffective Dilution had the values 4, 6, 6, and 4 for the wastewater, TiO_2/UV (pH 3), O_3 (pH 11), and O_3/UV treatment, respectively. Substance specific analyses indicated that aromatic organic structures were preferentially degraded partially to aliphatic oxidised polar compounds. In another study by Schrank *et al.* (2005) Fenton and H_2O_2/UV treatment methods were applied to a tannery wastewater pre-treated by alum coagulation. However, COD of wastewater was 130 mg/L. Maximum COD removal of 81 % was obtained with Fenton application (H_2O_2 = 500 mg/L, Fe^{2+} = 100 mg/L, pH = 3.5). TOC removal was 58 % with H_2O_2/UV application (H_2O_2 = 1020 mg/L for 2 hours reaction time). Qualitative analysis using instrumental methods indicated that compounds such as phenol, benzothiazole existing in the wastewater were dramatically reduced by the application of the oxidation processes.

Sauer *et al.* (2006) applied advanced oxidation to a real tannery wastewater pre-treated by alum coagulation and containing 801 mg/L COD and 27.6 mg N/L ammonia. Sulphide, chromium, and other metal content of the wastewater were less than 1.0 mg/L. Different dilutions of wastewater were treated by TiO_2/UV in a continuous reactor using 11 g/L TiO_2 at pH 2.5 and 1 hour residence time. COD removal increased with increasing dilution up to 65.7 % at ¼ dilution, in contrast BOD_5 reduced with increasing dilution. Ammonia removal was 22.3 % in undiluted sample and ammonia concentration was increased in the diluted sample due to hydrolysis of organic nitrogen. Another important finding was that toxicity of the samples, as measured by *Artemia Salina* mortality, increased after treatment and increased with increasing dilution. H_2O_2/UV treatment at 0.6 g/L H_2O_2 optimum dose provided around 60 % COD removal along with a relative decrease in aromaticity in the wastewater.

3.2.4 Oxidation applications to biologically treated effluents

3.2.4.1 Electrochemical oxidation and electrocoagulation

Szpyrkowicz *et al.* (1995) studied electrochemical treatment of tannery wastewater undergone aerobic biological treatment with COD values varying between 170–458 mg/L using Ti/Pt and Ti/Pt/Ir electrodes. For Ti/Pt electrodes current densities were between 2–4 A/dm^2, electrolysis times 25–45 minutes, and pH around neutral, COD removals with this system were 61–70 % while ammonia was totally eliminated. Ti/Pt/Ir electrode system worked at 2–3 A/dm^2 current density, 22–45 minutes electrolysis time and around neutral pH. In this system, COD removal was far from being satisfactory providing maximum 43 % efficiency while ammonia removals were high. In this study, the same treatment application was employed for a final effluent undergone nitrification and denitrification with 93–46 mg/L of COD and 12–24 mg/L of ammonia. Both electrode systems worked at 2 A/dm^2 and for 15–20 minutes electrolysis time and at near neutral pH, COD removal efficiencies of 60 % were obtained. Szpyrkowicz *et al.* (2001) applied electrolysis process also to an anaerobically treated effluent of tanning wastewater. Ti/Pt-Ir anode and stainless steel cathode were used at current density of 4 A/dm^2 for one hour duration. With wastewater COD varying between 245–419 mg/L and chloride concentration over 5000 mg/L, COD removal did not reach 50 %, while sulphide and ammonia were totally eliminated. Rao *et al.* (2001) used Ti/Pt, Ti/PbO_2, and Ti/MnO_2 anodes and a Ti cathode for the electrochemical treatment of aerobic biologically treated effluent. COD and NH_3 contents of the wastewater were 480–550 and 100–110 mg N/L, respectively. Ti/Pt electrode at 5.56 A/dm^2 provided 62 % COD removal in 4 hours. Ammonia removals were around 90 %. Ti/PbO_2 and Ti/MnO_2 electrodes provided over 70 % COD removals in 4 hours operation time at 1.67 and 0.612 A/dm^2 current densities, respectively. Panizza and Cerisola (2004) studied electrolysis with Ti/PbO_2 and Ti/TiRuO_2 as anodes to evaluate the treatability of a biologically treated tannery effluent using synthetically prepared sample (SPS) made up of tannic acid and inorganic salts. Although, COD concentrations of SPS were too high for biological treatment effluent being over 2000 mg/L, near complete removals were obtained with both electrodes. Rodrigues *et al.* (2008) aimed to recover and reuse of tannery wastewater using a photoelectrochemical and electrodialysis combination. Electrochemical system comprised of an electrolysis system and a UV light source embedded in the center of the reactor. The electrodes were DSA (70 TiO_2/30 RuO_2) and the system worked at 36 mA/cm^2 for 24 hours. Biologically treated wastewater containing 150 mg/L COD was used in the experiments. High COD removal approaching 90 % was obtained with the operation of the system at pH 7.6.

3.2.4.2 Ozonation

Jochimsen and Jekel (1997) applied ozonation to a biologically pretreated tannery wastewater to improve the quality of effluent, particularly, its biodegradability. They used UV-removal parameter for evaluation. The optimal specific ozone dose was found 1 to 3 g O_3/g DOC. They noted that ozonation led to a relative increase of the low molecular weight DOC fraction of the effluent. Doğruel *et al.* (2004) studied ozonation of biologically treated tannery wastewater. Ozonation was applied to several stages of activated sludge treatment. The first stage biologically treated effluent corresponded to the COD removal where readily biodegradable portion of COD was completely depleted. Ozonation of this sample (ozone feeding at the rate of 42.8 mg O_3/min with 84 % utilised ozone), at neutral pH, reduced COD to 495 mg/L in 30 minutes providing 41 % COD removal. The second stage effluent represented the level of treatment where the wastewater had only half of its initial rapidly hydrolysable COD content. Ozonation of this effluent applying the same procedure employed for the first sample, COD removal in 30 minutes was 34 % with utilised ozone of 59 %. The third stage sample corresponded to a level where the wastewater was assumed to be solely inert in nature. Ozonation of this sample with the same procedure yielded 47 % COD removal with 80 % ozone utilised. It was concluded that the application of ozone after the complete degradation of readily degradable COD component was the most promising alternative in terms of efficiency and increased the biodegradability after ozonation. The same conclusion was obtained in a further study by Doğruel *et al.* (2006) with the measurements of COD fractions. Di Iaconi *et al.* (2009) used aerobic sequencing batch biofilter granular reactor where they introduced an ozonation cycle. It was concluded that addition of ozonation improved quality of the effluent.

Tünay *et al.* (2006) employed advanced oxidation processes for polishing treatment of a biologically treated tannery effluent. The COD of the effluent varied 103–112 mg/L. Ozonation with a low dose of 10 mg O_3/min provided 50 % reduction in COD in 90 minutes. During ozonation pH decreased from 7.77 down to pH 5.5. The same experiment with the 50 mg/L Fe^{3+} addition as catalyst yielded near 70 % COD removal. Ozonation at pH 13 provided up to 90 % COD removal. O_3/H_2O_2 application at O_3/H_2O_2: 2 mol/1 mol ratio did not further increase the COD removal efficiency and around 70 % reductions were obtained.

3.2.5 Colour removal by chemical oxidation

There are a few studies for the removal of colour in tannery wastewaters as well as in synthetic samples prepared using dye stuffs that are commonly used in leather dyeing. Sauer *et al.* (2006) studied decolourisation of Direct Black 38 using advanced oxidation methods. Direct photolysis by UV (lamb emission 300–410 nm with maximum emission at 355 nm) provided 70 % colour removal together with only 8 % COD removal. Colour removals with H_2O_2/UV, TiO_2/UV and TiO_2/H_2O_2/UV applications (residence time of 3 hours, initial dye concentration of 300 mg/L, pH = 2.5, initial H_2O_2 = 20 mg/L) were over 80 %. In these experiments COD removal was less than 20 % for H_2O_2 but 68–82 % for the other methods. Ammonia removal in all experiments including direct photolysis was over 80 %. Preethi *et al.* (2009) investigated colour removal using direct brown azo dye effluent and tannery effluent by ozonation. Complete colour removals were obtained at pH 11 in 60 minutes with ozone dose of 500 ppm. Colour removal in the tannery effluent was also over 90 % at pH 11, but with 5000 ppm ozone dose. Srinivasan *et al.* (2009) applied ozone to remove colour of Sandopel Brown BRR dye in synthetically prepared solution. Decolourization efficiencies over 80 % were obtained at pH 4, 7, 9 and 11 within 30 minutes using 3 g/h ozone feed in the 2 L reactor at 30 mg/L dye concentration. Increasing the dye concentration up to 360 mg/L increased the time required to 120 minutes for decolourization efficiencies over 90 %.

3.3 CONCLUDING REMARKS

Oxidation of sulphide by air oxygen is a well defined, economic and suitable method for segregated beamhouse wastewaters. Although, the complex mechanism of the process has still been searched, practical experience about the process ensures a sound basis for process design. Soluble manganese salts function well as a catalyst and due to advantages of cost and non-toxic nature, provide a good alternative. The critical point in the design of the process is aeration and mixing. Particularly, mixing is of importance in dispersing the oxygen and catalyst homogeneously. Hydrogen peroxide works well for sulphide oxidation but it is costly and requires pH adjustment to avoid excessive doses. Therefore, it can be considered as a means of polishing treatment following air oxidation. Electrochemical techniques are promising treatment methods, however, further studies are needed to solve operating problems such as pH control, and deposition of elemental sulphur on the electrodes. Among the other oxidants peroxymonosulphate may also be evaluated as a promising candidate.

Chemical oxidation, in addition to sulphide removal, has the potential for use in several stages to solve many of the complex problems that tannery wastewaters exhibit. Among them, increase of biodegradability of organics, control of toxic organics, colour removal, and residual COD removal as well as in-plant control and recovery applications are of importance. Electrochemical treatment is again a method of considerable potential. Its application to raw wastewaters, however may pose problems such as chromium oxidation, formation of sulphur deposits, and operation problems due to high suspended matter content of the wastewater. Pretreated wastewater, namely the wastewater free from sulphide, chromium and excess suspended matter, however, may be handled by electrochemical oxidation and electrocoagulation. This application may be an alternative to biological treatment or can be combined with biological treatment. Use of ozone for oxidation may not be economical if this application preferentially removes biodegradable organics and having limited effect on inert or slowly degradable organics in biological treatment. However, its use as an intermediate step within biological treatment may be considered provided that this application solves residual COD problem. Use of advanced oxidation methods considering the quality and strength of the secondary effluent and power of oxidants is a viable alternative.

REFERENCES

Abel, E. (1956) Autooxidation in the sulfur group. *Monatsh. Chem.* **87**, 498–502.

Ahmad, N., Maitra, S., Dutta, B.K. and Ahmad, F. (2009) Remediation of sulfidic wastewater by catalytic oxidation with hydrogen peroxide. *J. Environ. Sci.-China* **21**, 1735–1740.

Almgren, T. and Hangström, I. (1974) The oxidation rate of sulphide in sea water. *Water Res.* **8**, 395–400.

Ateş, E., Orhon, D. and Tünay, O., (1997) Characterization of tannery wastewaters for pretreatment-selected case studies. *Water Sci. Technol.* **36**, 217–223.

Bes-Piá, A., Cuartas-Uribe, B., Mendoza-Roca, J.A., Galiana-Aleixandre, M.V., Iborra-Clar, M.I. and Alcaina-Miranda, M.I. (2008) Pickling wastewater reclamation by means of nanofiltration. *Desalination* **221**, 225–233.

Betterton, E.A. and Hoffmann, M. R. (1990) Kinetics and mechanism of the oxidation of aqueous hydrogen sulfide by peroxymonosulfate. *Environ.Sci.Technol.* **24**, 1819–1824.

Bouzid, J., Aydi, W., Ksibi, M. and Eleuch, B. (2008) Use of ozonation process for the oxidation of limiting float wastewater stream. *Environ. Eng. Sci.* **25**, 1139–1147.

Boyce, S.D., Hoffmann, M.R., Hong, P.A. and Moberty L.M. (1983) Catalysis of the autoxidation of aquated sulfur dioxide by homogeneous metal- phthalocyanine complexes. *Environ. Sci. Technol.* **17**, 602–611.

Buisman, C., Uspeert, P. Janssen, A. and Lettinga, G. (1990) Kinetics of chemical and biological sulphide oxidation in aqueous solutions. *Water Res.* **24**, 667–671.

Cadena, F. and Peters, R.W. (1988) Evaluation of chemical oxidizers for hydrogen sulfide control. *J. Water Pollut. Con. F.* **60**, 1259–1263.

Chen, K.Y. and Morris, J.C. (1970) Oxidation of aqueous sulphide by O_2: 1. General characteristics and catalytic influences. In Proc. of *Fifth International Water Pollution Research Conference*, p. III, 32/1–32/17. In *Advances in Water Research*, (ed. S. H. Jenkins), vol. II., Pergamon Press, New York, U.S.A.

Costa, C.R., Botta, C.M.R., Espindola, E.L.G. and Olivi, P. (2008) Electrochemical treatment of tannery wastewater using DSA electrodes. *J. Hazard. Mater.* **153**, 616–627.

Dalai, A.K. Majumdar, A. and Tollefson, E.L. (1999) Low temperature catalytic oxidation of hydrogen sulphide in sour produced wastewater using activated carbon catalysts. *Environ. Sci. Technol.* **33**, 2241–2246.

Di Iaconi, C., Ramadoni, R. and Lopez, A. (2009) The effect of ozone on tannery wastewater biological treatment at demonstrative scale. *Bioresource Technol.* **100**, 6121–6124.

Doğruel, S., Genceli, E.A., Babuna, F.G. and Orhon, D. (2004) Ozonation of nonbiodegradable organics in tannery wastewater. *J. Environ. Sci. Heal. A* **39**, 1705–1715.

Doğruel, S., Genceli, E.A., Babuna, F.G. and Orhon, D. (2006) An investigation on the optimal location of ozonation within biological treatment for a tannery wastewater. *J. Chem. Technol. Biot.* **81**, 1877–1885.

Dutta, P.K., Rabaey, K., Yuan, Z. and Keller, J. (2008) Spontaneous electrochemical removal of aqueous sulphide. *Water Res.* **42**, 4965–4975.

EC (2003) Integrated Pollution Prevention and Control, Draft Reference Document on Best Available Techniques in the Tanning of Hides and Skins, European Commission.

EC (2009) Integrated Pollution Prevention and Control, Draft Reference Document on Best Available Techniques in the Tanning of Hides and Skins, European Commission.

Elsheikh, M.A-S. (2009) Tannery wastewater pre-treatment. *Water Sci. Technol.* **60**, 433–440.

Espinoza-Quinones, F.R., Fornari, M.M.T., Monedes, A.N., Palacio, S.M., Trigueros, D.E.G., Borba, F.H. and Krumov, A.D. (2009) Electrocoagulation efficiency of the tannery effluent treatment using aluminium electrodes. *Water Sci. Technol.* **60**, 2173–2184.

Filpponen, H., Guerra, A., Hai, A., Lucia, L.A. and Argyropoulos, D.S. (2006) Spectral monitoring of the formation and degradation of polysulfide ions in alkaline conditions. *Ind. Eng. Chem. Res.* **45**, 7388–7392.

Fischer, H., Schulz, Ekloff, G. and Wöhrle, D. (1997) Oxidation of aqueous sulphide solutions by dioxygen part 1: autoxidation reaction. *Chem. Eng. Technol.* **20**, 462–468.

Floqi, T., Vezi, D. and Malollari, I. (2007) Identification and evaluation of water pollution from Albanian tanneries. *Desalination* **213**, 56–64.

Germann, H.P., (1995) Oxidation and precipitation. *Leather* **June**, 56–59.

Haydar, S. and Aziz, J.A. (2009) Characterization and treatability studies of tannery wastewater using chemically enhanced primary treatment (CEPT)- A case study of Saddiq Leather Works. *J. Hazard. Mat.* **163**, 1076–1083.

Hoffmann, M.R. (1977) Kinetics and mechanism of the oxidation of hydrogen sulfide by hydrogen peroxide in acidic solution. *Environ. Sci. Technol.* **11**, 61–66.

Hoffmann, M.R. and Lim, B.C. (1979) Kinetics and mechanism of the oxidation of sulphide by oxygen: catalysis by homogenous metal-phthalocyanine complexes. *Environ. Sci. Technol.* **13**, 1406–1414.

Jian, S., Wenyi, T. and Wuyong, C. (2010) Studies on the application of ultrasound in leather enzymatic unhairing. *Ultrason. Sonochem.* **17**, 376–382.

Jochimsen, J.C. and Jekel, M.R. (1997) Partial oxidation effects during the combined oxidative and biological treatment of separated streams of tannery wastewater. *Water Sci. Technol.* **35**, 337–345.

Jochimsen, J.C., Schenk, H., Jekel, M.R. and Hegemann, W. (1997) Combined oxidative and biological treatment for separated streams of tannery wastewater. *Water Sci. Technol.* **36**, 209–216.

Joseph, K. and Nithya, N. (2009) Material flows in the life cycle of leather. *J. Clean. Prod.* **17**, 676–682.

Kabdaşlı, N. I. and Tünay, O. (1992). Pollution Control Approaches in Leather Tanning Industry. In Proc. of the Third Industrial Pollution Control Symposium'92, İstanbul, 231–241, .

Kabdaşlı, I., Tünay, O. and Orhon, D. (1993) The treatability of chromium tannery wastes. *Water Sci. Technol.* **28**, 97–105.

Kabdaşlı, I., Tünay O. and Orhon, D. (1999) Wastewater control and management in a leather tanning district. *Water Sci. Technol.* **40**, 261–267.

Kabdaşlı, I., Tünay, O., Çetin, M.Ş. and Ölmez, T. (2002) Assessment of magnesium ammonium phosphate precipitation for the treatment of leather tanning industry wastewaters. *Water Sci. Technol.* **46**, 231–239.

Kamyshny, A.J., Goifman, A., Gun, J., Rizkov, D. and Lev, O. (2004) Equilibrium distribution of polysulfide ions in aqueous solutions at 25°C: A new approach for the study of polysulfides equilibria. *Environ. Sci. Technol.* **38**, 6633–6644.

Kamyshny, A.J., Gun, J., Rizkov, D., Voitsekovski, T. and Lev, O. (2007) Equilibrium distribution of polysulfide ions in aqueous solutions at different temperatures by rapid single phase derivatization. *Environ. Sci. Technol.* **41**, 2395–2400.

Khudaish, E.A. and Al-Hinai, A.T. (2006) The catalytic activity of vanadium pentoxide film modified electrode on the electrochemical oxidation of hydrogen sulfide in alkaline solutions. *J. Electroanal. Chem.* **587**, 108–114.

Kleinjan, W.E., Keizer, A. and Janssen, A.J.H. (2005) Kinetics of the chemical oxidation of polysulfide anions in aqueous solution. *Water Res.* **39**, 4093–4100.

Kotronarou, A. and Hoffmann, M.R. (1991a) Catalytic autoxidation of hydrogen sulphide in wastewater. *Environ. Sci. Technol.* **25**, 1153–1160.

Kotronarou, A. and Hoffmann, M.R. (1991b) Peroxymonosulfate: An alternative to hydrogen peroxide for the control of hydrogen sulfide. *J. Water Pollut. Con. F.* **63**, 965–970.

Kotronarou, A., Mills, G. and Hoffmann, M.R. (1992) Oxidation of hydrogen sulfide in aqueous solution bu ultrasonic radiation. *Environ. Sci. Technol.* **26**, 2420–2428.

Kurt,U., Apaydin, O. and Gonullu, M.T. (2007) Reduction of COD in wastewater from an organized tannery industrial region by electro-Fenton process. *J. Hazard. Mater.* **143**, 33–40.

Li, S., Li, J., Yi, J. and Shan, Z. (2010) Cleaner beam house processes trial on cattle sofa leather. *J. Clean. Prod.* **18**, 471–477.

Licht, S. and Davis, J. (1997) Disproportionation of aqueous sulphur and sulphide: kinetics of polysulfide decomposition. *J. Phys. Chem. B* **101**, 2540–2545.

Linkous, C.A., Huang, C. and Fowler, J.R. (2004) UV photochemical oxidation of aqueous sodium sulfide to produce hydrogen and sulfur. *J. Photochem. Photobiol. A: Chem.* **168**, 153–160.

Martin, J.L. and Rubin, A.J. (1972) Removal of sulfides by catalytic oxygenation in alkaline media. *J. Water Pollut. Con. F.* **49**(12), 2445–2452.

Millero, F.J., Hubinger, S., Fernandez, M. and Garnett, S. (1987) Oxidation of H_2S in seawater as a function of temperature, pH, and ionic strength. *Environ. Sci. Technol.* **21**, 439–443.

Millero, F.J., LeFerriere, A., Fernandez, M., Hubinger, S. and Hershey, J.H. (1989) Oxidation of H_2S with H_2O_2 in natural waters. *Environ. Sci. Technol.* **23**, 209–213.

Murugananthan, M., Raju, G.B. and Prabhakar, S. (2004) Removal of sulphide, sulfate and sulfite ions by electro coagulation. *J. Hazard. Mat.* **B109**, 37–44.

Nacheva, P.M. (1989) Catalytic air oxidation of sulphides at high concentrate sodium solutions. Seminar on Waste-Water Management, Munich (Federal Republic of Germany), 21–25 May, ENVWA/SEM.4/R.14.

Nazer, D.W., Al-Saed, R.M. and Siebel, M.A. (2006) Reducing the environmental impact of the unhairing-liming process in the leather tanning industry. *J. Clean. Prod.* **14**, 65–74.

Nielsen, A.H., Vollertsen, J. and Hvitved- Jacobsen, T. (2003) Determination of kinetics and stoichiometry of chemical sulphide oxidation in wastewater of sewer networks. *Environ. Sci. Technol.* **37**, 3853–3858.

O'Brien, D.J. and Birkner, F.B. (1977) Kinetics of oxygenation of reduced sulphur species in aqueous solution. *Environ. Sci. Technol.* **11**, 1114–1120.

Ouejhani, A., Hellal, F., Dachraoui, M., Lalleve, G. and Fauuvarque, J.F. (2008) Application of Doehlert matrix to the study of electrochemical oxidation of Cr(III) to Cr(IV) in order to recover chromium from wastewater tanning baths. *J. Hazard. Mat.* **157**, 423–431.

Ölmez, T., Kabdaşlı, I., Tünay, O., Kınacı, Eryılmaz, G. and Yılmazoğlu, D. (2000) Treatability alternatives for sulphide and sulphate indigo dyeing baths. *Fresenius Environ. Bull.* **9**, 306–313.

Panizza, M. and Cerisola, G. (2004) Electrochemical oxidation as a final treatment of synthetic tannery wastewater. *Environ. Sci. Technol.* **38**, 5470–5475.

Paschke, R.A., Hwang, Y.S. and Johnson, D.W. (1977) Catalytic oxidation of sulphide wastewater from a cellulose sponge-making operation. *J. Water Pollut. Con. F.* **49**, 2445–2452.

Preethi, V., Kalyani, K.S.P., Iyappan, K., Srinivasakannan, C., Balasubramaniam, N. and Vedaraman, N. (2009) Ozonation of tannery effluent for removal of COD and color. *J. Hazard. Mat.* **166**,150–154.

Purkait, M.K., Kumar, V. D. and Maity, D. (2009) Treatment of leather plant effluent using NF followed by RO and permeate flux prediction using artificial neural network. *Chem. En. J.* **151**, 275–285.

Rao, N.N., Somasekhar, S.N., Kaul, S.N. and Szpyrkowicz, L. (2001) Electrochemical oxidation of tannery wastewater. *J. Chem. Technol. Biot.* **76**, 1124–1131.

Rodrigues, M.A.S., Amado, F.D.R., Xavier, J.L.N., Streit, K.F., Bernardes, A.M. and Ferreira, J.Z. (2008) Application of photoelectrochemical-electrodialysis treatment for the recovery and reuse of water from tannery effluents. *J. Clean. Prod.* **16**, 605–611.

Sabate, J., Cervera-March, S., Simarro, R. and Gimenez, J. (1990) Photocatalytic production of hydrogen from sulphide and sulfite waste streams: A kinetic model for reactions occurring in illuminated suspensions of Cds. *Chem. Eng. Sci.* **45**, 3089–3096.

Saravanabhavan, S., Thanikaivelan, P., Rao, J.R., Nair, B.U. and Ramasami, T. (2006) Reversing the conventional leather processing sequence for cleaner leather production. *Environ. Sci. Technol.* **40**, 1069–1075.

Särkka, H., Kuhmonen, K., Vepsalainen, M., Pulliainen, M., Selin, J., Rantala, P., Kukkamaki, E. and Sillanpaa, M. (2009) Electrochemical oxidation of sulphides in paper mill wastewater by using mixed oxide anodes. *Environ. Technol.* **30**, 885–892.

Sauer, T.P., Casaril, L.C., Obernizer, A.L.B., Jose, H.J. and Moreira, R.F.P.M. (2006) Advanced oxidation processes applied to tannery wastewater containing Direct Black 38- Elimination and degradation kinetics. *J. Hazard. Mater.* **B135**, 274–279.

Schrank, S.G., Jose, H.J., Moreira, R.F.P.M. and Schroder, H.F. (2004) Elucidation of the behaviour of tannery wastewater under advanced oxidation conditions. *Chemosphere* **56**, 411–423.

Schrank, S.G., Jose, H.J., Moreira, R.F.P.M. and Schroder, H.F. (2005) Applicability of Fenton and H_2O_2/UV reactions in the treatment of tannery wastewaters. *Chemosphere* **60**, 644–655.

Sharma, V.K., Smith, J.O. and Millero, F.J. (1997) Ferrate(VI) oxidation of hydrogen sulfide. *Environ. Sci. Technol.* **31**, 2486–2491.

Sirajuddin, Kakakhel, L., Lutfullah, G., Bhanger, M.I., Shah, A. and Niaz, A. (2007) Electrolytic recovery of chromium salts from tannery wastewater. *J. Hazard. Mater.* **148**, 560–565.

Sivakumar, V., Swaminathan, G., Rao, P.G., Muralidharan, C., Mandal, A.B. and Ramasami, T. (2009) Use of ultrasound in leather processing Industry: Effect of sonication on substrate and substances - New insights. *Ultrason Sonochem.* Doi:10.1016/j.ultsonch.2009.10.012.

Sreeram, K.J. and Ramasami, T. (2003) Sustaining tanning process through conservation, recovery and better utilization of chromium. *Resour. Conserv. Recy.* **38**, 185–212.

Srinivasan, S.V., Rema, T., Chitra, K., Balakameswari, K.S., Suthanthararajan, R., Maheswari, B.U., Ravindranath and Rajamani, S. (2009) Decolourisation of leather dye by ozonation. *Desalination* **235**, 88–92.

Steudel, R. (1996) Mechanism for the formation of elemental sulphur from aqueous sulfide in chemical and microbiological desulfurization processes. *Ind. Eng. Chem. Res.* **35**, 1417–1423.

Szpyrkowicz, L., Naumczyk, J. and Zilio-Grandi, F. (1995) Electrochemical treatment of tannery wastewater using Ti/Pt and Ti/Pt/Ir electrodes. *Water Res.* **29**, 517–524.

Szpyrkowicz, L., Kelsall, G.H., Kaul, S.N. and Faveri, M.D. (2001) Performance of electrochemical reactor for treatment of tannery wastewaters. *Chem. Eng. Sci.* **56**, 1579–1586.

Şengil, İ.A., Kulaç, S. and Özacar, M. (2009) Treatment of tannery liming drum wastewater by electrocoagulation. *J. Hazard. Mater.* **167**, 940–946.

Thanikaivelan, P., Rao, J.R., Nair, B.U. and Ramasami, T. (2002) Zero Discharge tanning: a shift from chemical to biocatalytic leather processing. *Environ. Sci. Technol.* **36**, 4187–4194.

Thanikaivelan, P., Rao, J.R., Nair, B.U. and Ramasami, T. (2005) Recent trends in leather making: processes, problems, and pathways. *Crit. Rev. Env. Sci. Tech.* **35,** 37–79.

Tünay, O., Orhon, D. and Kabdasli, I. (1994) Pretreatment requirements for leather tanning industry wastewaters. *Water Sci. Technol.* **29**, 121–128.

Tünay, O., Kabdaşli, I., Orhon, D. and Ates, E. (1995) Characterization and pollution profile of leather tanning industry in Turkey. *Water Sci. Technol.* **32**, 1–9.

Tünay, O., Kabdaşlı, I., Başeğmezler, B. and Kösebay, O. (1998) The factors affecting sulphide oxidation by air oxygen. *Fresenius Environ. Bull.* **7**, 223–230.

Tünay, O., Kabdaşlı, I., Orhon, D. and Cansever, G. (1999) Use and minimization of water in leather tanning processes. *Water Sci. Technol.* **40**, 237–244.

Tünay, O., Yelmez, Z.B., Ölmez, T. and Kabdaşlı, I. (2006) Residual COD reduction in biologically treated leather tanning effluents by advanced treatment processes. *J. Am. Leather Chem. As.* **101**, 117–160.

UN (1997) Economic and Social Commission for Western Asia, Environmentally Sound Technologies in the Tanning Industry, United Nations, New York, U.S.A.

US EPA, (1982) Development Document for Effluent Limitations Guidelines and Standards, Leather Tanning and Finishing Point Source Category, EPA 440/1–79/016, Effluent Guidelines Division, Office of Water and Waste Management, U.S. Environmental Protection Agency, Washington, D.C., United States.

Valeika, V., Beleska, K. and Valeikiene, V. (2006) Oxidation of sulphides in tannery wastewater by use of manganese (IV) oxide. *Polish. J. Environ. Stud.* **15**, 623–629.

Vlyssides, A.G. and Israilides, C.J. (1997) Detoxification of tannery waste liquors with an electrolysis system. *Environ. Pollut.* **97**, 147–152.

Waterston, K., Bejan, Dorin and Bunce, N .J. (2007) Electrochemical oxidation of sulphide ion at a boron-doped diamond anode. *J. Appl. Electrochem.* **37**, 367–373.

Wilmot, P.D., Cadee, K., Katinic, J.J. and Kavanagh, B.V. (1988) kinetics of sulphide oxidation by dissolved oxygen. *J. Water Pollut. Con. F.* **60**, 1264–1270.

Zengin, G., Ölmez, T., Doğruel, S., Kabdaşlı, I. and Tünay, O. (2002) Assessment of source-based nitrogen removal alternatives in leather tanning industry wastewater. *Water Sci. Technol.* **45**, 205–215.

Chapter 4

Metal finishing industry

4.1 METAL FINISHING INDUSTRY

4.1.1 Overview

Metal finishing industry facilities employ a broad range of processes in order to form and shape the raw materials, to alter the appearance and properties of the material surface, and to construct the products. Metals and alloys in the form of raw stock are used as starting materials to product a wide range of metal components in the industry. The facilities covered by the metal finishing category are encompassed Standard Industrial Classification (SIC) Major Groups 34 through 39 listed in Table 4.1 (US EPA 1983). The facilities tabulated in Table 4.1 that are not exclusively in the metal finishing category cover coil coating, batteries manufacturing, porcelain enamelling, photographic equipment and supplies, electrical and electronic components, iron and steel, aluminium and aluminium alloys, copper and copper alloys, and shipbuilding.

The fabricated metal production (SIC 34) facilities manufacture a wide range of metal components, tools, and equipments such as shipping containers, cutlers, hand tools, heating equipments *etc*. The facilities listed under Major Group 35 are the productions of engines, turbines, farm and garden machinery and equipments,

Table 4.1 Facilities covered by metal finishing category

SIC	Facilities
34	Fabricated metal products exc. machinery&transportation equipment
35	Machinery except electrical
36	Electrical and electronic machinery, equipment, and supplies
37	Transportation equipments
38	Measuring, analysing, and controlling instruments
39	Miscellaneous manufacturing industries

industrial machinery and equipments, office machineries, computing and accounting machines, and so on. The manufacturing of electric transmission and distribution equipments, electrical industrial apparatuses, electric lighting and wiring equipments, radio and television receiving equipments, communication equipments, and electronic components and accessories are the industrial branches tabulated under Major Group 36. Transportation Equipment Major Group (SIC 37) covers the production of aircraft and its parts, motor vehicles and their equipments, railroad equipments, motorcycles, bicycles and their parts, guided missiles and space vehicles and their parts, and building and repairing of ship and boats. The facilities classified in Major Group 38 are engaged in production of watches, clocks and goods of photographic, medical and optical. Jewelleries, the producers of silverware, plated ware, musical instruments, the materials of office and artists, costume jewellery are inclusive of Major Group 39. With this profile, the metal finishing industry is among the most common industrial activities in the world.

Although, the processes employed in the metal finishing industry are similar, the capacities of the facilities vary in a wide range. According to their capacities, and the number and type of operations, the facilities are divided into two main sectors: job shops and captives (US EPA 1984). Job shops accomplish fewer operations, while captives have a complex range of operations. Captive facilities perform the production processes on in-house manufactured parts, whereas job shops work on the parts produced by others or do not own more than 50 % of materials being processed. Captive facilities may also hire the job shops as subcontractors. Job shops are more flexible in operations to respond to the varying demands of costumers, whereas captive facilities are more specialised in their productions (Tünay *et al.* 2004).

4.1.2 Production processes

In the metal finishing facilities, the operations applied start with materials in the form of bars, wires, plates, sheets, castings, or rod *etc.*, and can proceed to the most sophisticated surface finishing applications. The operations which are employed in manufacturing, rebuilding, or maintenance of metal components, products, vehicles, equipments, or machines can be grouped as follows (US EPA 2003):

- metal shaping;
- surface preparation;
- metal deposition;
- organic material deposition;
- surface finishing;
- assembly;
- dry dock;
- printed wiring board operations.

Metal shaping is a mechanical operation that is applied in order to alter the form of raw materials into intermediate or final products. Impact or pressure deformation, machining or grinding are typical examples of metal shaping processes. Surface preparation (*e.g.*, alkaline cleaning, acid treatment or barrel finishing) covers the mechanical and chemical operations to remove undesired materials or to alter the physical and chemical properties of the part surface. Deposition is a coating process that may be applied to the part surface by chemical or physical means using either a metal or organic material. This step precedes surface finishing and final assembly operations. Surface finishing (*e.g.*, chemical conversion coating) protects and seals the surfaces from wear and corrosion by chemical means. Assembly is accomplished throughout the manufacturing, reconstructing, or maintenance operation. Final shaping (*e.g.*, drilling and grinding) and surface preparation (*e.g.*, alkaline cleaning) operations are occasionally employed during this step. Dry dock operations cover only ship and boat facilities. Printed wiring board operations such as solder levelling, carbon black deposition are specialised by manufacturing or rebuilding/maintenance of wiring boards.

The above mentioned processes are accomplished by applying forty six unit operations listed in Table 4.2 (US EPA 1983). Depending on the complexity of the product, types of material processed and facility, these unit operations are employed in many different combinations and sequences. For instance, nearly all unit operations are used in the manufacturing of a complex product, while only a unit operation is employed in the production of a simple one. While preliminary operations such as disassembly, cleaning, or degreasing may not be performed at manufacturing facilities, the repairment, rebuilding or maintenance of products necessitates these preliminary operations.

Table 4.2 Unit operations performed in metal finishing industry (US EPA 1983)

Unit operations	
➢ Electroplating	➢ Electrochemical machining
➢ Electroless plating	➢ Electron beam machining
➢ Anodising	➢ Laser beam machining
➢ Coating	➢ Plasma arc machining
➢ Cleaning	➢ Ultrasonic machining
➢ Machining	➢ Sintering
➢ Etching and chemical milling	➢ Laminating
➢ Grinding	➢ Hot dip coating
➢ Polishing	➢ Sputtering
➢ Barrel finishing	➢ Vapour plating
➢ Burnishing	➢ Thermal infusion
➢ Impact deformation	➢ Salt bath descaling
➢ Pressure deformation	➢ Solvent degreasing
➢ Shearing	➢ Paint stripping
➢ Heat treating	➢ Painting
➢ Thermal cutting	➢ Electrostatic painting
➢ Welding	➢ Electropainting
➢ Brazing	➢ Vacuum metalizing
➢ Soldering	➢ Assembly
➢ Flame spraying	➢ Calibration
➢ Sand blasting	➢ Testing
➢ Abrasive jet machining	➢ Mechanical plating
➢ Electrical discharge machining	➢ Printed circuit board manufacturing

4.1.3 Wastewater sources

In the metal finishing industry, there are a great number of wastewater sources in parallel with the number of unit operations together with other activities producing wastewater. To provide ease of evaluation, these wastewater sources can be classified as follows (US EPA 1983):

- the exhausted and spent baths dumps;
- rinsing;
- spills, leakages, and drips;
- cooling and lubrication;
- quenching;
- spray booths;
- testing;
- washing;
- in plant treatment and recovery.

During the unit operations, plating baths, cleaning baths, rinse baths and other baths are periodically replenished until they are exhausted or spent. Although, low amount wastewaters are generated from these dumps, they are main wastewater sources of the metal finishing industry facilities and create the greatest part of pollution load (Tünay *et al.* 2004). Hence, they play a determining role on the selection of wastewater treatment system units and their design parameters. The workpieces are rinsed off to remove the fluid or solid films deposited on their surfaces during the unit operations. A large amount of wastewater with low pollutant concentrations occurs from the rinsing steps. Owing to high volume, pollution loads originating from rinses may become important. The total volume of rinsing wastewaters generally determines the wastewater treatment system capacity of the facilities. Spills, leaks, and drips of process solutions may happen by accident during the operations. These sources may contribute to the total wastewater. Control and reduction of these sources can be managed by in-plant pollution prevention practices. Some metal removal operations necessitate the use of coolants and lubricants in the form of free oils, emulsified oils, and grease. The residues originating from these steps are washed off by washing or rinsing operations. After intense heat applications such as heat treating, welding, or hot dip coating, workpieces are quenched and cooled in aqueous solutions to reach the desired properties or to enable their handlings. Quenching and cooling solutions are classified as process wastewaters since they become contaminated. Spray painting operation is sometimes accomplished in spray booths with water curtain to capture the paint overspray. The paint collected in this water is separated by skimmers or ultrafilters and the water is reused in the curtain. This water is occasionally discharged. During testing operations (*e.g.*, leak, pressure, and performance testing), a large amount of wastewater is generated. Washing wastewaters occur from the cleaning of workpieces processed, equipments used during the operations such as tanks, filters or pumps, and floor cleaning. Washing is a common application in all operations even for zero discharge processes. Wastewater generation is possible from some of in-plant treatment and recovery applications being integral part of many unit operations. Filter backwash of painting systems, concentrates of membrane and ultrafiltration applications are well-known examples (Tünay *et al.* 2004).

In the metal finishing facilities, unit operations generating a large amount of wastewater encompass electroplating, electroless plating, anodising, conversion coating, etching, cleaning, machining, grinding, tumbling, heat treating, welding, sand blasting, salt bath descaling, paint stripping, painting, electrostatic painting, electropainting, testing, mechanical plating, and printed circuit board manufacturing. Unit operations with zero discharge cover electron beam machining, plasma arc machining, laser beam machining, ultrasonic machining, sintering, sputtering, vapour plating, thermal infusion, vacuum metalizing, and calibration (US EPA 1983).

The amount of wastewater originating from the metal finishing activities vary in a wide range depending on unit operations employed, production type, capacity, and size of the facility. Pollution prevention practices applied in the facilities are also effective in reduction of water usage. As mentioned above, a large amount of wastewater is discharged from some unit operations, whereas some of them are dry processes. In integrated and large facilities such as automotive, the amount of daily wastewater generation rarely exceeds 2000 m^3, while the amount of the wastewater flow remains in the range of 0–400 m^3/day for large and medium-size activities (Tünay *et al.* 2004). In a detailed wastewater control and management survey study conducted on a large-size metal finishing facility producing steel wheel, tractor, engine assembly and spring, total amount of daily process wastewater flow was reported as 425 m^3 (Orhon *et al.* 2000). The contributions of steel wheel production, tractor production, engine assembly shop and spring production flows to the total process wastewater were determined as 34.4 %, 38.9 %, 24.2 %, and 2.5 %, respectively. Total wastewater generation together with domestic wastewater (385 m^3/day) and cooling water discharged (112 m^3/day) was 922 m^3/day in the plant investigated.

4.1.4 Wastewater characterisation

Due to a wide range of process activities and the great variety of chemicals and materials used in the unit operations, there is a long list of pollutants existing in the metal finishing industry wastewaters. The list covers metals, cyanide, organic wastes, total suspended solids, and other inorganic pollutants such as sulphate, chloride, phosphates, fluoride and ammonia (US EPA 1995; EC 2006).

Common and precious metals are important groups originating from the metal finishing industry facilities. These groups include over thirty metals. These metals are classified with respect to their toxic effects on the environment and/or human health. Some of them (*e.g.*, iron, magnesium, manganese, and titanium) are sorted as non-toxic and are not regulated as they are effectively treated or recovered by suitable processes. The metals having adverse effects are classed as toxic and are regulated by imposing stringent discharge limitations. Depending on their speciation cadmium, lead, nickel, hexavalent chromium, copper, and zinc are of most concern for their environmental and/or health effects. Therefore, their uses in the industry have been restricted by European

Commission (EC) regulations (EC 2006). Addition to these six metals, silver has also been recognised and regulated as a toxic metal existing at high concentrations in the metal finishing industry wastewaters by US EPA (1983; 1984).

Cyanide is a well-known hazardous compound but is commonly used as a complexing agent for cadmium and precious metals plating and for other solutions such as copper and zinc baths. In the form of strongly complexed metal such as nickel cyanide complex species, cyanide creates several treatment problems. As cyanide concentrations may reach several thousand mg/L in stream segregated flows of the metal finishing industry, this pollutant is regulated by EC and US EPA.

Organic wastes resulting from the metal finishing industry operations vary in a wide spectrum and are principally classified into two main groups: toxic organics, and oil and grease. The origins of toxic organics are complicated. Hence, for ease of evaluation, they can be sorted out into three groups: solvents; complexing agents; and surfactants. Some degreasing methods necessitate organic solvent usage. These applications are the primary sources of priority pollutants in metal finishing industry. Since spent organic solvents contain priority pollutants, they are segregated to haul or reclaim on-site (US EPA 1984). Therefore, their controls rely on in-plant control practices. Complexing agents such as ethylene diamine tetraacetic acid (EDTA), tartaric acid, nitrilo triacetic acid (NTA) *etc.* are commonly used in the plating baths. They retain metals in solution as complexes. Some of them such as EDTA have low biodegradability. As the function of complexing agents is to prevent metals from separating out of solution, the flows containing complexed metals necessitate specialised control and treatment processes. Hence, these flows are conceived as a separate group on the basis of metal removal processes. Surfactants are also widely used in many operations such as degreasing, wetting metal surfaces and assisting other operations, and as brighteners by promoting finely divided metal deposition employed in metal finishing industry (EC 2006). Some surfactants have low biodegradability and their biodegradation by-products may have adverse effects. Owing to the bio-accumulative and toxic characteristics, some surfactants such as nonyl phenyl ethoxylate have been listed as priority pollutants and their usages are banned by EC regulations (EC 2006). Oil and grease is a traditional pollutant parameter causing several aesthetic and material transport problems in receiving media and having toxic effects on aquatic organisms. As oil and grease is found at significant concentrations in the metal finishing wastewaters, it has been commonly regulated. Total suspended solids is a conventional parameter used for the control of organic and inorganic toxic pollutant discharges as well as the performance of the treatment. Hence, this parameter has also been limited by US EPA and EC regulations (US EPA 1983; EC 2006).

The main pollutant groups most commonly present in wastewaters generated by the unit operations employed in the metal finishing industry are tabulated in Table 4.3 (US EPA 1983). The table includes thirty unit operations bearing more than one pollutant groups. The other unit operations which are not covered in the table, generate only common metals bearing wastewaters.

Flow segregation is the basic approach used for the treatment, control and management of the metal finishing industry wastewater. It serves to conduct an appropriate treatment application for each class of waste adequately. The main reason for segregation is that combination of some waste classes may result in failure of some of the treatment applications. Furthermore, the mixture of wastewater groups brings about health hazards and economic losses. Within this context, seven waste groups corresponding to wastewater streams have been defined by US EPA on the basis of characters and treatment requirements of wastewaters (US EPA 1983; 1984). These groups are given as follows:

- common metals;
- precious metals;
- complexed metals;
- hexavalent chromium;
- cyanide;
- oils;
- solvents.

The main components of the common metals bearing wastewaters are the metals and acids. Metal concentrations present in the flows vary from several mg/L to several hundred mg/L.

The major constituents of precious metals bearing wastewaters are gold and silver. They are much more commonly used than other precious metals such as palladium, platinum, osmium, ruthenium, and rhodium in the metal finishing industry facilities. Due to the high cost of precious metals, the precious metals bearing flows are wasted only after efficient recovery applications. In the recovered flows, gold concentration reduces in a few mg/L while silver concentration may reach as high as one hundred mg/L.

Table 4.3 Waste characteristic distribution by unit operation (US EPA 1983)

Potential wastewater pollutants → Unit operations ↓	Common metals	Precious metals	Complexed metals	Chromium (VI)	Cyanide	Oils	Solvents
Electroplating	√	√		√	√		
Electroless plating	√	√	√		√		
Anodising	√	√		√	√		
Conversion coating	√	√		√	√		
Etching (Chemical Milling)	√	√	√	√	√		
Cleaning	√	√	√	√	√	√	√
Machining	√					√	
Grinding	√					√	√
Polishing	√					√	
Tumbling	√			√	√	√	
Burnishing	√	√			√	√	
Impact Deformation	√					√	
Pressure Deformation	√					√	
Shearing	√					√	
Heat Treating	√				√	√	
Thermal Cutting	√					√	
Abrasive Jet Machining	√					√	
Electrical discharge machining	√					√	
Electrochemical machining	√				√		√
Salt Bath Descaling	√					√	
Solvent Degreasing	√						√
Paint Stripping	√					√	√
Painting	√			√			√
Electrostatic Painting	√			√			√
Electropainting	√			√			√
Assembly	√					√	√
Calibration	√					√	
Testing	√					√	
Mechanical Plating	√			√			
Printed Circuit Board Manufacture	√	√	√	√	√	√	√

Complexed metals bearing flows are generated from a number of unit operations employed in the metal finishing industry. The main sources of these flows are the electroless and immersion plating baths and their subsequent rinses.

Nickel, copper and tin are the most commonly used metals in these operations. The complexed metals bearing wastewaters must be segregated and treated so as to prevent further complexing of metals present in the free form in the other flows (US EPA 1983).

Wastewaters with high cyanide concentrations commonly originate in electroplating and heat treatment operations. Many other unit operations, as listed in Table 4.3, contribute to cyanide wastes. Cyanide concentrations may reach up to 500 mg/L. The flows containing cyanide are subjected to stream segregation for treatment before being combined with other wastewaters.

Hexavalent chromium bearing flows are generated in many unit operations such as electroplating, anodising, conversion coating, and acid treatment. The hexavalent chromium may be present at high concentrations in these flows. Owing to its high toxicity, hexavalent chromium bearing flows are strictly segregated and undergone separate pretreatment applications.

In the metal finishing industry, oily wastewater flows consist of a mixture of free and emulsified oils, grease, and other assorted organics. These flows are characterised as concentrated and diluted streams. Oil concentrations may reach several hundred thousand mg/L in the concentrated flows. The selection of treatment methods of oily wastewaters is based on the concentration levels of the wastes and their physical state as well.

Since they are classified as hazardous wastes by EC and US EPA regulations, spent solvents are separately collected to reclaim or dispose off as a hazardous waste. Therefore, they are not considered as a wastewater flow.

4.1.5 In-plant control and clean technologies

US EPA has described pollution prevention as any practice which minimises the amount of any hazardous substance, pollutant, or contaminant entering the waste stream or otherwise released to the environment prior to recycling, treatment, or disposal, and reduces or eliminates the hazards to public health and the environment (US EPA 1992; 1997; 2003). The issue of the integrated pollution prevention and control (IPPC) for the metal finishing industry has been also well defined in "Best Available Techniques (BAT-EC) Reference Document (BREF), Surface Treatment of Metals and Plastics (STM)" by EC (2006). Similar to those of US EPA, the implementations of IPPC described in BREF for STM encompass the effective management systems, efficient raw material, energy and water usage, the substitution by less harmful substance, as well as minimisation, recovery and recycling of wastewater and wastes (EC 2006).

The pollution prevention applications are based on source control. Source control is a process of reducing the pollution and health hazards that originate from a certain production by taking appropriate measures from the selection of production method to the good operating practices. Every effort for pollution prevention ultimately, at least legally, targets to avoid banned applications, to satisfy occupational health measures, and to fulfil discharge standards for water pollution control, within the economical constraints of the facilities and the market (Tünay *et al.* 2004).

Waste reduction practices include several activities such as training, awareness and participation of the employees, the improvement of process schedules, and the improvement of procedures used in the processes. The specific approaches proposed for the waste minimisation for the metal finishing industry facilities encompass source reduction and recycling/resource recovery (US EPA 1992; EC 2006). Source reduction technologies are designed to minimise the amount of waste initially generated. Many source reduction practices necessitate only simple housekeeping changes or minor in-plant process modification.

Source reduction techniques are also applied to process baths and rinsing operations. At the process bath level, source reduction can be realised by material substitution, extending bath life, and drag-out reduction. Within this context, the substitution of toxic plating solution with non or less toxic constituents are the recommended practices. The replacement of hexavalent chromium with trivalent chromium is a good example of the material substitution. In so doing, drag-out is minimised and an extra treatment step necessary to reduce the chromium from hexavalent to trivalent state before precipitation is eliminated. Similar to hexavalent chromium replacement, the use of non-cyanide plating baths omits cyanide oxidation step. The substitutions of EDTA and the other strong complexing agents with weaker and biodegradable ones such as those based on gluconic acid are also recommended as BAT-EC to avoid their usage and to eliminate the extra treatment step required for their removals. Rinse water reduction can be achieved by improving rinse efficiency, and rinse water flow control. Spray rinsing, rinse water agitation, increased contact time, rinse elimination, and counter flow multiple tank rinsing can be used to minimise the volume of rinse water and to enhance the rinsing system efficiency.

Resource recovery and recycling technologies either directly use waste from one process as raw materials for another process or recover valuable materials from a waste stream before disposal (US EPA 1992; EC 2006). Some spent process baths and much more rinse water can be reused in other operations, and process ingredients can be recovered from rinse water in order to return to process baths or to sell. Multiple-use rinse water is a common reuse application. After rinse water becomes too contaminated for its original purpose, it may be used for other rinse processes. Rinse water can be recycled in a closed loop or open loop system. While treated effluent is returned to the rinse system in the closed loop operation, it is also reused in the rinse system in open loop operation but the rinsing is completed by final rinse being fed by fresh water to ensure high quality rinsing. Metal and their salts can also be recovered from spent baths and rinse water by suitable treatment methods in order to return to baths as makeup and to sell or return to suppliers. Evaporation, reverse osmosis, ion exchange, electrolytic recovery and electrodialysis can be applied individually or in combination for the recovery of metals and their salts.

4.1.6 Conventional end-of-pipe wastewater treatment

As mentioned in Subsection 4.1.4, general treatment approach for the wastewaters originating from the metal finishing facilities is commonly based on the flow segregation. The flow segregation allows the reduction of hexavalent chromium, the destruction of cyanide, the recovery of precious metals, and the removal and/or recovery of oils prior to the removal of common metals. Stream segregation minimises the flow rate of wastewater to be treated in each component and accordingly reduces the cost of these pretreatment applications. A general flow segregation and wastewater treatment scheme are depicted in Figure 4.1 (US EPA 1983).

The segregated flows are subjected to their specific pretreatment applications before combining with the other wastewaters for end-of-pipe treatment. As shown in Figure 4.1, complexed metals bearing wastewaters are segregated and separately treated, solvents are hauled or claimed. The segregated flows containing more than one pollutant such as cyanide and hexavalent chromium bearing streams, and oily wastewaters receive the specialised treatment before passing on to common metal removal. A typical treatment scheme is terminated by metal removal units for the metal finishing industry wastewaters.

Physical and physico-chemical treatment methods based on chemical precipitation, oxidation and separation technologies are commonly used for the treatment of the metal finishing industry wastewater. Table 4.4 highlights the potential treatment methods applied to segregated flows, and covers resource recovery/recycling techniques used in-plant control (updated upon US EPA (1983)). This table also includes promising novel techniques compiled from the related literature.

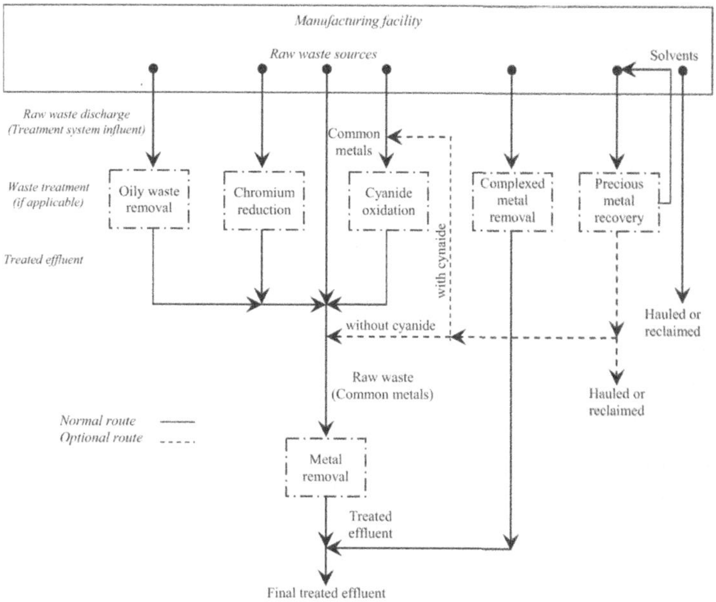

Figure 4.1 A typical wastewater treatment scheme based on flow segregation for the metal finishing industry

"*The best available technologies economically achievable* (BAT-US)" have been described by US EPA for the treatment and control of process wastewater produced in the metal finishing industry facilities (US EPA 1983). The factors taken into consideration in assessing BAT-US cover the cost of achieving BAT-US effluent reductions, the age of equipment and facilities involved, the processes employed, potential process changes, and non-water quality environmental impacts including energy requirements.

In BAT-US definition for the metal finishing industry, treatment systems are described under three options. "*Option 1*" includes emulsion breaking and oils separation by skimming for oily flows, cyanide oxidation, hexavalent chromium reduction prior to hydroxide precipitation and sedimentation for common metal removal; complexed metals bearing flow segregation and separate treatment by high pH hydroxide precipitation and sedimentation; and sludge treatment by drying beds. "*Option 2*" treatment system encompasses all of "*Option 1*" together with polishing filtration. "*Option 3*" treatment system covers "*Option 1*" end-of-pipe treatment system plus the addition of in-plant controls for cadmium

Table 4.4 Potential treatment methods for segregated flows

Potential wastewater pollutants → Treatment methods ↓	Common metals	Precious metals	Complexed metals	Chromium (VI)	Cyanide	Oils	Solvents	Recyclying/recovery
Carbon adsorption	√	√	√			√	√	
Centrifugation						√	√	
Chemical reduction			√	√				
Coalescing						√		
Diatomaceous earth filtration	√	√	√				√	
Electrocoagulation	√		√					
Electrochemical oxidation					√			
Electrochemical reduction				√				
Electrochemical regeneration				√				√
Electrodialysis				√				√
Electrolytic recovery	√	√						√
Emulsion breaking						√		
Evaporation	√	√		√	√			√
Flotation	√					√		
Granular bed filtration	√	√	√					
High pH precipitation			√					
Hydroxide precipitation	√	√	√					
Insoluble starch xanthate	√	√	√					
Ion exchange	√	√	√	√				√
Membrane filtration	√	√	√					
Oxidation by Cl_2 or H_2O_2 or O_3					√			
Oxidation by O_3/UV or H_2O_2/UV					√			
Peat adsorption	√	√	√					
Pressure filtration	√	√	√			√		
Resin adsorption						√	√	
Reverse osmosis	√	√	√			√	√	√
Sedimentation	√	√	√					
Skimming						√		
Sulphide precipitation	√	√	√	√				
Ultrafiltration	√	√	√			√	√	√

including evaporative recovery, ion exchange, and recovery rinses to minimise toxicity. The similar treatment systems for process wastewaters and recovery/recycling techniques used for in-plant control practices are also defined in BREF published for STM by EC (EC 2006). The BAT treatment systems recommended by both US EPA and EC comprise the unit processes that are proven and effective techniques such as precipitation, filtration, and sedimentation. Information about these unit processes is extensive and readily accessible in the related literature (EC 2006; US EPA 1983; 1984; 1997; 2003; Tünay *et al.* 2004) and need not to be reiterated here, since the intention of this book is to provide more detailed information about the oxidation and reduction processes used in treatment of the metal finishing industry wastewaters. Therefore, the oxidation and reduction processes are introduced and discussed in more detail in the following subsections.

4.2 OXIDATION AND REDUCTION PROCESSES

As mentioned in the previous subsections, the oxidation and reduction processes are commonly used for the treatment and recovery/recycling practices of the metal finishing industry wastes. Since implementations of these processes are strongly related to pollutant type to be treated or recovered/recycled, these processes are introduced and discussed on the basis of pollutant type in this chapter. Within this context; the pollutants to be treated by oxidation or reduction process are classified as follows:

- cyanide;
- hexavalent chromium;
- complexed metals;
- precious metals;

In the following subsections, special emphasis is given to well-proven and common treatment methods in terms of their efficiencies, design criteria, and costs, while promising novel techniques are introduced considering their basic features.

4.2.1 Cyanide destruction

4.2.1.1 Alkaline chlorination

Alkaline chlorination is the oldest and most widely recognized cyanide destruction method based on chemical oxidation. The first industrial applications of this method are in the treatment of cyanide plating flows generated from the metal finishing industry facilities (Smith and Mudder 1991). This method has already been recommended for the treatment of cyanide bearing wastewaters of the metal finishing industry as a BAT by both US EPA and EC since it has proved to be a very efficient one (US EPA 1983; 1997; 2003; EC 2006).

The destruction of free cyanide and some weakly and moderately strong complexed metal cyanides, such as copper, nickel and cadmium cyanide, is achieved by the application of alkaline chlorination method. Chlorine can be used as elemental, sodium hypochlorite (NaOCl), or calcium hypochlorite ($Ca(OCl_2)$) form as the oxidant. The destruction of cyanide takes place in two stages under alkaline conditions at ambient temperatures. In the first stage reaction, cyanide oxidises to the intermediate cyanogen chloride (CNCl) according to one of the following reactions depending on the chlorine compound used;

$$NaCN + Cl_2 \rightarrow CNCl + NaCl \tag{4.1}$$

$$NaCN + NaOCl + H_2O \rightarrow CNCl + 2NaOH \tag{4.2}$$

$$2NaCN + Ca(OCl)_2 + 2H_2O \rightarrow 2CNCl + Ca(OH)_2 + 2NaOH \tag{4.3}$$

Since cyanogen chloride is highly toxic and readily volatilises, it requires to be converted into a harmless compound as soon as it is formed. Cyanogen chloride is rapidly hydrolysed to cyanate at high pH values (>10) and its hydrolysis rate decreases at lower pH values. The hydrolysis process is represented as follows;

$$CNCl + 2NaOH \rightarrow NaCNO + NaCl + H_2O \tag{4.4}$$

The conversion of cyanide to cyanate by oxidation completes within 10–30 minutes at pH 8.5–9.0. When solution pH increases to 10–11, the complete conversion is accomplished within 5–7 minutes at ambient conditions. Hence, first stage reaction is always performed by addition of an alkali agent such as sodium hydroxide or lime to ensure pH values above 10.5 to provide safe working conditions.

In the second stage, cyanate is hydrolysed to ammonia and carbonate in the presence of excess chlorine or hypochlorite as a catalyst according to the following reaction;

$$3Cl_2 + 4H_2O + 2NaCNO \rightarrow 3Cl_2 + (NH_4)_2CO_3 + Na_2CO_3 \tag{4.5}$$

The above reaction takes place several hours at alkali pH values and completes in reasonable contact times at pH 8.5–9.0. In spite of the accelerator effect of chlorine, the hydrolysis reaction requires at least 1 or 1.5 hours of contact time. In the presence of excess free chlorine, ammonia reacts further through the breakpoint chlorination process to form nitrogen gas:

$$3Cl_2 + 6NaOH + (NH_4)_2CO_3 + Na_2CO_3 \rightarrow N_2 + 2NaHCO_3 + 6NaCl + 6H_2O \tag{4.6}$$

The second stage reaction can also be accomplished at acid pH values lower than 2.5 within 5 minutes. It is represented by the below reaction;

$$2NaCNO + H_2SO_4 + 4H_2O \rightarrow (NH_4)_2SO_4 + 2NaHCO_3 \tag{4.7}$$

On the other hand, in spite of short reaction time, the above application is rarely used owing to relatively high cost and neutralisation requirement of the effluent (Tünay *et al.* 2004).

The oxidation of cyanide bearing wastewaters by alkaline chlorination is a widely used treatment method in the facilities utilising cyanide in cleaning and plating baths owing to some advantages as compared to the other removal techniques. The advantages of this process for handling cyanide bearing wastewaters are listed as operation at ambient temperature, suitability for automatic control, and relatively low cost. Disadvantages include the failure in the treatment of some strong metal cyanide complexes, the necessity for careful pH control, possible chemical interference in treating combined wastewaters, and the potential hazard of storing and handling chlorine gas unless sodium or calcium hypochlorite is used, and possibility of formation of chlorinated organic compounds (US EPA 2003).

High cyanide removal efficiencies (higher than 99 %) can be achieved by alkaline chlorination provided that operating conditions such as chlorine dose, operation pH, and contact time are carefully optimised (US EPA 1983). Free cyanide and weakly complexed metal cyanide concentrations are typically reduced to less than 0.2 mg/L and 1 mg/L, respectively (Ghosh *et al.* 2006a). On the other hand, some strong cyanide metal complexes such as ferro-cyanide and cobalt-cyanide are not oxidised by alkaline chlorination processes (Tünay *et al.* 2004). Oxidation of nickel-cyanide requires longer contact time for the case of complete destruction of cyanide. Chlorine demand increases beyond breakpoint to destroy chloramines for ammonia containing wastewaters. Though, chloramines are able to oxidise cyanide, the oxidation rate of insoluble metal cyanides by chloramines is too slow resulting in incomplete destruction in reasonable contact times.

Theoretical chlorine demands are 2.73 g and 4.1 g per gram of cyanide for the first and second stages of alkaline chlorination process, respectively. In most cases, higher chlorine dose may be required owing to other chlorine-consuming constituents such as ammonia, organic matter, and oxidisable metals like cuprous copper. The required sodium hydroxide doses are 1.13 g and 1.02 g per g chlorine gas applied for the first and second stages of the process.

Alkaline chlorination process can be operated either in a batch or continuous operation mode. Treatment system is composed of a balancing tank followed by two reaction tanks for continuous mode operation. Each tank is equipped with an electronic controller to monitor and maintain the required pH and oxidation-reduction potential (ORP). Chlorine or sodium hypochlorite is metered to the first reaction tank as required to maintain the ORP in the range of 350 to 400 mV, and meanwhile aqueous sodium hydroxide (50 %) is fed to the tank to keep the pH of approximately 11. After oxidising the cyanides to cyanates in the first reaction tank, the reaction solution are passed to the second tank in which the reaction conditions are maintained to hydrolyse cyanate to ammonia and carbonate. The required ORP and pH for this conversion process are 600 mV and 8 to 9, respectively. Each reaction tank is equipped with a mechanical mixer designed to provide slow mixing conditions (approximately one turnover per minute) (US EPA 1983; 2003). The required reaction times are minimum 30 minutes and 1.5 hours for the first and second stage reactions, respectively (Tünay *et al.* 2004). The batch operation takes place in two tanks. Cyanide bearing flows are collected over a specified time period in one tank and the collected wastewater is treated in the other tank. In the case of frequent dumps, another tank can serve to equalise the flows. When this holding tank fills up, the wastewater is delivered to the treatment tank (US EPA 2003).

Capital costs for batch and continuous alkaline chlorination systems to destroy cyanide prior to chemical precipitation and sedimentation have been estimated by US EPA (2003) as approximately €21,650 (on 1996 cost basis; $28,862) for flows less than 3.78 L/min. For the cyanide bearing wastewater flows varying between 3.78 L/min and 757 L/min, a flow-dependent capital cost (CC) estimation formula has been developed for batch and continuous alkaline chlorination systems by US EPA (2003). Using this CC formula, capital costs for a 500 L/min system can be calculated as €56,500. US EPA (2003) has also proposed a formula being function of flow rate, operating hours per day and days per year in order to estimate annual costs including operation & maintenance, labour and materials, energy costs, and chemicals (*e.g.*, sodium hydroxide, sulphuric acid, sodium hypochlorite). Typical operation and maintenance costs have also been reported to be in the range of €3.75 and €5.25 per kilogram cyanide oxidised by Ghosh *et al.* (2006a).

4.2.1.2 Ozonation

Recently, ozonation is becoming more and more common application for the oxidation of cyanide, particularly in European countries. Free cyanide is effectively oxidised by ozonation. This method has also proven to be a successful oxidation process in the treatment of some metal cyanide complexes such as zinc, cadmium, silver, copper and nickel cyanides, but ferri and cobalt cyanide complexes are resistant to ozonation (US EPA 1983; Ghosh *et al.* 2006a). Ozone oxidises cyanide to cyanate under alkaline condition, as shown by the reaction below:

$$CN^- + O_3 \rightarrow CNO^- + O_2 \tag{4.8}$$

The oxidation of cyanide to cyanate is a rapid process and completes within 15 minutes at pH values of 9.0 to 10.0. If ozone supply is prolonged to yield an excess dose in the solution, the cyanate can be oxidised to ammonia and bicarbonate (or carbonate depending on solution pH) by the reaction:

$$CNO^- + O_3 + 2H_2O \rightarrow NH_3 + HCO_3^- + 1.5O_2 \tag{4.9}$$

The conversion presented by Eq. (4.9) is a slow process and requires the prolonged ozonation application for complete oxidation unless it is accelerated by a homogenous catalyst such as copper (Patterson 1985; Gurol and Holden 1988), a heterogeneous catalyst such as nickel and cobalt oxides (Udrea and Avramescu 2004), promoted by ultraviolet (UV) radiation (Kumar and Bose 2005; Mudliar *et al.* 2009), or combination of UV with H_2O_2 (Kepa *et al.* 2008) to form hydroxyl ($HO^\bullet$) radicals. Hence, the prolonged ozonation implementation in order to supply excess ozone concentration is not economically feasible (US EPA 1983).

Alternatively, cyanate can be hydrolysed to carbonate and ammonia under highly alkaline conditions and thereafter, ammonia is oxidised to nitrate. These reactions are shown as below (Parga *et al.* 2003);

$$CNO^- + H_2O + OH^- \rightarrow NH_3 + CO_3^{2-} \tag{4.10}$$

$$NH_3 + 4O_3 \rightarrow NO_3^- + H^+ + H_2O + O_2 \tag{4.11}$$

The major disadvantage of the above alternative conversion is the formation of nitrate which may require further treatment step.

Because of the above mentioned disadvantages, in practice, other economically applicable processes such as biological oxidation following the first stage ozonation are preferred for further destruction of cyanates (Tünay *et al.* 2004).

Some advantages of ozonation for handling wastewaters are that it is stronger than chlorine, it is well suited to automatic control, its on-site generation eliminates chemicals procurement, no dissolved solid contributes to treated effluents, and it oxidises organic compounds to relatively less toxic, short-chain organic acids, ketones, and aldehydes (US EPA 1983; 2003). The disadvantages of ozonation process are high capital costs, possible chemical interference in the treatment of combined wastes, and energy requirement of 15 to 22 kWh per kilogram of ozone generated. Additionally, cyanide is not economically destructed beyond the cyanate form (US EPA 1983; 2003).

Since ozone is supplied as a gas which is bubbled through the wastewater, the oxidation rate is limited by mass transfer of ozone, and temperature. Oxidation rate increases with ozone feeding rate during the cyanide ozonation process and cyanide concentration decreases linearly with time. According to literature findings cyanide oxidation is represented by the first order reaction kinetics with respect to ozone and zero order with respect to cyanide (Carrillo-Pedroza *et al.* 2000; Parga *et al.* 2003; Barriga-Ordonez *et al.* 2006).

Optimum ozone dose has been reported as 1.8 g to 2.2 g ozone per gram of cyanide for the partial oxidation of cyanide to cyanate by US EPA (1983) and Barriga-Ordonez *et al.* (2006), and 4.6 g to 5.0 g per gram of cyanide for complete oxidation based on excess ozone application by US EPA (1983). Under the optimised operation conditions of ozonation application, cyanide can be oxidised to below detection limit (0.2 μg/L) in electroplating wastewaters.

A typical ozonation system consists of an ozone generator, gas diffusion system, a closed reaction tank equipped with mixer, and off-gas control system to collect the release of unreacted ozone. Ozone must be generated on-site and delivered directly to the reaction tank.

Ozonation is an economically impractical process for wastewater flows higher than 1.9 m^3/min. Capital costs for a typical 1.9 m^3/min ozonation system for treating cyanide bearing wastewaters have been reported as approximately €656,000, with typical operation and maintenance costs of €1.5 per kilogram cyanide oxidised by Ghosh *et al.* (2006a).

4.2.1.3 Oxidation by H_2O_2

Oxidation by hydrogen peroxide is a well-established cyanide destruction method for wastewaters generated from the metal finishing and mining industries. Although, this method and particularly its modifications seem to be a promising technique in the treatment of cyanide bearing wastewaters, owing to their high operation costs, their full scale applications are limited (EC 2006). Cyanate is the first intermediate product of cyanide oxidation by hydrogen peroxide, and then cyanate oxidises to ammonium and carbonate ions in the presence of excess hydrogen peroxide (Goldstein 1976; Young and Jordan 1995; Sarla *et al.* 2004) by the following reactions:

$$CN^- + H_2O_2 \rightarrow CNO^- + H_2O \tag{4.12}$$

$$CNO^- + 2H_2O \xrightarrow{H_2O_2} NH_4^+ + CO_3^{2-} \tag{4.13}$$

Optimum pH range is between 9.5 and 10.5 for the Eq. (4.12) (Hartinger 1994). Equation (4.13) proceeds very slowly under alkaline conditions and accelerates with decreasing pH (US EPA 2000). Due to disproportionation reaction of hydrogen peroxide into oxygen and water, the loss of hydrogen peroxide is one of the problems associated with long reaction time (Pak and Chang 1997). Presence of other oxidisable organic and inorganic matters competing for hydrogen peroxide is another factor resulting in high hydrogen peroxide consumption. These factors lead to the requirement of excess dose of hydrogen peroxide varying between 200 and 450 % (Smith and Mudder 1991). To overcome the problem associated with long reaction time, the cyanide oxidation by hydrogen peroxide can be accelerated by the addition of a homogenous metal catalyst such as copper, zinc and cadmium (Sarla *et al.* 2004), heterogeneous catalyst such as Ru/MgO (Pak and Chang 1997), or activated carbon prepared from olive stone (Yeddou *et al.* 2010). Since the copper-catalysed hydrogen peroxide oxidation processes, being proprietary, have been commonly employed as cyanide destruction methods in the treatment of mining industry wastes, their process descriptions, operation conditions, and design criteria will be addressed in Chapter 8.

One of the copper-catalysed hydrogen peroxide oxidation processes, known as Kastone Process, is also applicable to cyanide bearing wastewaters, especially those from cyanide-zinc and cyanide-cadmium electroplating (US EPA 1983). By applying this process, cyanide is oxidised to cyanate and metals (*e.g.* cadmium or zinc) are precipitated as hydroxides. This process is performed by the addition of a proprietary formulation (made up of 41 % hydrogen peroxide with formaldehyde and catalyst) at 49–54°C and alkaline pH values 10.5 to 11.8 under vigorous mixing conditions. The reaction completes in about one hour. Two tanks equipped with heaters, and air spargers or mechanical stirrers require for conversion step. While conversion reactions take place in one tank, the second tank may serve to collect the wastewaters. These tanks can be operated in batch or continuous mode. A settling tank or a filtration system is used to separate the metal hydroxides. The major advantage of this process is that the precipitated metals can be recovered without requiring an additional treatment step for metal precipitation, and together with cyanide, zinc or cadmium can be reduced below 0.1 mg/L. The main drawback of this process is that it is only able to oxidise cyanide to cyanate, the addition of formaldehyde may bring about treated wastewater with high organic matter values, and operating costs are expensive due to energy expenditures for heating. Chemical costs are cheaper than those for oxidation with hypochlorites, and close to those for alkaline chlorination (US EPA 1984).

A limited number of studies have been published on the cyanide oxidation by hydrogen peroxide accelerated by the addition of heterogeneous catalysts or activated carbon. Pak and Chang (1997) performed a series of experiments to find out the usability of Ru/MgO as a heterogeneous catalyst for the oxidation of cyanide by hydrogen peroxide at lab-scale. It has been reported that the oxidation rate for this system depends on operation conditions and the optimum operation conditions were determined as initial H_2O_2 concentration of 600 mg/L, Ru/MgO dose of 1 g/L, pH 7.0, and temperature higher than 50°C for complete oxidation of 300 mg/L cyanide. Recently, cyanide oxidation by hydrogen peroxide in the presence of activated carbon prepared from olive stone has been studied by Yeddou *et al.* (2010). The results obtained from their lab-scale experiments have been confirmed the catalytic effect of activated carbon on the process as well as effectiveness of this system on the complete conversion of cyanide to cyanate.

4.2.1.4 Sulphur dioxide/air process

The INCO process developed and patented by the International Nickel Company of Canada utilises mixtures of sulphur dioxide and oxygen in air, in the presence of copper to oxidise cyanide. This process has been proven to be an effective and successful treatment technique for the removal of all forms of cyanide including the stable iron complexes from clarified barren, decant solution, and heap leach rinse solution of mines (Smith and Mudder 1991). Hence, it is widely used in the treatment of mining industry wastes, the INCO process will be introduced in Chapter 8.

4.2.1.5 Oxidation by chlorine dioxide

There has been recent attention to the substitution of chlorine with chlorine dioxide in the treatment of cyanide bearing wastewaters by the oxidation process. This oxidant may become competitive with chlorine if suitable chlorine dioxide generators are developed for this purpose.

Chlorine dioxide yields an instantaneous conversion of cyanide to cyanate and chlorite according to the following reaction (Noack and Iacovielo 1994);

$$CN^- + 2ClO_2 + H_2O \rightarrow CNO^- + 2ClO_2^- + 2H^+ \tag{4.14}$$

In the presence of cuprous copper being in the form of copper cyanide complex, the above reaction proceeds in two stages: (i) rapid oxidation of free and complex cyanide to cyanate with formation of chlorite and (ii) a slower cuprous copper catalysed oxidation of cyanide by chlorite. This reaction completes within 10 minutes in the presence of 25 mg/L cuprous copper.

Recently, the use of chlorine dioxide in a gas-sparged hydrocyclone has been tested in a bench and pilot scale applications to destruct the cyanide wastes by Parga *et al.* (2003). Their results indicated that the complete destruction of free cyanide could be achieved at a wide range of pH varying between 2.57 and 11.23 while complexed cyanides destruction was stringently dependent on the reaction pH. The highest complexed cyanide destruction efficiency was obtained as 78.8 % at pH 11.23 in their study.

4.2.1.6 Hexavalent ferrate oxidation

Recently, hexavalent ferrate (FeO_4^{2-}) suggested as an environmentally-friendly oxidant (a green chemical), has been nominated as a candidate for the treatment of cyanide, thiocyanate, and some metal cyanide complexes such as cuprous copper-cyanide, and zinc-cyanide (Sharma 2004; Sharma *et al.* 1998; 2005; 2008; Yngard *et al.* 2007; Lee and Tiwari 2009). Hexavalent ferrate is a powerful oxidant, and under acidic conditions, its redox potential (E^o: 2.2 V) is the highest of any oxidant used in wastewater treatment (Sharma 2004). Ferric hydroxide which is a by-product formed during the reduction of hexavalent ferrate to ferric iron, serves as a potential coagulant and adsorbent for metal removal. Hence, the application of hexavalent ferrate plays a multifunctional role in wastewater treatment (Sharma 2004; Lee and Tiwari 2009). The reactions of cyanides with hexavalent ferrate can be presented as follows (Sharma *et al.* 1998; 2005; 2008; Sharma 2004; Yngard *et al.* 2007; Lee and Tiwari 2009):

$$CN^- + HFeO_4^- + H_2O \rightarrow CNO^- + Fe(OH)_3 + OH^- + 0.25O_2 \quad (4.15)$$

$$SCN^- + HFeO_4^- + 5H_2O \rightarrow CNO^- + 4Fe(OH)_3 + 2OH^- + SO_4^{2-} + O_2 \quad (4.16)$$

$$CuCN_4^{3-} + 5HFeO_4^- + 8H_2O \rightarrow 4CNO^- + 5Fe(OH)_3 + 6OH^- + Cu^{2+} + 1.5O_2 \quad (4.17)$$

Sharma *et al.* (2008) have suggested that the further reaction of cyanates produced in the oxidation of cyanides yields nitrite which is further oxidised by hexavalent ferrate to nitrate. On the other hand, the results given in the related scientific literature indicates that the conversion rate of cyanate to nitrate depends on the operation pH and this conversion partly occurs at pH 7.5 while complete conversion is obtained at pH 9 and 12.0 (Sharma *et al.* 1998; 2005).

Hexavalent ferrate completely and rapidly oxidises cyanides to cyanate in seconds to minutes. On the other hand, its oxidation rate strictly depends on the form of cyanides to be oxidised (*e.g.* free cyanide, weakly or moderately strong complexed metal cyanides), initial cyanide concentration, solution pH, and temperature. In a study of Sharma *et al.* (1998), the treatment of an electroplating rinse water sample containing free cyanide of 24 mg/L (928 μM) by hexavalent ferrate was explored so as to determine suitable operation condition at lab-scale. It was found that an excess dose of higher than 2.3 times the molar amount of hexavalent ferrate was required for complete oxidisation of cyanide of 6 mg/L (233 μM) at pH 8.6. When cyanide concentration increased to 16.4 mg/L (630 μM) the conversion rate was limited to 90 % at FeO_4^{2-}/CN^-: 2/1 (on the molar basis) at pH 8.6, 9.1, and 9.8.

Except the study conducted by Sharma *et al.* (1998), in the other related scientific researches, synthetically prepared samples were used to mimic the real industrial wastewaters originating from gold mining and metal finishing industries (Sharma 2004; Sharma *et al.* 2005; 2008; Yngard *et al.* 2007; Lee and Tiwari 2009). In these researches, oxidation of free cyanide and metal complexed cyanides such as zinc, copper, and nickel cyanides by hexavalent ferrate at varying operation conditions was examined under lab-scale conditions and this method has been suggested as an applicable and effective treatment method.

4.2.1.7 Photochemical and photocatalytic oxidation methods

As mentioned in Chapter 1, photochemical oxidation processes involve generation of $HO^{\bullet}$ through UV photolysis of conventional oxidants such as ozone and hydrogen peroxide. In the photocatalytic oxidation processes, homogenous or heterogeneous catalysts are used to destruct pollutants by means of UV light-induced redox reactions. Although, these processes have been recognized as successful methods in the oxidation of cyanides, in practice, they are only economically feasible for the treatment of low flow rates. Additionally, these processes require special reactor designs for UV illumination. Therefore, their full scale applications are limited.

A limited number of research focused photochemical or photocatalytic oxidation of cyanide performed using real metal finishing industry wastewaters in the scientific literature (Mudliar *et al.* 2009). Other studies have been conducted using synthetically prepared samples (SPS) under lab-scale conditions (Augugliaro *et al.* 1999; Chiang *et al.* 2002; Dabrowski *et al.* 2002; Hernández-Alonso *et al.* 2002; Barakat *et al.* 2004; Christoskova and Stoyanova 2009). The results obtained from these studies are summarised as follows.

Photochemical and photocatalytic oxidations of cyanide (SPS; 100 mg/L; 3.85 mM) were studied in an annular type batch reactor with 25 W low-pressure UV lamp in the presence of hydrogen peroxide using with/without Cu^{2+} as catalyst at pH 10 by Sarla *et al.* (2004). While complete oxidation of cyanide to cyanate was obtained at an optimum H_2O_2 dose of 35.5 mM in 40 minutes for UV/H_2O_2 system, the addition of 19 mg/L Cu^2 to this system remarkably accelerated the reaction rate and cyanide was completely oxidised to cyanate in 9 minutes. Pseudo-first order reaction rates were calculated as 0.099 min^{-1} and 0.7035 min^{-1} for the UV/H_2O_2 and $UV/H_2O_2/Cu^{2+}$ systems, respectively. The prolonged reaction time (2 hours) yielded further oxidation of cyanate to bicarbonate and nitrogen.

Mudliar *et al.* (2009) investigated destruction of cyanide from an automobile industry wastewater by UV/O_3, UV/H_2O_2, and $UV/H_2O_2/O_3$ using a double walled sieve photoreactor with 8 W UV lamp. The wastewater had cyanide of 250.76 ± 8.6 mg/L, iron concentration of 212.84 ± 13.8mg/L, cadmium of 119.17 ± 9.5 mg/L, nickel of 78.22 ± 9.2 mg/L, lead of 53.79 ± 4.4 mg/L, and pH 12 to13. The wastewater was diluted at varying ratios (1/25–15/25) to determine the effect of initial cyanide concentration on the processes performance. For all tested processes, an increase in initial cyanide concentration accelerated cyanide oxidation rate, decreased specific energy consumption, and extended reaction time. Optimum doses were determined as an H_2O_2 of 116.8 g/L and initial O_3 of 5 mg/L for all samples. Cyanide oxidation of undiluted sample completed in 2.51 hours yielding conversion efficiency of 99 % and the lowest specific energy consumption was obtained as 67 ± 5 kWh/kg CN treated for the $UV/H_2O_2/O_3$ system.

Until now, a large number of studies dealing with oxidation of cyanide by semiconductor-mediated heterogeneous photocatalysis have been published in the scientific literature. These studies are mostly carried out using SPS containing free cyanide at varying concentrations between 12 and 100 mg/L. The catalytic efficiencies of various types of semiconductor heterogeneous catalysts such as titanium dioxide (Augugliaro *et al.* 1999; Dabrowski *et al.* 2002; Hernández-Alonso *et al.* 2002; Barakat *et al.* 2004; Bozzi *et al.* 2004), manganese modified cobalt oxide (Christoskova and Stoyanova 2009), and copper doped titanium dioxide (Chiang *et al.* 2002) have been tested and demonstrated at varying doses of catalysts (1–10 g/L) under lab-scale conditions. These applications yielded complete conversion of cyanide to cyanate. In some cases, nitrate formation at low stoichiometric ratios implying partial oxidation of cyanate was also observed.

4.2.1.8 Other methods

Electrochemical oxidation (EC) is an effective treatment method for the removal of free cyanide and weakly complexed metal cyanides from the concentrated cyanide bearing streams. On the other hand, this process fails to treat strongly complexed metal cyanides such as iron and cobalt cyanides.

Cyanide is completely oxidised to cyanate, and then further oxidation of cyanate takes place to yield carbon dioxide, nitrogen, and ammonia during EC application under alkaline conditions (pH >10). Cyanide concentration reduces from a few hundred mg/L to less than 1.0 mg/L by the application of this process (US EPA 1983). The treated effluents sometimes bear cyanate in the case of its incomplete oxidation. EC application can also be conducted on at high temperatures (52–93°C) to accelerate the oxidation reactions (Ghosh *et al.* 2006b). The required units of EC process are a reactor equipped with anode and cathode, a power supply, a holding tank, a pump and a proper ventilation system. EC can be operated in a batch or continuous mode. The key parameters affecting removal performance as well as specific energy consumption are the applied current density (A/m^2), electrolyte concentrations, reaction time, and operation temperature. Optimisation based on operation parameters is the best approach to maximise process performance and minimise specific energy consumption.

High temperature alkaline chlorination process (HTACP) is another alternative treatment method for cyanide bearing wastes originating from metal finishing and mining industry. Free cyanide, weakly and strongly complexed metal cyanides can be effectively removed from waste streams and sludges by the application of this method. HTACP which is a modification of alkaline chlorination is performed at elevated temperatures varying between 140 and 180°C instead of ambient temperatures. Application bases of this process are similar to alkaline chlorination (explained in Subsection 4.2.1.1). HTACP takes place in three stages: (i) liberation of free cyanide from complexed metal cyanides by means of their dissociations; (ii) oxidation of free cyanide to cyanate (iii) oxidation of cyanate to nitrogen, carbon dioxide and water. Some disadvantages of this process can be listed as relatively high operation and maintenance costs, the prolonged reaction times up to 9 hours for the case of sludge treatment, and in some cases incomplete cyanide oxidation because of association of cyanide with metal hydroxides precipitating during the first stage of process (Ghosh *et al.* 2006b). Full scale implementation of HTACP has been limited owing to relatively high operation costs.

4.2.2 Hexavalent chromium reduction

4.2.2.1 Chemical reduction by the reduced sulphur compounds

Reduction of hexavalent chromium to trivalent chromium by means of chemical reducing agents and subsequent hydroxide precipitation of trivalent chromium have already been recommended as BATs by both US EPA and EC, as these processes have been fully proven techniques based on years of experience (US EPA 1983; 1997; 2003; EC 2006).

In aqueous solution, hexavalent chromium exists as oxyanions such as chromates ($HCrO_4^-$ and CrO_4^{2-}), or dichromate ($Cr_2O_7^{2-}$) species depending on pH. Chemical reduction of hexavalent chromium is accomplished using reduced sulphur compounds as reduction agents under acidic conditions. Gaseous sulphur dioxide (SO_2), sodium bisulphite ($NaHSO_3$), and sodium metabisulphite ($Na_2S_2O_5$) are used as reducing agents (US EPA 1983). Among them, gaseous sulphur dioxide is the most commonly preferred reduction agent, particularly for continuous treatment systems, because of its low cost. When gaseous sulphur dioxide is fed into water, it hydrolyses to sulphurous acid to yield sulphite:

$$3SO_2 + 3H_2O \rightarrow 3H_2SO_3 \tag{4.18}$$

And then sulphite reduces hexavalent chromium to trivalent chromium according to the following reaction;

$$3SO_3^{2-} + 2HCrO_4^- + 8H^+ \rightarrow 2Cr^{3+} + 3SO_4^{2-} + 5H_2O \tag{4.19}$$

Hexavalent chromium reduction using sodium bisulphite and sodium metabisulphite is chemically similar to Eq. (4.19). The reduction rates accelerate with decreasing pH values and the fastest reduction rate is obtained at the pH range 1.8 to 2.0. The reduction process completes in 5–10 minutes at this pH level. On the other hand, the reduction rate is too slow at pH levels above 5 (Taylor and Qasim 1982; US EPA 1983; 2003). Considering that the other oxidising substances such as dissolved oxygen and ferric iron consume the reducing agents, an excess dose of reducing agents are always used to ensure the complete reduction of hexavalent chromium. Thus, minimum 10 mg/L of remaining sulphite is required after the complete reaction (Tünay *et al.* 2004).

Some advantages of chemical reduction of hexavalent chromium reduction by the reduced sulphur compounds, particularly sulphur dioxide, can be listed as: the process is well suited to automatic control; operation is well defined and straightforward; and the equipment is readily obtainable from many suppliers (US EPA 1983; 2003). Additionally, there is no sludge production during the reduction process. The major disadvantage of the process is that the cost of treatment chemicals may be significant for the high concentrations of hexavalent chromium. Furthermore, storage and handling of sulphur dioxide is hazardous.

The treatment system consists of an equalisation tank with 2 hours retention time, and two reaction tanks connected in series. The reaction tanks are equipped with mixers designed to provide approximately one turnover per minute, pH and ORP meters, and chemical feed systems. Retention time for each reaction tank is about 45 minutes. After equalisation, the wastewater is passed on the reaction tanks. Sulphuric acid is fed into the tanks to keep a pH level of approximately 2 and a reducing agent is metered the tanks to maintain the ORP within the range of 250 to 300 mV. Hexavalent chromium concentration can be reduced to less than 0.05 mg/L by this application (US EPA 1983; 2003).

Capital costs for batch and continuous systems to reduce hexavalent chromium to trivalent chromium by sodium metabisulphite prior to chemical precipitation and sedimentation have been determined by US EPA as approximately €15,670 (on 1996 cost basis; $20,892) for hexavalent chromium bearing flows less than 3.78 L/min (US EPA 2003). Using CC formula proposed by US EPA (2003) for flows varying between 3.78 L/min and 170 L/min, capital costs for a 500 L/min system can be calculated as €44,100.

4.2.2.2 Chemical reduction by iron and iron compounds

Ferrous iron is also an effective reducing agent used in the chemical reduction of hexavalent chromium. The capability of ferrous iron ($FeSO_4$) in the reduction of hexavalent chromium has been well documented (Eary and Rai 1988; Fenford and Li 1996; Sedlak and Chang 1997; Pettine *et al.* 1998; Schlautman and Han 2001; Chang 2003). The reduction of bichromate by ferrous iron can be represented by the following general overall reaction:

$$3Fe^{2+} + H_xCrO_4^{x-2} + iH_2O \rightarrow Cr(OH)_y^{3-y} + 3Fe(OH)_z^{3-z} + jH^+ \quad (4.20)$$

where the possible hydrolysed species are defined by $x = 0$ to 2, y and $z = 0$ to 4, $i = \{y + 3z\}$, and $j = \{x + y + 3z\}$ (Fendorf and Li 1996). This reaction is rapid with complete reduction occurring within 1-2 minutes at the stoichiometric dose (Fe^{2+}/Cr^{6+}: 3/1 on the molar basis) (Eary and Rai 1988). The reduction rate of above mentioned reaction is a function of pH. It decreases over the pH range of 1.5–4.5 and increases from 5.0 to 8.7, while remains approximately constant in the pH range of 4.5–5.0 (Pettine *et al.* 1998). Dissolved oxygen may have a negative effect on the reduction of hexavalent chromium at high pH values greater than 8.0 and very low hexavalent chromium concentrations less than 10 μM because of possible oxidation of ferrous iron by dissolved oxygen to ferric iron (Fendorf and Li 1996). The reduction of hexavalent chromium by ferrous iron brings about rapid precipitation of ferric iron and/or chromium hydroxides in a pH range of 5.0–11.0 (Eary and Rai 1988). Hexavalent chromium concentration can be reduced to under detectable level (2 μg/L) by ferrous sulphate reduction (Chang 2003).

Hexavalent chromium can also be effectively removed by means of ferrous sulphide used mainly as a precipitation agent for metal sulphide precipitation. In this application, hexavalent chromium is reduced to trivalent state by sulphide and ferrous iron which are oxidised to elemental sulphur and ferric iron, and precipitates as chromium hydroxide (US EPA 1983; Bhattacharyya *et al.*, 1979). Therefore, the application of metal sulphide precipitation plays a multifunctional role in the treatment of hexavalent chromium. This process can be represented according to the following reaction:

$$2FeS + Cr_2O_7^{2-} + 7H_2O \rightarrow 2S^0 + 2Cr(OH)_3 + 2Fe(OH)_3 + 2OH^- \quad (4.21)$$

The above reaction is accomplished using ferrous sulphide slurry obtained by mixing ferrous sulphate and sodium hydrosulphite. Solution pH must be maintained above 8 throughout the reaction to prevent the generation of toxic hydrogen sulphide gas. Since hydroxyl ions are produced during the process, pH adjustment may be required prior to discharge of the treated effluent.

During the last decade, there has been great interest in the remediation of groundwater contaminated with heavy metals by means of permeable reactive barrier using zero-valent iron. Therefore, researches have been focused on applicability of different forms of zero-valent iron such as nanoparticle (Wu *et al.* 2009), powder (Hou *et al.* 2008; Zhou *et al.* 2008), granular (Lo *et al.* 2006), sphere (El-Shazly *et al.* 2005; Fiúza *et al.* 2010), and scrap (Gheju and Iovi 2006; Gheju *et al.* 2008) in the reduction of hexavalent chromium to trivalent state. The results demonstrated that hexavalent chromium can be effectively removed by zero-valent iron and removal efficiency of the process depends on several parameters such as type and the specific surface area of zero-valent iron, initial hexavalent chromium concentration, presence of foreign ions such as carbonate, divalent metal cations, and complexing agents, and reaction conditions *e.g.* pH, zero-valent iron dosage, temperature, reaction time, and mixing rate. In the view of the success in the remediation of groundwater supported by laboratory and field tests, zero-valent iron has been nominated as a candidate reducing agent for the treatment of hexavalent chromium bearing wastewaters originating from the metal finishing industry (Chang 2003; 2005; Chen *et al.* 2008). Chang (2003; 2005) has defined the responsible reactions for reduction of chromate by zero-valent iron, at initial pH values of 3.0 to 10.0, as:

$$Fe^0 + CrO_4^{2-} + 4H_2O \rightarrow Cr(OH)_3 + Fe(OH)_3 + 2OH^- \quad (4.22)$$

$$xCr(OH)_3 + (1-x)Fe(OH)_3 \rightarrow (Cr_xFe_{1-x})(OH)_3 \quad (4.23)$$

The reactions given by Eqs. (4.22) and (4.23) mainly take place on the solid surface of zero-valent iron. Removal of hexavalent chromium by zero-valent iron is through chromate adsorption, reduction, and subsequent precipitation of $Fe(OH)_3$ and $Cr(OH)_3$, or $(Cr_xFe_{1-x})(OH)_3$) ((oxy)hydroxide phase). The results obtained from Chang's studies (2003; 2005) indicated that (i) the hexavalent chromium reduction efficiency was strictly pH-dependent and increased with decreasing pH; (ii) the adsorption/reduction efficiency was limited by the transport process (for the case of powder iron) or the diffusion (for the case of iron wires) of hexavalent chromium from bulk solution or wastewater to the iron surface, and because of these limitations proper mixing and sufficient contact time were crucial in achieving higher removal efficiencies; (iii) due to hydroxyl ion production, solution pH was increased through the reaction resulting in formation of solid phase; and (iv) the total chromium concentration could be reduced from 20 mg/L to 1 mg/L within 3–4 hours in batch system. Consequently, this process has been suggested by Chang (2003; 2005) as a feasible and effective method, particularly, to treat metal cleaning wastewaters containing low amount of hexavalent chromium.

In another study, the treatability of electroplating wastewater with high chromate concentration (534 mg/L) was investigated using wasted cast iron by using a plug flow reactor at acidic pH values (<1.8) under lab-scale conditions (Chen *et al.* 2008). Their results indicated that process efficiency depended on hydraulic retention time and initial pH of the solution. A decrease in pH or an increase in hydraulic retention times resulted in faster chromate reductions and longer breakthrough times. Complete chromate reductions were achieved at pH values 1.3 to 1.7 in hydraulic retention times of 26–98 minutes. Solution pH markedly increased through the process.

4.2.2.3 Electrochemical oxidation and generation

During last few decades, electrochemical treatment techniques involving combinations of coagulation, adsorption, oxidation, and flotation processes have been successfully applied to metal finishing industry wastewaters in order to treat hexavalent chromium. In these researches, various types of electrode materials such as stainless steel (Ölmez-Hanci 2009), iron (Arroyo *et al.* 2009) polypyrrole-modified stainless steel (Tian and Yang 2007), aluminium (Adhoum *et al.* 2004) gold, glassy carbon, boron doped diamond anode (Welch *et al.* 2005), reticulated vitreous carbon and its polyaniline modification (Ruotolo *et al.* 2006), iron and copper-iron bimetallic (Lugo-Lugo *et al.* 2010), and lead-nickel (US EPA 1983) were used as an anode and cathode in the shape of either rode or plate. Results demonstrated that these techniques were very efficient in reducing hexavalent chromium to trivalent chromium from metal finishing industry wastewaters under the optimum operation conditions in batch or continuous mode.

In electrochemical techniques, the applied electric current induces oxidation/reduction coupled redox reactions upon electrode surfaces as mentioned in Chapter 1. Additionally, simultaneous coagulation and adsorption take place as a consequence of in-situ metal hydroxide formation when stainless steel, iron or aluminium sacrificed electrodes are used. The inherent advantages of these techniques over well-known physicochemical processes have been also discussed in Chapter 1.

When inert electrodes such as gold, glassy carbon, boron doped diamond *etc.* are used to reduce hexavalent chromium, the responsible removal mechanisms of electrochemical treatment are oxidation/reduction coupled redox reaction favouring acidic pH conditions (< 3). Hexavalent chromium is completely reduced to trivalent state by electrochemical techniques if current density and pH are selected properly (Welch *et al.* 2005; Ruotolo *et al.* 2006; Tian and Yang 2007; Lugo-Lugo *et al.* 2010). For electrochemical techniques using the sacrificed electrodes, widely called as electrocoagulation, reduction/oxidation, coagulation, adsorption, and flotation processes are the responsible removal mechanisms. In this technique, while sacrificed anode is releasing free metal ions into the solution, hexavalent chromium and water are reduced in the cathode. In addition, ferrous iron produced in anode through electrocoagulation process can directly reduce hexavalent chromium to trivalent state for the case of stainless steel and iron electrodes. Because of hydroxyl ion production in cathode, the solution pH increases and trivalent chromium is removed by means of co-precipitation of $Cr(OH)_3$ and $Fe(OH)_3$ or $(Cr_xFe_{1-x})(OH)_3$. As mentioned in the literature, current density applied and initial as well as final pH achieved at the end of electrocoagulation application are important operation parameters significantly affecting the process performance and operating cost (Adhoum *et al.* 2004; Arroyo *et al.* 2009; Ölmez-Hanci 2009; Lugo-Lugo *et al.* 2010). Increasing the electrolyte (NaCl) concentration accelerates hexavalent chromium removal, decreases the power consumption and shortens the reaction time (Adhoum *et al.* 2004; Arroyo *et al.* 2009; Ölmez-Hanci 2009). In addition, chloride may also generate chlorine/hypochlorite serving as an oxidising agent during electrocoagulation process at a proper pH range (3–4<pH<11) and for an appropriate electrode material *e.g.* iron or stainless steel (Adhoum *et al.* 2004; Arroyo *et al.* 2009). Therefore, optimisation of operation parameters such as the applied current density, initial pH, and supporting electrolyte concentration is crucial to maximise electrocoagulation process performance and minimise operating costs (Ölmez-Hanci 2009). Under optimised operation conditions, both hexavalent and trivalent chromium can be completely removed by electrocoagulation processes within 20–70 minutes (Arroyo *et al.* 2009; Ölmez-Hanci 2009; Lugo-Lugo *et al.* 2010).

4.2.3 Complexed metal bearing wastewater treatment

As mentioned earlier, complexed metal bearing wastewaters must be segregated and separately treated to prevent further complexing of metals present in the free form in the other flows. In these wastewaters, the metals are weakly or strongly tied up or complexed by inorganic or organic ligands (complex formers) whose function is to keep them in solution. When metals are bound by weak complex formers such as succinic acid, acetic acid *etc.*, they are successfully treated by hydroxide precipitation under regular treatment conditions without requiring any pretreatment step (Kabdaşlı 1990; Tünay and Kabdaşlı 1994). On the other hand, since strong complex formers are mostly chelates that bind the metals with more than one hand to form stable structures *i.e.* strongly complexed, due to their high stabilities these types of complexed metals cannot be efficiently treated under conventional hydroxide precipitation conditions. Therefore, they call for the modified hydroxide precipitation or specialised pretreatment application (Kabdaşlı *et al.* 2009). The main obstacle of the modified hydroxide precipitation is that complex formers which may be harmful in many ways are simultaneously released into the environment. Via application of a specialised pretreatment the complexing agents can be completely destroyed or converted into forms that do not interfere with subsequent conventional hydroxide precipitation. The specialised pretreatment methods which are applicable for complexed metal bearing wastewaters can be classified as reduction of complexed metals to zero-valent state, electrochemical processes, advanced oxidation processes, sonochemical destruction, and physical separation techniques. These methods, except physical separation techniques, are discussed as follows.

4.2.3.1 Chemical reduction

Reduction of complexed metals to zero-valent metals can be accomplished by electrolytic recovery method or using reducing agents such as sodium borohydride, dithiocarbamate, and hydrazine (US EPA 2003). Since electrolytic recovery method is generally applied to precious metal bearing flows as an in-plant control practice, it will be discussed in the following subsection.

Sodium borohydride ($NaBH_4$) has been widely used to reduce a variety of cations to metallic state to recover precious and heavy metals as well as to remove heavy metals from metal finishing industry wastewaters (Lu *et al.* 1997). It has been demonstrated that the process is sensitive to the reaction conditions such as initial metal concentration, and mixing efficiency. The stoichiometry of borohydride depends on the experimental conditions, the electronegativity of the metal and catalysis on borohydride, reaction pH, and the form of the metal salt in solution (Lu *et al.* 1997). The following independent reactions representing the reduction of the cobalt by sodium borohydride depending on pH are proposed by Shen *et al.* (1993):

$$4Co^{2+} + BH_4^- + 8OH^- \rightarrow 4Co^0 + BO_2^- + 6H_2O \qquad (4.24)$$

$$4Co^{2+} + 2BH_4^- + 6OH^- \rightarrow 2Co_2B + 6H_2O + H_2 \tag{4.25}$$

One of the above reactions takes place depending on working pH. Lu *et al.* (1997) reported that cobalt ions transform to Co_2B according to Eq. (4.25) at a pH range of 5–8. On the other hand, in the related literature this pH range was reported as lower than 10.0–12.0 for Ni_2B whose formation can be written similar to Eq. (4.25) (Tünay *et al.* 1997; Ying *et al.* 1987).

Applicability of sodium borohydride reduction to nickel-EDTA and nickel-acetic acid complexes was also tested at pH range of 10–12 by Tünay and co-workers (1997). Results showed that at the stoichiometric sodium borohydride doses and at pH 12.0 (i) nickel concentration could be reduced from 6000 mg/L to 45.5 mg/L for nickel-EDTA system; (ii) complete nickel removal could be obtained for nickel-acetic acid system; and (iii) bivalent nickel converted to colloidal zero-valent nickel during the process at all tested pH values (10–12).

Dithiocarbamate (DTC) is an effective reducing agent used in the treatment of complexed metal bearing flows. It can reduce complexed metals to zero-valent metals in stoichiometric ratio to the metals present ($(M^{m+})/(DTC^{2-})$: 1/1 on the molar basis). On the other hand, if used incorrectly, DTC compounds which are a class of pesticides, may create several problems in the aquatic systems. In addition to this disadvantage, large amount of sludge is produced by DTC reduction and the process is expensive because of high cost of DTC (Chen and Lim 2002; US EPA 2003). The capital costs associated with DTC systems are equal to those of hexavalent chromium reduction systems using sodium metabisulphite given in Subsection 4.2.2.1 (US EPA 2003).

Hydrazine (N_2H_4) can effectively reduce various metal cations (M^{2+}) to zero-valent state (M^0) at alkaline conditions (pH >11) according to the following reaction (Chen and Lim 2002):

$$2M^{2+} + N_2H_4 \rightarrow 2M^0 + N_2 + 4H^+ \tag{4.26}$$

One mole hydrazine is enough to reduce two mole divalent metals. For the case of monovalent cations the reaction stoichiometry is 4 mole M^+ to 1 mole N_2H_4. Atmospheric oxygen may decompose dilute solutions of hydrazine into nitrogen. Chen and Lim (2002) found that a 0.05 M hydrazine solution was decomposed to extent of 20 % at the end of 16 hours exposure to the atmosphere. They also reported that (i) hydrazine was an effective reducing agent for the recovery of silver and copper; (ii) presence of humic acids did not hinder the recovery of copper, but rather, improved process efficiency whereas humic acids thwarted the recovery of silver; and (iii) at high initial metal concentrations, more finer particles were produced, and the seeding and ageing process had no positive effect on particle size.

The major operational problem faced in the reduction of complexed metals by reducing agent to zero-valent is the separation of fine particles formed upon process. This separation problem can be overcome using suitable separation techniques such as gravity clarification, centrifugation, or coagulation prior to sedimentation/filtration.

4.2.3.2 Other methods

Recently, it has been demonstrated that the electrocoagulation process using stainless steel electrodes has a high capacity in simultaneously removing both heavy metals and organic complex formers from a nickel and zinc plating effluents (Kabdaşlı *et al.* 2009). Under the optimised operation conditions, complete metal removals (nickel and zinc) and about 66 % TOC abatement were achieved. The results of another study performed by Kabdaşlı *et al.* (2010) indicated that combination of electrocoagulation with Fenton's reagent could enhance organic matter removal which almost equalled complex former removal. In these studies, heavy metal removal mechanism has been described as breaking the structure of complex former via oxidation, followed by hydroxide precipitation, and adsorption or entrapment of the metals on freshly produced ferric hydroxide flocs.

Recently, many researchers have also studied heterogeneous photocatalysis of complexed metals over titanium dioxide as an alternative treatment method. The process efficiency was tested for various metal-complex systems such as hexavalent chromium-citric acid (Meichry *et al.* 2007), copper-ferric-EDTA (Park *et al.* 2006), nickel-EDTA (Madden *et al.* 1997; Salama and Berk 2005), copper-EDTA (Madden *et al.* 1997), lead-EDTA (Madden *et al.* 1997; Vohra and Davis 2000), and cadmium-EDTA (Madden *et al.* 1997; Davis and Green 1999) at varying reaction conditions in lab-scale. It has been demonstrated that high removal efficiencies in terms of metals and complex formers could be achieved by the process.

4.2.4 Recovery of precious and common metals

A number of methods are available to recover common metals as well as precious metals such as gold and silver from the metal finishing industry wastewaters. The recovery methods can be listed as electrolytic metal recovery, electrodialysis, ion exchange, and evaporation. Among these techniques, only electrolytic metal recovery involves oxidation/reduction process and other methods are out of scope of this book. Hence, only electrolytic metal recovery will be introduced in this subsection.

Electrolytic metal recovery is one of the most common methods employed to recover dissolved metals from metal finishing wastewaters. This process is able to recover 90 to 95 % of the available metals in gold, silver, copper, zinc, cadmium, tin, and solder alloy platings (US EPA 1997; 2003). The most common applications of electrolytic metal recovery encompass gold cyanide plating, silver cyanide plating, brass cyanide plating, acid copper plating, cadmium cyanide plating, copper cyanide plating, and zinc cyanide plating (US EPA 1997). Electrolytic metal recovery can also be used in electroless plating operations, even though its application is not as straightforward due to the presence of other components such as complex formers, reducing agents, and stabilisers. Hexavalent chromium cannot be recovered by electrowinning. This method is the most effective in the case of high metal concentrations and is not feasible or applicable to continuous rinses with lower metal concentrations because of the extended operation time. Electrolytic metal recovery is applied to the static rinses following plating operations by circulating the bath solution through the recovery tank or to the exhausted plating baths before the end-of-pipe wastewater treatment. Furthermore, acidic, metal-rich, and cation regenerates originating from ion exchangers are well suited to electrolytic recovery.

A typical electrolytic metal recovery system comprises of an electrolytic cell equipped with two electrodes (an anode and a cathode) placed in the solution, a transfer pump, and a DC power supply (rectifier). Distance between the anode and the cathode is closely fixed. While ions in the solution are moving toward the charged electrodes, the dissolved metals are reduced and deposited on cathode. The deposited metals are removed either by mechanical or chemical means to refine, recycle or dispose of (US EPA 1997). The capacity of recovery system depends on total surface area of cathode and the efficiency of rectifier. A cathode area can range from 0.09 m^2 to 9.9 m^2 or larger. The cathodes used in electrolytic metal recovery system are classified as: (i) flat-plate; (ii) expanded metal, wire mesh, or reticulate plate; and (iii) porous or woven carbon and graphite plate (US EPA 1997). Flat-plate cathodes with the lowest surface area are used only for the flow containing high concentration of metal (1000–20,000 mg/L). Reticulate cathodes with a metalized woven fibre design are typically used for recovering metals with lower concentrations (<100 mg/L). Carbon and graphite cathodes with the highest surface area are usually used where metal concentration is below 1,000 mg/L (US EPA 2003).

The advantages of electrolytic metal recovery are that its application results in saleable non-hazardous products; and cleaning and maintenance of recovery unit are required low labour. Some disadvantages of this method are that, in some cases, energy costs may exceed the recovery value of the metal; electrode replacement costs may be significant for units using disposable cathodes, particularly for high metal recovery rates; it is required a ventilation system for the case of flows containing hydrochloric acid or other compounds containing chloride because of in-situ generation of chlorine gas; and strong oxidising agents such as nitric acid or fluoroboric acid shorten usability of anodes (US EPA 1997; 2003).

Capital cost of electrolytic recovery system is moderately low and varies between €6,000 and €11,250 (on the basis of 1995 cost) depending on electrode types for recovering 500 grams of metal per day (US EPA 1997). Several factors such as metal concentration recovered, rectifier efficiency, and electrode materials and their replacement periods play determining roles on operating costs. The most significant components of operating costs are electricity, electrode replacement and maintenance costs.

4.3 CONCLUDING REMARKS

The metal finishing industry is one of the largest industrial activities using a wide range of chemicals in the world. If not properly managed, the chemicals used in the metal finishing operations may adversely impact public health and the environment. The best management policy proposed by environmental protection authorities such as US EPA and EC includes (i) efficient raw material, energy and water usage; (ii) the substitution by less harmful chemical in the production; and (iii) minimisation, recovery and recycling of wastewater and wastes. During last few decades, there has been a great effort in substitutions of hazardous chemicals such as hexavalent chromium, cyanide, and cadmium by less harmful and environmentally-friendly substances. On the other hand, in some cases, these substitution efforts fail by economical constrains and/or unsatisfactory product quality. Capital, operation and maintenance costs as well as achievable treatment levels are important key factors for wastewater and waste minimisation, recovery and recycling implementations. When these factors are taken into account, stream segregation seems to be the most feasible approach in the treatment of metal finishing industry wastewaters with different character in the view of maximisation of removal performance and reduction of treatment cost. Within this context, the application of oxidation/reduction techniques to stream segregated flows of the metal finishing industry can be evaluated as follows.

Oxidation/reduction techniques used in the treatment of cyanide, hexavalent chromium and precious metals bearing wastewaters are well-established and straightforward methods. Among them, alkaline chlorination for the destruction of cyanide, chemical reduction by reduced sulphur compounds for reduction of hexavalent chromium and electrolytic metal recovery for precious metals are the most widely used treatment methods in the metal finishing industry facilities because of moderately low costs. When properly applied, satisfactorily removal efficiencies ensuring discharge standards can be achieved by these methods. Some oxidation/reduction techniques introduced in this chapter, particularly those used in the treatment of complexed metals, are sensitive to reaction conditions and their misapplications result in unsatisfactory pollutant removal efficiencies. Some methods yield partial oxidation of pollutant or certain by-products such as cyanate, ammonia, nitrate and sulphate which necessitate an additional treatment step prior to discharge to receiving media. Some processes are expensive to operate and/or energy-intensive methods and hence they are considered economically applicable only for the source-based treatment or the treatment of flows generated in small volumes. Some techniques such as photochemical and photocatalytic oxidations are required specific reactor designs. Application of some techniques such as electrocoagulation produces treatment sludge and brings on an additional sludge treatment costs. Therefore, optimisation of operation conditions is a crucial step to minimise sludge production and operating costs as well as maximise the process performance. Besides these limitations and economical constrains, all oxidation/reduction techniques introduced in this chapter are effective treatment methods in the removal of specific waste groups of the metal finishing industry under the defined operation conditions. At any rate, pilot-scale or at least lab-scale studies are essential steps on a case by case to select the most suitable treatment techniques and to determine their specific design criteria and operation conditions in the implementations of these treatment methods.

REFERENCES

Adhoum, N., Monser, L., Bellakhal, N. and Belgaied J.E. (2004) Treatment of electroplating wastewater containing Cu^{2+}, Zn^{2+} and Cr(VI) by electrocoagulation. *J. Hazard. Mater.* **B112**, 207–213.

Arroyo, M.G., Pérez-Herranz, V., Montañéz, M.T., García-Antón, J. and Guiñón, J.L. (2009) Effect of pH and chloride concentration on the removal of hexavalent chromium in a batch electrocoagulation reactor. *J. Hazard. Mater.* **169**, 1127–1133.

Augugliaro, V., Gálvez, J.B., Vázquez, J.C., López, E.G., Loddo, V., Muñoz M.J.L., Rodrígez, S. M., Marcì, G., Palmisano, L., Schiavello, M and Ruiz, J. S. (1999) Photocatalytic oxidation of cyanide in aqueous TiO_2 suspensions irradiated by sunlight in mild and strong oxidant conditions. *Catal. Today* **54**, 245–253.

Barakat, M.A., Chen, Y.T. and Huang, C.P. (2004) Removal of toxic cyanide and Cu (II) ions from water by illuminated TiO_2 catalyst. *Appl. Catal. B* **53**, 13–20.

Barriga-Ordonez, F., Nava-Alonso, F. and Uribe-Salas, A. (2006) Cyanide oxidation by ozone in a steady-state flow bubble column, *Minerals Eng.* **19**, 117–122.

Bhattacharyya, D., Jumawan, Jr., A.B. and Grieves, R.B. (1979) Separation of toxic heavy metals by sulfide precipitation. *Sep. Sci. Technol.* **14**, 441–452.

Bozzi, A., Guasaquillo, I. and Kiwi, J. (2004) Accelerated removal of cyanides from industrial effluents by supported TiO_2 photo-catalysts, *Appl. Catal. B* **51**, 203–211.

Carrillo-Pedroza, F.R., Nava-Alonso, F. and Uribe-Salas, A. (2000) Cyanide oxidation by ozone in cyanidation tailings: Reaction kinetics. *Minerals Eng.* **13**(5), 541–548.

Chang, L.Y. (2003) Alternative chromium reduction and heavy metal precipitation methods for industrial wastewater. *Environ. Progress* **22**(3), 174–182.

Chang, L.Y. (2005) Chromate reduction in wastewater at different pH levels using thin iron wires-A laboratory study. *Environ. Progress* **24**(3), 305–316.

Chen, S.S., Hsu, B.C. and Hung, L.W. (2008) Chromate reduction waste iron from electroplating wastewater using plug flow reactor. *J. Hazard. Mater.* **152**, 1092–1097.

Chen, J.P. and Lim, L.L. (2002) Key factors in chemical reduction by hydrazine for recovery of precious metals, *Chemosphere* **49**, 363–370.

Chiang, K., Amal, R. and Tran, T. (2002) Photocatalytic degradation of cyanide using titanium dioxide modified with copper oxide. *Adv. in Environ. Res.* **6**, 471–485.

Christoskova, St. and Stoyanova, M. (2009) Catalytic oxidation of cyanide in a aqueous phase over individual and manganese-modified cobalt oxide systems. *J. Hazard. Mater.* **165**, 690–695.

Dabrowski, B., Zaleska, A., Janczarek, M., Hupka, J. and Miller, J.D. (2002) Photo-oxidation of dissolved cyanide using TiO_2 catalyst. *J. Photochem. Photobiol. A: Chem.* **151,** 201–205.

Davis, A.P. and Green D.G. (1997) Photocatalytic oxidation of cadmium-EDTA with titanium dioxide. *Environ. Sci. Technol.* **33**, 609–617.

Eary, L.E. and Rai, D. (1988) Chromate removal from aqueous waste by reduction with ferrous iron. *Environ. Sci. Technol.* **22**, 972–977.

EC (2006) Integrated Pollution Prevention and Control Reference Document on Best Available Techniques for the Surface Treatment of Metals and Plastics, European Commission.

El-Shazly, A.H., Mubarak, A.A. and Konsowa, A.H. (2005) Hexavalent chromium reduction using a fixed bed of scrap bearing iron spheres, *Desalination* **185**, 307–316.

Fendorf, S.E. and Li, G. (1996) Kinetics of chromate reduction by ferrous iron. *Environ. Sci. Technol.* **30**, 1614–1617.

Fiúza, A., Silva, A., Carvalho, G., de la Fuente, A.V. and Delerue-Matos, C. (2010) Heterogeneous kinetics of the reduction of chromium (VI) by elemental iron. *J. Hazard. Mater.* **174**, 1042–1047.

Gheju, M. and Iovi, A. (2006) Kinetics of hexavalent chromium reduction by scrap iron. *J. Hazard. Mater.* **B135**, 66–73.

Gheju, M., Iovi, A. and Balcu, I. (2008) Hexavalent chromium reduction with scrap iron in continuous-flow system part 1: effect of feed solution pH, *J. Hazard. Mater.* **153**, 655–662.

Ghosh, R.S., Thheis, T.L., Smith, J.R. and Whong-Chong, G.M. (2006a) Ambient temperature oxidation technologies for treatment of cyanide. In *Cyanide in Water and Soil* (eds. R.S. Ghosh, D.A. Dzombak and G.M. Whong-Chong), pp. 517–570, Taylor&Francis Group, LLC, CRC Press, London, United Kingdom (ISBN 978-1-56670-66-7).

Ghosh, R.S., Smith, J.R. and Whong-Chong, G.M. (2006b) Thermal and high temperature oxidation technologies for treatment of cyanide. In *Cyanide in Water and Soil* (eds. R.S. Ghosh, D.A. Dzombak and G.M. Whong-Chong), pp 439–451, Taylor&Francis Group, LLC, CRC Press, London, United Kingdom (ISBN 978-1-56670-66-7).

Goldstein, M. (1976) Economics of treating cyanide wastes. *Pollut. Eng.* **8**, 36–38.

Gurol, M.D. and Holden, T.E. (1988) The effect of copper and iron complexation on removal of cyanide by ozone, *Ind. Eng. Chem. Res.* **27**, 1157–1162.

Hernández-Alonso, M., Coronado, J., Maira, A.J., Soria, J., Loddo, V. and Augugliaro, V. (2002) Ozone enhance activity of aqueous titanium dioxide suspensions for photocatalytic oxidation of free cyanide ions. *Appl. Cat. B:Environ.* **39**, 257–269.

Hartinger, L. (1994) *Handbook of Effluent Treatment and Recycling for the Metal Finishing Industry*, Finishing Publications Ltd & ASM International, 2nd ed., Herts, United Kingdom.

Hou, M., Wan, H., Liu, T., Fan, Y., Liu, X. and Wang, X. (2008) The effect of different divalent cations on the reduction of hexavalent chromium by zerovalent iron. *Appl. Cat. B: Environ.* **84**, 170–175.

Kabdaşlı, N.I. (1990) Complexed metal treatment by hydroxide precipitation. M.Sc. thesis, Environ. Egn. Dept. Institute of Sciences, İstanbul Technical University. (in Turkish).

Kabdaşlı, I., Arslan, T., Ölmez T., Arslan-Alaton, I. and Tünay, O. (2009) Complexing agent and heavy metal removals from metal plating effluents by electrocoagulation with stainless steel electrodes. *J. Hazard. Mater.*, **165**, 838–845.

Kabdaşlı, I., Arslan, T., Arslan-Alaton, I., Ölmez T. and Tünay, O. (2010) Organic matter and heavy metal removals from complexed metal plating effluent by the combined electrocoagulation/Fenton process. *Wat. Sci. Technol.* **61**, 2617–2624.

Kepa, U., Stanczyk-Manazek, E. and Stepniak, L. (2008) The use of the advanced oxidation process in the ozone + hydrogen peroxide system for the removal of cyanide from water. *Desalination* **223**, 187–193.

Kumar, R. and Bose, P. (2005) Modeling free and copper-complexed cyanide degradation in a continuous flow mixed ozone conductor. *Ind. Eng. Chem. Res.* **44**, 776–788.

Lee, S.M. and Tiwari, D. (2009) Application of ferrate (VI) of industrial wastes containing metal-complexed cyanide: A green treatment. *J. Environ. Sci.* **21**, 1347–1352.

Lo, I.M.C., Lam, C.S.C. and Lai, K.C.K. (2006). Hardness and carbonate effects on the reactivity of zero-valent iron for Cr (VI) removal. *Wat. Res.* **40**, 595–605.

Lu, J., Dreisinger, D.B. and Cooper, W.C. (1997) Cobalt precipitation by reduction with sodium borohydride. *Hydrometallurgy* **45**, 305–322.

Lugo-Lugo, V., Barrera-Díaz, C., Bilyeu, B., Balderas-Hernández, P., Ureña-Nuñez, F. and Sánchez-Mendieta, V. (2010) Cr (VI) reduction in wastewater using bimetallic galvanic reactor. *J. Hazard. Mater.* **176**, 418–425.

Madden, T.H., Datye, A.K. and Fulton, M. (1997) Oxidation of metal-EDTA complexes by TiO_2 photocatalysis. *Environ. Sci. Technol.* **31**, 3475–3481.

Meichtry, J.M., Brusa, M., Mailhot, G, Grela, M.A. and Litter, M.I. (2007) Heterogeneous photocatalysis of Cr (VI) in the presence of citric acid over TiO_2 particles: Relevance of Cr (VI)-citrate complexes. *Appl. Cat. B: Eniron.* **71**, 101–107.

Mudliar, R., Umare, S.S., Ramteke, D.S. and Wate, S.R. (2009) Energy efficient-Advanced oxidation process for treatment of cyanide containing automobile industry wastewater. *J. Hazard. Mater.* **164**, 1474–1479.

Noack, M. and Iacoviello, A.S. (1994) The chemistry of chlorine dioxide in industrial and wastewater treatment applications. In *Chemical Oxidation Technologies for the Nineties* (eds. W.W. Echenfelder, A.R. Bowers and J.A. Roth), Technomic Publishing Company Inc., Lancaster, United States.

Orhon, D., Germirli Babuna, F., Kabdaşlı, I., Dulkadiroğlu, H., Doğruel, S. and Ölmez, T. (2000) *Water Consumption, Amount and Characterization of Wastewater*. Research and Development Foundation Project Report, Project No: 2001/01/23, Uzel Makine San. A.Ş., ITU, Istanbul, Turkey (in Turkish).

Ölmez-Hanci, T. (2009) The optimization of Cr (VI) reduction and removal by electrocoagulation using response surface methodology. *J. Hazard. Mater.* **162**, 1371–1378.

Pak, D. and Chang, W. (1997) Oxidation of aqueous cyanide solution using hydrogen peroxide in the presence of heterogeneous catalyst. *Environ. Technol.* **18**, 557–561.

Parga J.R., Shukla, S.S. and Carillo-Pedroza, F.R. (2003) Destruction of cyanide waste solutions using chlorine dioxide, ozone and titania sol, *Waste Man.* **23**, 183–191

Park, E.H., Jung, J. and Chung H.H. (2006) Simultaneous oxidation of EDTA and reduction of metal ions in mixed Cu(II)/Fe(III)-EDTA system by TiO_2 photocatalysis. *Chemosphere* **64**, 432–436.

Patterson, J.W. (1985) *Industrial Wastewater Technology*, 2nd edn, Butterworth-Heinemann, Boston, USA, p. 115.

Pettine, M., D'Ottone, L., Campanella, L., Millero, F.J. and Passino, R. (1998) The reduction of chromium (IV) by iron (II) in aqueous solutions. *Geochim. Cosmochim. Acta* **62**, 1509–1519.

Ruotolo, L.A.M., Santos-Júnior, D.S. and Gubulin, J.C. (2006) Electrochemical treatment of effluents containing Cr(VI). Influence of pH and current on the kinetic. *Water Res.* **40**, 1555–1560.

Sarla, M, Pandit, M., Tyagi, D.K. and Kapoor J.C. (2004) Oxidation of cyanide in aqueous solution by chemical and photochemical process. *J. Hazard. Mater.* **B116**, 49–56.

Schlautman, M.A. and Han, I. (2001) Effects of pH and dissolved oxygen on the reduction of hexavalent chromium by dissolved ferrous iron in poorly buffered aqueous systems. *Water Res.* **35**, 1534–1546.

Sedlak, D.L. and Chan, P.G. (1997) Reduction of hexavalent chromium by ferrous iron, *Geochim. Cosmochim. Acta* **61**, 2185–2192.

Sharma, V.K., Rivera, W., Smith, J.O. and O'Brandon, B. (1998) Ferrate (VI) oxidation of aqueous cyanide. *Environ. Sci. Technol.* **32**, 2608–2613.

Sharma, V.K. (2004) Use of iron (VI) and iron in water and wastewater treatment. *Water Sci. Technol.* **49**, 69–74.

Sharma, V.K., Burnett, C.R., Yngard, R.A. and Cabelli, D.E. (2005) Iron (VI) and iron (V) oxidation of copper (I) cyanide. *Environ. Sci. Technol.* **39**, 3849–3854.

Sharma, V.K., Yngard, R.A., Cabelli, D.E. and Baum, J.C. (2008) Ferrate (VI) and ferrate (V) oxidation of cyanide, thiocyanate, and copper (I) cyanide. *Rad. Phys. Chem.* **77**, 761–767.

Shen, J., Li, Z., Yan, Q. and Chen, Y. (1993) Reactions of bivalent metal ions with borohydride in aqueous solution for the preparation of ultrafine amorphous alloy particles. *J. Phys. Chem.* **97**, 8504–8511.

Salama, P. and Berk, D. (2005) Photocatalytic oxidation of Ni-EDTA in a well-mixed reactor. *Ind. Eng. Chem. Res.* **44**, 7071–7077.

Smith, A. and Mudder, T.I. (1991) *The Chemistry and Treatment of Cyanidation Wastes*, Mining Journal Books Ltd., London, England, (ISBN 0-900117-51-69).

Tian, Y. and Yang, F. (2007) Reduction of hexavalent chromium by polypyrrole-modified steel mesh electrode. *J. Cleaner Prod.* **15**, 1415–1418.

Taylor, C.R. and Qasim, S.R. (1982) More economical treatment of chromium-bearing wastes. *Proc. of 37th Industrial Waste Conference.* Purdue University, West Lafayette, IN, USA, pp. 189–196.

Tünay, O. and Kabdaşlı, N.I. (1994) Hydroxide precipitation of complexed metals. *Wat. Res.* **28**, 2117–2124.

Tünay, O., Alp, K., Kabdaşlı, N.I., Çeviker, S. and Eremektar, G. (1997) Heavy metal treatment by chemical reduction. *Fresenius Envir. Bull.* **6**, 278–283.

Tünay, O., Kabdaşlı, I. and Hung, Y.T. (2004) Treatment of metal finishing wastes. In *Handbook of Industrial Hazardous Waste Treatment*, 2nd edn (eds. L.K. Wang, Y.T. Hung, H.H. Lo and C. Yapijakis), Marcel Dekker Inc., Revised and Expanded, pp. 289–359, (ISBN 0-8247-4114-5).

Udrea, I. and Avramescu, S. (2004) Catalytic oxidation of SCN^- and CN^- ions from aqueous solutions. *Environ. Technol.* **25**, 1131–1141.

US EPA (1983) *Development Document for Effluent Limitation Guidelines New Source Performance Standards for the Metal Finishing Point Source Category*, Effluent Guidelines Division Office of Water Regulations and Standards, US Environmental Protection Agency, Washington D.C., United States.

US EPA (1984) *Guidance Manual for Electroplating and Metal Finishing Pretreatment Standards*, EPA-440/1-84/091g, Effluent Guidelines Division and Permit Division, U.S Environmental Protection Agency, Washington D.C., United States.

US EPA, (1992) *Guides to Pollution Prevention, The Metal Finishing Industry*, EPA/625/R-92/011, Risk Reduction Engineering Laboratory, Center for Environmental Research Information, Office of Research and Development, US Environmental Protection Agency, Cincinnati, United States.

US EPA (1995) *EPA Office of Compliance Sector Notebook Project.* Profile of the Fabricated Metal Product Industry, EPA/310-R-007, Office of Compliance, Office of Enforcement and Compliance Assurance, US Environmental Protection Agency, Washington D.C., United States.

US EPA (1997) *Pollution Prevention for the Metal Finishing Industry, A Manual for Pollution Prevention Technical Assistance Providers*, EPA/742/B-97/005, Office of Pollution Prevention, US Environmental Protection Agency, Washington D.C., United States.

US EPA (2000) *Managing Cyanide in Metal Finishing*, Capsule Report, EPA 625/R-99/009, Office of Research and Development, US Environmental Protection Agency, Cincinnati, United States.

US EPA (2003) *Development Document for the Final Effluent Limitation Guidelines and Standards for the Metal Products and Machinery Point Source Category*, EPA-821-B-03-001, Office of Water, US Environmental Protection Agency, Washington D.C., United States.

Vohra, M.S. and Davis, A.P. (2000) TiO_2-assisted photocatalysis of lead-EDTA. *Water Res.* **34**, 952–964.

Welch, C., Nekrassova, O. and Compton, R.G. (2005) Reduction of hexavalent chromium at solid electrodes in acidic media: Reaction mechanism and analytic applications. *Talanta* **65**, 74–80.

Wu, Y., Zhang, J., Tong, Y. and Xu, X. (2009) Chromium (VI) reduction in aqueous solutions by Fe_3O_4-stabilized Fe^0 nanoparticles. *J. Hazard. Mater.* **172**, 1640–1645.

Yeddou, A. R., Nadjemi, B., Halet, F., Ould-Dris, A. and Capart, R. (2010) Removal of cyanide in aqueous solution by oxidation with hydrogen peroxide in presence of activated carbon prepared from olive stone. *Min. Eng.* **23**, 32–39.

Ying., W., Bonk, R.R. and Tucker, M.F. (1987) Precipitation treatment of spent electroless nickel plating baths. In *Proc. of Thirty Seventh Industrial Waste Conf.* West Lafayette.

Yngard, R., Damrongsiri, S., Osathaphan, K. and Sharma, V.K. (2007) Ferrate (VI) oxidation of zinc-cyanide complex. *Chemosphere* **69**, 729.735.

Young, C.A. and Jordan, T.S. (1995) Cyanide remediation: Current and past technologies. In *Proc. of the Tenth Annual Conference on Hazardous Waste Research,* Kansas, USA, pp. 104–129.

Zhou, H., He, Y., Lan, Y., Mao, J. and Chen, S. (2008) Influence of complex reagents on removal of chromium (VI) by zero-valent iron. *Chemosphere* **72**, 870–874.

Chapter 5

Pharmaceutical industry

5.1 INTRODUCTION

The pharmaceuticals industry is a subcategory of the organic chemicals industry and many of the cleaner production practices in chemical manufacturing apply to the pharmaceuticals production. The laboratory operations of pharmaceutical companies are similar to those in medical research organizations and hospitals. The industry is characterised by large plants with highly advanced technology and sophisticated instrumentation for analytical procedures (US EPA 1996). Major cleaner production improvements usually come from redesigning processes or recycling of major waste streams such as solvents (US EPA 1996). Case studies are the best sources of information for such solutions. But low-cost options such as improved housekeeping, dry cleaning, and solvent substitution maintenance can also offer significant savings and reduce waste emissions. Cleaner production and pollution guides for chemical industries, for hospitals and medical research organizations and centres should be also considered. A lot of leading pharmaceutical companies have put their Environment Health and Safety reports on their websites that provide very useful references. However, it is often very difficult to obtain process-specific, quantitative information because process details are a major competitive secrecy issue in this sector.

Increasing competition from both domestic and foreign firms, as well as from the generic drug market has forced mergers between the larger and medium-scaled pharmaceutical companies. Such mergers and alliances allow companies to draw upon each others' research expertise, bring products to the market more rapidly and fast once they are approved by Food and Drug Administration (FDA). For instance, the US pharmaceuticals industry accounts for about one-third of all pharmaceuticals marketed worldwide, followed by Europe (27 %) and Japan (22 %). Other major U.S.A. trading partners are Canada and Mexico that together imported 15 % of all U.S.A. exports (US Department of Commerce 1992). The largest importer of US pharmaceuticals is the European Community (EC). Although Japan still remains one of the largest importers of U.S. pharmaceuticals, Japanese pharmaceutical companies are investing heavily in their research and development (R&D), thereby reducing Japan's import share recently.

5.2 PHARMACEUTICAL PRODUCTS AND PROCESSES

5.2.1 Pharmaceutical products

According to the Standard Industrial Classification (SIC) code established by the US Office of Management and Budget (OMB) to track the flow of goods and services within this sector is "283" (SIC 1987). OMB is going to change the classification code system to a system based on similar production processes called North American Industrial Classification System (NAICS) very soon. The industry is further categorised by four 4-digit SIC codes consisting of (i) Medicals and botanicals (SIC 2833); (ii) Pharmaceutical preparations (SIC 2834); (iii) In vivo and in vitro diagnostic substances (SIC 2835); and (iv) Biological products, except diagnostics (SIC 2836). According to the U.S. Census of Manufacturers in 1992, the categories coded 2833 and 2834 account for 64 % of establishments in the U.S. and 81 % of the value of shipments in the industry (US EPA 1997). As defined by its SIC code, the pharmaceuticals industry (SIC 283) consists of establishments that are primarily involved in fabricating or processing medicinal chemicals and pharmaceutical products. The industry also includes establishments that formulate pharmaceutical products and are involved in grinding, grading and milling of botanical products. Although this industry is part of the Chemicals and Allied Products category (SIC 28), it differs significantly from the rest of the chemicals industry in its processes and regulatory requirements (US EPA 2004). For example, pharmaceutical industries use more batch operations and since bulk manufacturing operations involve extracting relatively small, highly concentrated quantities of active ingredients from large volumes of raw materials, the production yield of this industry is relatively low. The pharmaceutical industry is also subjected to an extensive regulatory oversight as compared with other industrial sectors. According to US EPA, any drug may be marketed only after approval by the U.S. Federal FDA. The drug development process averages 15 years (US EPA 2004). Following the discovery of a new drug that appears to have efficacy in treating illness, pre-clinical tests and clinical trials are conducted. Because of high approval costs and times, effective patent protection, that are valid for 20 years, is an essential component in the decision to invest in drug development and marketing. While the International Conference on Harmonization is proposing harmonised rules for drug registration and approval for Europe, Japan and the Unites States, each country retains its approval system. In other countries, especially developing countries, the issue of adequate patent protection is a central concern of pharmaceutical manufacturers (PhRMA 1997). The competition for discovering new drugs and bringing them to the market is extremely high. Hence, a significant amount of the industry's sales are reinvested into R&D. The Pharmaceutical Research and Manufacturers of America (PhRMA) estimates that over 21 % of total sales will be reinvested into R&D by its members (PhRMA 1997). The pharmaceutical industry manufactures bulk substance pharmaceutical intermediates and active ingredients which are further processed into finished products.

5.2.1.1 Medicals and botanicals (SIC 2833)

Companies in this category are primarily engaged in (i) manufacturing bulk organic and inorganic medicinal chemicals and their derivatives, and (ii) processing bulk botanical drugs and herbs. The industry is made up of facilities that manufacture products of natural origin, hormonal products, and basic vitamins, besides active medicinal principals (*i.e.* alkaloids) from botanical drugs and herbs (US EPA 1997). These substances are used as active ingredients for the Pharmaceutical Preparations industry category.

5.2.1.2 Pharmaceutical preparations (SIC 2834)

This category is made up of companies that manufacture, fabricate and process raw materials into pharmaceutical preparations for human and veterinary uses. Finished products are sold in various dosage forms including tablets, capsules, ointments, solutions, suspensions and powders. These are (i) preparations aimed for use mainly by dental, medical, or veterinary professionals, and (ii) those aimed for use by patients and the general public by prescriptions or over-the-counter (US EPA 1997).

5.2.1.3 In vivo and in vitro diagnostic substances (SIC 2835)

This category includes facilities that produce in vivo (tested inside a living organism) and in vitro (tested outside of a living organism) diagnostic substances. Chemical, biological and radioactive substances are produced being used in diagnosing and monitoring health.

5.2.1.4 Biological products, diagnostics excluded (SIC 2836)

Bacterial and virus vaccines, toxoids, serums, plasmas and other blood derivatives for human and veterinary use other than those classified under SIC 2835. According to 1992 US Census of Manufacturers, US Department of Commerce, Bureau of Census 1992 and 1995, the number of establishments, companies, and value of shipments are being highest in the category "Pharmaceutical Preparations" and account for 49 %, 49 % and 75 % of the total numbers and amounts, respectively (U.S. Department of Commerce 1992; 1995). The percentage values of the other categories are more evenly distributed. The same category sector has also a somewhat higher proportion of large facilities.

5.2.2 Pharmaceutical processes

5.2.2.1 Research and development

New drug development involves four principal phases (i) pre-clinical research; (ii) clinical research and development; (iii) review of new drug application (NDA); and (iv) post-marketing surveillance. Pre-clinical research phase involves extensive laboratory and animal tests on the clinical trial material (CTM) to determine whether the compound is biologically active and safe. The average time to complete this phase is 6 years. Clinical research and development takes another 6 years in the average; in this stage the developed drug is progressively tested on people. Its first phase involves healthy volunteers. In the second phase its effectiveness is studied. In the third phase, the drug is used in clinics and hospitals. Meanwhile its manufacturing is optimised and improved. The NDA includes data to demonstrate that the drug is safe and effective for use under the conditions described in its labelling. The median total approval time for NDAs is 15 months. Each step in the manufacturing process as well as its quality and ingredients of chemicals used in each production step must be specified in the NDA and approved by the FDA. Once approved, changes require a new approval procedure by the FDA. FDA approval may take several years to obtain depending on the nature of the change; some changes even require new clinical tests. Pharmaceutical development includes the evaluation of synthetic methods on a larger scale and the assessment of various ways of formulating the product to optimise its delivery (US EPA 1997).

5.2.2.2 *The conversion of organic and natural substances into bulk pharmaceutical substances or ingredients through fermentation, extraction, and/or chemical synthesis*

Pharmaceutical substances typically consist of structurally complex organic compounds manufactured via a series of intermediate steps and under precise conditions. These substances are manufactured by (i) chemical synthesis; (ii) fermentation; (iii) isolation/recovery from natural sources; or (iv) a combination of these processes. Most pharmaceuticals are produced in batch processes. When the same equipment is used for manufacturing different intermediates or different bulk substances, the equipment is throughout cleaned and validated prior to its reuse. The specific methods and materials (*e.g.* deionised water, water, steam, detergents and/or organic solvents) used to clean the equipment depend on the ability of the cleaning process to remove residues of raw materials, intermediates, precursors, degradation products and isomers (FDA 1996).

Chemical synthesis (production range: antibiotics, antihistamines, cardiovascular agents, vitamins and hormones, *etc.*): Co-products such as salts may be recovered and reused on-site, or sold for reuse. Reactors that can also be used as mixers, heaters, holding tanks, crystallisers and evaporators due their flexibility, are often attached to a process condenser to recover solvents from processes and operations.

Separation processes employed by the pharmaceutical industry are extraction, decanting, centrifugation, crystallisation, purification (most often: re-crystallisation and washing with additional solvents) and drying to isolate, and separate the active ingredients/substances from the product bulk.

Extraction of biological and natural products (production range: insulin, vaccines, haematological agents, antineoplastic agents, *etc.*): involves the isolation of an active ingredient from natural sources such as plants, roots, parasitic fungi or animal glands. Active ingredients are recovered by precipitation, purification and solvent extraction.

Fermentation (production range: steroids, antibiotics, certain food additives including vitamins): In fermentation, microorganisms (*e.g.* bacteria, yeast or fungi) are typically inoculated in a liquid broth supplemented with nutrients that are acclimated to specific environmental conditions (pH, temperature, oxygen, *etc.*) to achieve their rapid growth. The fermentation process begins with seed preparation to generate enough inoculum for the production fermentor. Fermentation may last for hours to weeks. A fermentation broth is produced that is thereafter subjected to product recovery via filtration, solvent extraction, precipitation, ion exchange or adsorption on a resin.

5.2.2.3 The formulation, mixing and compounding of final pharmaceutical products

The primary objective in this process is to convert the manufactured bulk substances into a final, usable form. Common dosage forms of pharmaceutical products include powders, pellets, tablets, capsules, liquids, creams and ointments as well as aerosols, patches and injectable dosages. Formulation and packaging is performed in accordance with "*Good Manufacturing Practices*" (GMP) that are regulated by the FDA to assure that the drug meets the safety requirements as well as the quality and purity characteristics that it represents to possess (US EPA 1997).

5.3 RAW MATERIAL INPUTS AND POLLUTANT OUTPUTS

5.3.1 Chemical synthesis

Spent solvents (benzene, chloroform, toluene, methylene chloride, methanol, ethylene glycol, xylenes, hydrochloric acid, *etc.*): VOC emissions, acid gases (halogenated) acids, sulphur dioxide, nitrous oxides, wastewater containing solvents, catalysts, reactants, wet scrubber wastewater, equipment cleaning wastewater resulting in a high BOD, COD and TSS effluent at a pH ranging between 1 to 11.

5.3.2 Natural product extraction

Plants, roots, animal tissues, extraction solvents (*e.g.* ammonia, chloroform, phenol, toluene): Solvent vapours, VOCs, spent solvents resulting in a wastewater having a low BOD, COD, TSS and a pH of 6 to 8.

5.3.3 Fermentation

Inoculum, sugars, starches, nutrients, phosphates, solvents including ethanol, methanol and acetone: Particulates, odoriferous gases, extraction solvent vapours, fermentation wastewater containing sugars, starches and nutrients with a high COD, BOD, TSS and a pH of 4 to 8.

5.3.4 Formulation

Drug and drug binders, sugars, syrup: tablet dusts and other particulates, drug residues from equipment cleaning, wash water typically contain low levels of COD, BOD, TSS and a pH at 6–8. Solid wastes are generally spent raw materials and reaction residues, filter cakes, particulates, rejected tablets, capsules and waste packages. In particular some solvents used in the chemical synthesis processes are priority pollutants classified under the US EPA CAA (benzene, cyanide, chloroform, chlorobenzene, phenol, toluene, methylene chloride, *etc.*) and/or hazardous air pollutants regulated under the CAA (acetonitrile, aniline, benzene, chloroform, chlorobenzene, 1,4-dioxane, ethylene glycol, formaldehyde, hexane, phenol, xylenes, *etc.*). Some of these solvents (acetonitrile, methylene chloride, *etc.*) are also used in the extraction of biological and natural products as well as in the fermentation processes. Process wastewater includes water used or formed during the reaction, water used to clean process equipment and floors and pump seal water. Non-process wastewater includes noncontact cooling water, noncontact ancillary water, sanitary wastewater, and other sources. Pharmaceutical manufacturing facilities generate wastewaters that generally contain significant amounts of the collective parameters BOD, COD, TSS and pH and other chemical constituents. Some chemicals being discharged are priority pollutants. The most frequently used solvents are methanol, ethanol, isopropanol and acetone. A lot of facilities use distillation either to recover a specific solvent from a process stream or to treat process streams containing solvents.

5.4 IN-PLANT CONTROL STRATEGIES

The best way to reduce pollution is to prevent it in the first place. Some companies have creatively implemented pollution prevention techniques that improve efficiency and increase profits while at the same time minimising environmental impacts. The bulk manufacturing processes of the pharmaceutical industry are characterised by a low ratio of finished product to raw material. Therefore, large quantities of residual waste are generated, especially in fermentation and natural product extraction. Chemical synthesis processes generate wastes containing hazardous spent solvents and reactants, combined with residual wastes such as reaction residues. Source reduction methods such as process modification, material substitution may not be easily implemented in the pharmaceutical industry as in other manufacturing sectors. This is because even a simple change in the production process of an existing product may need the approval of the FDA. In that case, the company needs to prove that the new drug is of the same or better quality as the old one, and the reformulation of the new drug will not adversely affect the identity, strength, quality, purity or bioavailability of the drug.

5.5. CONVENTIONAL TREATMENT METHODS

Most process water receives a treatment, either in-plant at the process unit or combined with other process effluents at a treatment plant facility prior to discharge to a permitted outfall. Treatment technologies most often being applied by the pharmaceutical industry are neutralisation and pH adjustment, biological treatment processes, primary clarification, multimedia filtration, secondary clarification, granular activated carbon and oxidation. Most commonly used aerobic systems are activated sludge besides aerated lagoons, rotating biological contactors (RBC) and trickling filters. Advanced biological treatment processes are also applied for nitrogen removal and coupled with activated carbon columns and other tertiary treatment processes to enhance micropollutants abatement. In most cases, biological treatment is not applicable for the effluents originating from the production of specific drugs including antibiotics (US EPA 1997).

The most common pollutants in chemical synthesis effluent are organic matter, forms of nitrogen, cyanide and TDS. Treatment of this kind of strong wastes poses problems even if a treatment combination bearing physicochemical and biological systems is employed. Toxics control further complicates treatment schemes and the application of more advanced chemical oxidative treatment technologies as well as stripping in addition to conventional biological treatment systems becomes a requirement. However, as mentioned before, various forms of biotreatment are still the most common ones. Small scale facilities, especially those producing strong effluent, exhibit a significant demand for balancing and dilution, for instance with municipal (domestic) wastewater. However, if treated effluent has to satisfy the discharge limit values for acceptance to receiving water bodies, a thorough treatment based on the intense evaluation of process and pollution profiles is needed. Kabdasli *et al.* (1999), conducted a detailed case study for a plant producing three different active pharmaceuticals, namely paracetamol, mephenoxalone and omeprazole. Paracetamol is a widely used analgesic and the capacity of the process under study was 1250 kg/batch with a batch taking 8 h/d. The wastewater source is the distillation equipment washing water, however, the washing is not carried out on every batch but every 10–20 batches. Hence, the amount of wastewater varies between 1–2 m^3/t of product. The production is 1 batch/day and 5 days/week. Mephenoxalone, a well-known miorelaxant was also produced batch-wise. The capacity of the plant is 55 kg product/batch (1 batch/day, 5 days/week). The amount of wastewater is 1.2–1.3 m^3/t product. Omeprazole is a widely used active material as anti-ulcergenic. The capacity of the process used in the production is 115 kg/batch and every batch is completed in around 20 days. The amount of wastewater was 4 m^3/t of product. The wastewater characterisation is as follows; paracetamol production effluent: 200–400 kg/batch, COD = 160–12,650 mg/L, TOC = 60–3940 mg/L, TKN = 5.6–570 mg/L, alkalinity = 105–145 mg $CaCO_3/L$, TDS = 1165–2295 mg/L, pH = 6.96–7.41; mephenoxalone production effluent: 240 kg/batch, COD = 3830–21,250 mg/L, TOC = 1330–9050 mg/L; TKN = 435–3315 mg/L, alkalinity = 150–485 mg $CaCO_3/L$, TDS = 2060–16215 mg/L, pH = 6.80-6.98; omeprazole production composite: 3979 kg/batch, COD = 37,500 mg/L; TOC = 14,050 mg/L; TKN = 226 mg/L; TDS = 83,350 mg/L; pH = 2.21. Wastewater production rates were 160 kg/kg product for paracetamol, 4.36 kg/kg product for mephenoxalone and 34.6 kg/kg product for omeprazole. Unlike paracetamol effluent, mephenoxalone and omeprazole wastewaters required dilution and pretreatment prior to biological processes. Chemical oxidation experiments were conducted for these effluents using NaOCl (50 %) and H_2O_2 (35 %) at pH 3–4 to evaluate their maximum oxidisability using conventional oxidants. Biological treatability as the core treatment process for pharmaceutical effluent was also tested at varying dilutions and with different combinations of wastewater as well as chemically pretreated samples. $FeSO_4.7H_2O$, $ZnCl_2$ and $FeCl_3$ were used as the catalysts in the chemical oxidation experiments. Pretreatment was made to remove zinc and iron via precipitation at pH 8–9. Biological treatability tests were performed in 5 litre batch reactors with activated sludge that was obtained as the seed source from a wastewater treatment facility of a paracetamol production plant. System performance of the chemical and biological treatment processes was followed in terms of TOC and soluble COD parameters, respectively. Biological treatment was applied to paracetamol wastewater without pretreatment at varying dilutions (1:2 to 1:6) and the wastewaters were found to be readily biodegradable even at low dilutions. Decreasing the F/M ratio from 0.2 to 0.09 g COD/g VSS did not affect COD removal efficiency that was around 70 % for the paracetamol effluent with the lowest COD. The inert COD fraction was about 4 % of the total soluble COD. Although the effluent was readily biodegradable, significant operational problems were encountered such as a dark brown coloured supernatant as compared to the colourless original sample. Another problem observed for this wastewater was foaming that was controlled by adding a defoaming agent. Sludge settling problems did not occur at the higher organic loadings. Mephenoxalone effluent was diluted at different ratios (1/10, 1/20 and 1/100) and fed to the bioreactor seeded with paracetamol acclimated sludge. No COD removal was obtained for up to 10 days. Chemical oxidation of this effluent resulted in not more than 25 % TOC removal and some samples could not even be oxidised with very high oxidant doses and in the presence of catalyst. No biodegradation was observed for the chemically pre-oxidised mephenoxalone effluent samples. Omeprazole wastewater was biodegraded at different dilutions and also subjected to chemical oxidation at high oxidant doses. H_2O_2 was found to be more effective and approximately 30 % TOC removal was obtained at the stoichiometric dosage while increasing the oxidant dose did not provide further removal. The pre-oxidised samples were subjected to activated sludge treatment at dilutions $\leq$ 25 %. Even for the 1:10 dilution, biodegradation was high. Higher F/M ratios of 1.14–1.29 g COD/g VSS did not diminish the treatment performance. Chemical pre-oxidation entirely changed the characteristics of the omeprazole samples and fairly high biodegradabilities were observed. No substrate inhibition was found for this effluent and its pretreated form. In conclusion, it could be demonstrated that paracetamol and chemically oxidised omeprazole could be effectively treated via activated sludge separately or in combination with up to 80 % COD abatement at reasonable F/M ratios. However, dilution was necessary for all treatment cases and operational problems such as bulking or dark colour occurred. Mephenoalone effluent was in general very hard to treat both via biological and chemical processes. It should be dealt with at a central plant or as a hazardous waste.

5.6 ADVANCED AND EMERGING TREATMENT TECHNOLOGIES

Aside from conventional unit operations and processes present at wastewater treatment facilities that often allow residual pharmaceuticals to pass through, chemical oxidation processes have been tested in laboratory-scale for the removal of pharmaceutical compounds. Potential treatment include ferrate (Sarmah *et al.* 2006), chlorine dioxide (Huber *et al.* 2005), UV photolysis (Pereira *et al.* 2007) as well as different advanced oxidation processes including titanium dioxide as the photocatalyst (Kaniou *et al.* 2005; Abellan *et al.* 2007). Numerous studies found ozonation and enhanced ozonation particularly effective, achieving over 90 % degradation for a wide variety of compounds. Example studies conducted with aqueous solutions and real effluent are presented in the forthcoming sections.

5.6.1 Ozone and combinations

Biological treatment has been employed as the principal unit process in most of the pharmaceutical manufacturing plants. However, some pharmaceutical formulation activities produce a complex and difficult to degrade effluents that cannot be effectively treated via conventional activated sludge treatment units as far as treatment performance and residual COD values are concerned. In such cases, pretreatment prior to main biological oxidation is required. In a study reported by Tünay *et al.* (2004), three pharmaceutical production groups have been selected for investigation considering their frequent use: liquid, laque tablet and antibiotic formulation effluents. The wastewater mainly originated from equipment and space washing activities during formulation of pharmaceutical products. Composite samples from liquid and laque tablet products were prepared by taking source-based wastewater samples from each washing step and mixing them on a flow-weighted manner. Antibiotic manufacturing wastewater samples were synthetically prepared using finished drugs and dissolving them in water by considering their effluent concentrations. The effluent characteristic were as follows; Liquid products: pH = 6.3; COD = 6760 mg/L; soluble COD = 6720 mg/L; BOD_5 = 1700 mg/L; Laque tablet products: pH = 7.4; COD = 4100 mg/L; soluble COD = 3530 mg/L; BOD_5 = 770 mg/L; Antibiotic I: pH = 7.2; COD = 1015 mg/L; soluble COD = 670 mg/L; BOD_5 = 130 mg/L; Antibiotic II: pH = 7.7; COD = 1060 mg/L; soluble COD = 1040 mg/L; BOD_5 = 30 mg/L. The organic matter of the liquid production effluent was mainly in the soluble form and the BOD/COD ratios revealed that the biodegradability of this effluent was highest. Antibiotic formulation effluent was a biologically inhibitory (toxic) effluent considering the extremely low biodegradability ratios. Therefore, pretreatment is needed to enhance biodegradability and decrease the toxicity of antibiotic formulation effluent. Ozonation and catalytic ozonation studies were carried out with laque and antibiotic formulation effluents. Biological treatability experiments were conducted in 5 litre capacity fill-and-draw reactors in which activated sludge provided from the wastewater treatment plant of the pharmaceutical company was used as the seed source. The performance of the bioreactors was evaluated in terms of soluble COD removal efficiency. Ozone was produced from dry air at a rate of 10 mg/min at an air flow rate of 0.57 m^3/h. The Pyrex glass ozone reactor had the following dimensions: Internal diameter = 4.5 cm; height = 120 cm. Catalytic ozonation experiments were carried out with manganese sulphate at a concentration of 750 mg/L at pH 2.5. Liquid production effluent was subjected to biotreatment at organic loading up to 0.7 g COD/(g VSS × day) that resulted in 85–87 % COD removal at all studied organic loadings. Laque production effluent also exhibited a high COD but the biological treatment efficiency for this effluent decreased from 65 % to only 40 % when the organic loading rate was increased from 0.009 to 0.016 g COD/(g VSS × day). Ozonation was applied for 180 min at a rate of 10 mg/min to remove the residual inert COD (1000 mg/L). COD removal obtained after 180 min was only 20 % and the BOD_5/COD ratio increased to 0.43. Although the effluent seemed to be more readily biodegradable, poor biological treatment efficiencies were observed; COD removal after biotreatment of the pre-ozonated sample was only 60 % at an organic loading rate of 0.016 g COD/(g VSS × day) and even decreased to 40 % when the organic loading rate was elevated to 0.100 g COD/(g VSS × day). Antibiotic formulation effluents could not be directly treated via activated sludge process and hence were firstly subjected to ozonation and catalytic ozonation for 90 min. COD removals of around 30 % were achieved after 90 min. However, the increase in the BOD_5/COD ratio was rather limited for Antibiotic Formulation Effluent Nr. II (ultimate BOD/COD = 0.077). Catalytic ozonation applied to Antibiotic Formulation Effluent Nr. I provided significant COD removal and an increase of the BOD_5/COD ratio to 0.27 (to 0.22 after mere ozonation).

Benitez *et al.* (2009) studied the ozonation of four different pharmaceuticals (the beta-blocker metoprolol (pK_a = 9.6), the nonsteroidal anti-inflammatory compound naproxen (pK_a = 4.2), the antibiotic amoxicillin (pK_a = 2.8 and 7.2) and the analgesic phenacetin (pK_a = 2.2) in ultra-pure water and the pH range of 2.5–9.0. The experiments allowed the determination of the apparent rate constants for the reactions between ozone with a mass flow rate of 16 mg/h and the selected compounds by competition kinetics with reference compounds of similar structures and known rate coefficients with ozone. The values obtained varied depending on the pH, and ranged between 239 and 1.27×10^4 M^{-1} s^{-1} for metoprolol; 2.62×10^4 and 2.97×10^5 M^{-1} s^{-1} for naproxen; 2.31×10^3 and 1.21×10^7 M^{-1} s^{-1} for amoxicillin; and 215 and 1.57×10^3 M^{-1} s^{-1} for phenacetin. As a conclusion, from the calculated rate constants at pH = 7, the following order of reactivity and half-lives for a residual ozone concentration of 10 μM (0.48 mg L^{-1}) in the liquid phase could be deduced; amoxicillin (0.05 s) > naproxen (0.26 s) > phenacetin (44 s) > metoprolol (49 s). Therefore, the selected pharmaceuticals would be rapidly oxidised during ozonation at circum neutral pH. Due to the acidic nature of these substances, the degree of dissociation of each pharmaceutical was determined at every pH of work, and the specific rate constants of the neutral and ionic species formed were evaluated. Additionally, the simultaneous ozonation of the pharmaceuticals plus a reference compound (p-CBA in all cases) in different water matrices was carried out by considering a groundwater (pH = 7.2; TOC<1 mg/L; UV_{254} = 0.013 cm^{-1}; COD = 1.7 mg/L; alkalinity = 388 mg/L $CaCO_3$; conductivity = 1846 μS/cm), a surface water from a public reservoir (pH = 7.1; TOC = 4.3 mg/L; UV_{254} = 0.118 cm^{-1}; COD = 13.8 mg/L; alkalinity = 30 mg/L $CaCO_3$; conductivity = 119 μS/cm), and three secondary effluents from municipal wastewater treatment plants (pH = 7.8–8.1; TOC = 13.2–22.9 mg/L; UV_{254} = 0.181–0.249 cm^{-1}; COD = 28.2–49.7 mg/L; alkalinity = 93–325 mg/L $CaCO_3$; conductivity = 637–905 μS/cm). Several conditions were maintained constant in all these experiments: T = 20°C, a buffered pH of 7 and initial concentrations of the pharmaceuticals and p-CBA of 1.0×10^{-6} M, while the initial dose of ozone was varied between 0.1 and 2 mg/L for the natural waters, and between 0.5 and 5 mg/L for the secondary effluents. In these ozonation experiments, amoxicillin, the most reactive compound towards ozone, was totally consumed for an ozone dose of 2 mg/L. Complete removal also occurred in the case of naproxen for an ozone dose of 5.0 mg/L, while some phenacetin and metoprolol amounts were still present in the water after total consumption of ozone doses of 5 mg/L. There was a higher pharmaceutical removal in the groundwater in relation to the reservoir water and a less removal in the wastewaters from the municipal treatment plants. Ozone doses higher than 5 mg/L had to be applied to ensure complete depletion of the selected pharmaceuticals, specially the more resistant phenacetin and metoprolol. The rate constants for the indirect reactions of the pollutant with hydroxyl radicals were also provided from a recent work being: $6.8 \times 10^9\ M^{-1}\ s^{-1}$ for metoprolol; $8.4 \times 10^9\ M^{-1}\ s^{-1}$ for naproxen; $7.9 \times 10^9\ M^{-1}\ s^{-1}$ for amoxicillin, and $4.0 \times 10^9\ M^{-1}\ s^{-1}$ for phenacetin. The contribution of the free radical pathway could also be determined by measuring the abatement of a $HO^{\bullet}$ probe compound, p-, whose rate constants with ozone and $HO^{\bullet}$ are known: $k_{O3,R} = 0.15\ M^{-1}\ s^{-1}$ (Yao and Haag 1991) and $k_{OH,R} = 5 \times 10^9\ M^{-1}\ s^{-1}$ (Buxton *et al.* 1988). Contribution of free radical reactions in the ozonation of the pharmaceuticals present in the different water systems was examined at two standard ozone doses: 0.5 mg/L for groundwater and reservoir water and 1.0 mg/L for the three effluents from the wastewater treatment plants. It could be concluded that for this ozone dose there is no contribution of the radical pathway for the elimination of amoxicillin and naproxen in the groundwater, as well as for amoxicillin in the reservoir water, being this elimination totally due to the direct ozone pathway. Thus, amoxicillin and naproxen are almost exclusively oxidised by the direct ozone pathway, due to their higher bimolecular rate constants with ozone; and on the contrary, metoprolol is mainly oxidised by the free radical pathway; and finally, phenacetin presents values of contribution lower or higher than 0.5 depending on the nature of the specific water considered. In addition, the lowest contribution of hydroxyl radicals to pharmaceuticals oxidation was obtained for the groundwater, since its low organic matter content did not favour ozone decomposition into hydroxyl radicals. On the contrary, higher indirect (HO^{c}) reaction contributions were found for the reservoir water, which can be explained by its lower alkalinity. This parameter can be controlled by the content of carbonate/bicarbonate ions which consume hydroxyl radicals.

Reverse osmosis (RO) is one of the most effective means for removing a wide range of micropollutants during water recycling. However, one significant disadvantage is the need to dispose the resultant RO concentrate (brine) (Van der Bruggen *et al.* 2003). A crucial class of pharmaceuticals are beta blockers, which are permanently discharged by municipal WWTPs (Bendz *et al.* 2005). Several studies focusing on the toxicological potential of beta blockers show that they could be of environmental relevance. Benner *et al.* (2008) studied the ozonation of beta blockers in RO concentrates. RO concentrate samples were obtained from a RO-membrane system that uses effluents of wastewater treatment plants (WWTP) as feed water for the production of drinking water. A number of different pharmaceuticals (*e.g.* antibiotics, contrast media, beta blockers) were found in the WWTP effluent as well as in the RO-concentrate. Overall, a concentration factor (feed: concentrate) of approximately 3–4 was measured. Beta blockers (acebutolol, atenolol, bisoprolol, celiprolol, metoprolol, propranolol, timolol) were found in the range of low ng/L to low μg/L. Because metoprolol and propranolol are classified as potentially toxic to aquatic organisms and all beta blocker molecules have moieties which are reactive towards ozone (amine groups, activated aromatic rings) it was tested whether ozonation can be applied for their degradation. Rate constants for the reaction of acebutolol, atenolol, metoprolol and propranolol with ozone and •OH radicals were determined. Briefly, rate constants for slowly reacting compounds were determined at room temperature (20–22°C) under pseudo first-order conditions with an excess of O_3 ($[O_3]_0$:[substrate]$_0$ ≥10) by measuring the reduction of the target concentration via high performance liquid chromatography (HPLC) equipped with UV detection. Glass bottles (250 mL) with a dispenser system were used as reaction vessels. By adding ozone stock solution (final concentration in reaction solution~10 μM) to the solution containing the compound to be investigated (~ 1 μM) and *tert*-butanol (~ 20 mM) as the radical scavenger, the kinetic runs were started. For higher rate constants (>2000 M^{-1} s^{-1}) a competition kinetics method was applied using cinnamic acid as the competitor (for instance, in the presence of propanolol). In general, the experiments were carried out at room temperature (20–22°C) in 50-mL vials at a pH ranging from 7.5 to 8.5 (for propranolol pH 3–8.5, 10–50 mM phosphate buffer) with solutions containing equal concentrations of beta blocker and reference compound as well as *t*-BuOH (~ 20 mM). HO• radical rate constants were determined by competition kinetics as well. Hydroxyl radicals were generated by UV-C photolysis of H_2O_2 (~2 mM) with a low-pressure mercury lamp in solutions containing the competitor substrate *para*-chlorobenzoic acid (~1 μM p-CBA) and the beta blockers (~1 μM). The rate constants were determined by accounting for direct photolysis of 4 % for acebutolol, 3 % for atenolol, 0 % for metoprolol and 8 % for propranolol after 30 min irradiation. Bromate was determined in the reaction samples. During the determination of second-order rate constant of the beta blockers with ozone, the pH dependence of the kinetic measurements was also considered. At pH 7, acebutolol, atenolol and metoprolol reacted with ozone with an apparent second-order rate constant (k_{O3}) of about 2000 M^{-1} s^{-1}, whereas propranolol reacted with ~10^5 M^{-1} s^{-1}. The major difference in the calculated rate coefficients is related to the naphthalene moiety in the structure of propanolol, while the other three beta blockers contain a phenyl group. For acebutolol, atenolol and metoprolol this oxidation pathway will play a more important role than in the case of propranolol, because their k_{O3} are lower. The rate constants for the reaction of the selected compounds with $HO^{\bullet}$ radicals were determined to be $0.5–1.0 \times 10^{10}$ M^{-1} s^{-1}. With a bromide level in the RO concentrate sample of about 1200 µg/L, there is a potential of elevated bromate concentrations after ozonation. The dose needed for a 90 % elimination of propranolol of ~ 8 mg O_3/L (metoprolol ~11 mg O_3/L) caused a bromate concentration of around 24 µg/L (35 µg/L). These were relatively low bromate concentrations for an effluent. Experiments with RO concentrate indicated that an ozone dose of only 5 mg/L resulted in a quantitative removal of propranolol in 0.8 s and 10 mg O_3/L oxidised 70 % of metoprolol in only 1.2 s. Tests with chlorinated and non-chlorinated WWTP effluent showed an increase of ozone stability but a decrease of hydroxyl radical exposure in the samples after chlorination. A reason for the increase of the ozone stability could be the reaction of chlorine with amine moieties of the organic matrix. This would lead to a formation of chloramines, which are no longer susceptible for a direct reaction with ozone. This may shift the oxidation processes towards direct ozone reactions and favours the degradation of compounds with high k_{O3}.

Effluents generated from the antibiotic formulation can be defined as wastewater with neutral pH (6–8), medium strength COD (200–2000 mg/L), low BOD (0–100 mg/L), and low VSS (100–400 mg/L). Hence, they can generally be categorised as ideal candidates for AOP. Balcioglu and Ötker (2003) reported the ozonation of three different synthetic pharmaceutical formulation effluents containing two human antibiotics and a veterinary antibiotic to enhance their biodegradability. Synthetic prepared samples (SPSs) were used throughout the study since the formulation wastewater contains only the finished product and the exact composition of the formulation wastewaters are known. SPSs of human antibiotic I (ceftriaxone sodium, cephalosporine group; Roche®) and human antibiotic II (penicillin VK, penicillin group; Aventis®) contain only active substance, whereas that of veterinary antibiotic (enrofloxacin, quinolone group; Bayer®) includes 10 % active substance and some inorganic additives. Depending upon the studies carried out on actual antibiotic formulation wastewaters initial COD values of SPSs were chosen between 250 and 1400 mg/L. Due to the limited solubility of veterinary antibiotic at pH 3 and 7, its formulation mixture with COD values up to 900 mg/L was used in the experiments. Ozonation experiments were performed in a 1500 mL capacity ozone bubble column for 1 h at semi-batch mode with counter-current recirculation of the liquid to the gas flow. The oxygen flow rate to the generator utilising dry and pure oxygen was maintained at 100 L/h and monitored with a rota meter incorporated into the ozone generator. The diffusion rate of the ozone+oxygen mixture introduced from the bottom of the reactor through a sintered glass diffusing plate was 2.96 g/L after 1 hour ozonation. For human antibiotic I sample, 50 % TOC and 74 % COD removals were achieved at an ozone input rate of 2.96 g/L after one hour ozonation. With the increment of the solution pH from 3 to neutral value, overall COD abatement was enhanced for all formulation wastewaters as expected. Since the oxidation potential of hydroxyl radicals is much higher than that of ozone molecule, direct molecular oxidation is slower than free radical oxidation and causes incomplete oxidation of organic compounds as observed in this study. However, the overall COD removal of human antibiotic II sample increased from 24 % at pH 3 to 69 % at pH 7 and remained almost unchanged (71 %) at pH 11. The effect of pH on the overall COD removal of other wastewaters was less pronounced. For veterinary antibiotic wastewater the lower values of overall COD removal (79 %) at pH 11 compared to that of pH 7 (88 %) could be explained by the fact that the veterinary antibiotic formulation contained inorganic additives, which may act as radical scavengers at pH 11. The overall aromaticity removal of human antibiotic II sample decreased by increasing the pH value from 3 to 11 since the increase in UV_{254} absorbance was more pronounced at higher pH values. When the initial COD was high, more intermediates were generated during the initial period of ozonation. Hence they consume more ozone either by indirect or direct reaction pathways. However, the total COD removal of human II antibiotic sample decreased while overall UV_{254} elimination exhibited an increasing trend by increasing the initial COD of the wastewater. Since higher UV_{254} absorbance values in the initial stages of ozonation were observed by increasing the initial COD value of sample it was expected that reaction products of human antibiotic II sample exerted a higher ozone demand. The contradiction between the obtained results for human II antibiotic sample and other wastes could be explained by the fact that the ratio of chemicals having absorbance values at UV_{254} to the organic chemicals exerting COD was low in antibiotic II wastewater. The results obtained by H_2O_2/O_3 process for human I and veterinary antibiotic samples indicated that during one hour ozonation the COD and absorbance results paralleled those without the added hydrogen peroxide. These similarities may be due to the high reactivity of the selected compounds to ozonation and the application of high ozone concentrations. However, for human antibiotic II, COD and aromaticity removals were enhanced from 69 % and 29 % to 95 % and 90 %, respectively, in the presence of 20 mM hydrogen peroxide at an ozone input rate of 2960 mg/L. These results confirmed that the oxidation of human antibiotic II proceeded mainly by hydroxyl radicals. In order to assess the effect of ozonation on the biodegradability of the effluents, BOD_5 measurements were conducted and the biodegradability was represented as the BOD_5/COD ratio of the samples. The BOD_5 value of untreated veterinary antibiotic wastewater (COD = 900 mg/L) was only 70 mg/L, whereas untreated human I and human II antibiotic samples were determined as nonbiodegradable (≈0). The initially low/negligible biodegradability of human antibiotic I levelled off, that of human antibiotic II and veterinary antibiotic increased with an increasing the ozonation time. While the BOD_5/COD ratio of veterinary antibiotic formulation sample increased from 0.077 to 0.380 at an applied ozone dosage of 2.96 g/L of wastewater, this ratio for human antibiotic I and human antibiotic II increased from 0 to 0.10 and 0.27, respectively. Moreover the results of this investigation showed that the ozonation process is capable of achieving high levels of COD and aromaticity removals at the natural pH values of the effluents.

Procaine Penicillin G (PPG) is a representative antibiotic formulation due to its low price, common use, and wide activity spectrum. The molecular weight of penicillin G potassium salt ($C_{16}H_{17}KN_2O_4S$) is 372.5 g/mol whereas procaine ($C_{13}H_{20}N_2O_2$) has a molecular weight of 236.3 g/mol. In a study conducted by Arslan-Alaton and Caglayan (2006), the effect of O_3 (applied dose = 1440 mg/h; pH 7 and pH 12) and H_2O_2/O_3 (applied dose = 1440 mg/h; pH 7; 10 mM H_2O_2) pretreatment of PPG formulation effluent (COD = 600 ± 9 mg/L; BOD_5 = 48 ± 3 mg/L) on its biodegradability and acute toxicity was investigated. For that purpose, ISO 8192 activated sludge inhibition and *Daphnia magna* acute toxicity tests were conducted. BOD_5 measurements that were carried out to pre-evaluate the biocompatibility of untreated, ozonated, and perozonated PPG effluent have shown that BOD_5 alone was not a very correct or useful tool to predict the actual biodegradability of PPG effluent. For the biodegradability experiments, four 2000 mL capacity bioreactors containing untreated, ozonated and perozonated PPG + synthetic domestic effluent mixtures (SWW) were run for 8 weeks with a hydraulic retention time of 24 h and a sludge age fixed as 10 days to keep an MLVSS at 3450–3500 mg/L typically encountered in activated sludge treatment facilities. The bioreactors fed with mixtures of simulated pharmaceutical effluent (SWW+PPG mixtures) were gradually acclimated to raw, ozonated, and perozonated PPG to finally constitute 30 % raw or pretreated PPG on volumetric basis in the wastewater samples. In other words, the COD contribution of PPG in the synthetic pharmaceutical wastewater mixture was ultimately 30 %. The initial CODs of SWW (COD_{SWWo}) and raw PPG effluent (COD_{PGGo}) were again selected as 600 mg/L to mimic typical penicillin formulation effluent. The COD removal obtained for the control reactor (fed with SWW only) after 24 h biological treatment was taken as the reference value for the calculation of total and fractional COD removal efficiencies after 24 h biotreatment. The COD removal efficiency increased from 37 % after 1 h ozonation at pH 7 without H_2O_2 to 76 % in the presence of 10 mM H_2O_2. Time-dependent changes in BOD_5 values clearly indicated that the BOD content of the PPG effluent remained nearly unchanged throughout the ozonation experiments. No significant increase in BOD_5 values was obtained when acclimated sludge was used for seeding the samples (data not shown) instead of non-acclimated sludge. Values were always a little higher when acclimated sludge was used as the seed source, although the difference was less than 15 %. Significant biodegradability ratios (> 0.2) could be achieved with the O_3/H_2O_2/pH 7 process and only after 1 h ozonation, corresponding to a specific ozone dose of 3 mg O_3 per mg of initial COD (COD_0). COD removal efficiency had to exceed 55 % which could only be obtained only via perozonation, resulting in 76 % COD removal. Due to the fact that PPG is not very stable and easily decomposes at alkaline and acidic pH, the changes in its acute toxicity after exposure to low pH (3) and high pH (12) have also been investigated. If toxicity reduction is observed upon alkali hydrolysis or acidic decomposition, treatment at alkaline or acidic pH should be preferred for the toxicity reduction in PPG effluent instead of ozonation or perozonation, which are far more costly treatments than pH adjustment. Exposure to alkaline or acidic pH for 24 h had only a slightly positive effect on PPG effluent toxicity. The acute toxicity of PPG effluent decreased appreciably after pretreatment with ozone at varying pH values and in the presence of H_2O_2, revealing that rather *"destructive"* pretreatment is required to ensure an appreciable detoxification. Among the oxidative pretreatment processes, ozonation at pH 7 (37 % COD abatement) and pH 12 (51 % COD abatement) appeared to have the same positive effect on PPG toxicity reduction compared to perozonation at pH 7. During the activated sludge inhibition experiments EC_{20} and EC_{50} values (in mg/L) were determined for all reaction samples. According to our experimental findings the order of increasing toxicity is as follows: raw PPG < pH 12/HD < O_3/pH 12 < O_3/pH 7 < O_3/H_2O_2/pH 7. Interestingly, a 13-fold increase in activated sludge inhibition occurred when PPG effluent was subjected to O_3/H_2O_2/pH 7 treatment, *i.e.*, the pretreatment process that was found to be most effective with regard to both COD abatement and BOD_5/COD ratio improvement. From these findings it can be concluded that the toxicity of an effluent cannot be predicted by BOD_5 measurements alone. However, the biological COD removal with acclimated activated sludge before and after chemical oxidation also has to be considered to decide which option is more feasible, namely, long-term PPG acclimation or pretreatment with ozone or perozone processes. From the practical point of view, it is more appropriate to ozonate PPG at pH 7, because the pH of penicillin formulation effluent is typically 6–8, *i.e.*, in the neutral pH range. Alkali hydrolysis does not remove the original compound and is acutely toxic toward *D. magna*. At the pharmaceutical plant from where the PPG formulation was obtained, 30 % of the total wastewater (Q = 150 m^3/h) originates from penicillin formulation and the remaining wastewater has a rather *"readily biodegradable"* character. In the batch activated sludge reactor fed with SWW only, the final COD was found as 90 mg/L after 24 h (initial COD =

600 mg/L; F/M = 0.17 gCOD/gMLVSS). The remaining COD is most probably attributable to the accumulation of microbial degradation products. In the case of untreated PPG+SWW mixed synthetic pharmaceutical effluent, no inhibition in COD removal of the SWW fraction of the total wastewater was observed. In addition, 17 % COD of the raw PPG fraction was also removed. This may be attributed to the short-term acclimation period that can be further extended to achieve more complete COD removal. A similar trend in COD abatement was found for the reactor fed with a mixture of SWW+ozonated PPG (from the O_3/pH 7 process). For this effluent, the final COD decreased to 190 mg/L instead of 240 mg/L obtained after 24 h activated sludge treatment of the raw PPG+SWW wastewater mixture. On the other hand, COD removal of the effluent mixture containing perozonated PPG (treated with the O_3/H_2O_2/pH 7 process) was seriously inhibited; *i.e.*, the perozonated PPG effluent fraction was not degraded at all, and at the same time COD removal of the SWW fraction was reduced from 79 to only 50 %, resulting in a final total COD of 260 mg/L instead of 210 mg/L for the wastewater that had an original COD of 470 mg/L (F/M= 0.13 g COD/g MLVSS). From the above indicated findings it is obvious that PPG effluent should be pretreated only with ozone at pH 7 when the final effluent COD must be below 200 mg/L to meet the Turkish national discharge limitation for ultimate discharge into a receiving water body. Another benefit of pretreatment is the fact that the acute toxicity of ozonated PPG effluent is significantly less than that of the untreated PPG effluent. This has also to be considered since a significant PPG fraction in the final biotreated pharmaceutical wastewater mixture will remain in the biologically treated effluent and enter natural water bodies.

Lin *et al.* (2009) examined the degradation of sulphonamide and macrolide antibiotics in spiked water and a pharmaceutical wastewater by ozonation under varied reaction conditions such as concentration, contact time, pH, and H_2O_2/O_3 molar ratio. For this purpose, aqueous sulphamethoxazole (SMX), sulphadimethoxine (SDM), sulphamethazine (SMT), erythromycin (ERM) and tylosin (TYL) solutions were synthetically prepared. The ozone reactor was made of stainless steel and formed by a cylinder with flanges secured by screws to a top plate and a bottom plate with an O-ring on each; the dimensions were of 13 cm in ID, 18.5 cm in OD, 40 cm in height, and 5 L in volume. An O_3/O_2 mixture was supplied by an O_3 generator (OZONIA CFS-1 2G) at 5.3 % O_3 (v/v) and 1.6 L/min. Experiments were performed with and without controlling the pH. Initial pH of the spike solution was 4.2–5.7 and it decreased to 2.4–3.4 without buffering agents. Process variables included contact time (0–1 h), pH (3–11), contaminant concentration (0–200 ppm) and antibiotic type (five compounds), as well as the H_2O_2 to O_3 molar ratio (0–20). Reaction rate constants of O_3 with individual compounds were not determined because of the O_3-limiting reaction conditions (*i.e.*, expected O_3 was less than 10 mg/L prior to full saturation and particularly so at the start of ozonation and it was less than the initial concentration of any of the target compounds at 200 mg/L) and because of concentration parameters changing concurrently throughout the reaction, *i.e.* dissolved O_3 concentration rising as ozonation started, target compound concentration continually decreasing as a result of degradation by O_3, and pH decreasing due to hydrolysis of O_3 which in turn determined the attainable steady-state concentration of O_3. At an O_3 application rate of 0.17 g O_3/min the removal of SMX, SDM, SMT, ERM and TYL was 93, 96, 95, > 99, and > 99 %, respectively, within 10 min of ozonation and higher than 99 % for all compounds after 20 min ozonation. The UV absorbance at 254 nm *vs.* time profiles showed clear decreases in all solutions corresponding to degradation of parent and organic fragments throughout ozonation, except for the solution of ERM that absorbs little UV light for the absence of unsaturated C-C bonds in contrast to all others. Degradation experiments for a mixed solution were performed and the obtained kinetic profiles were used to obtain pseudo first-order rate constants for each compound. The rate constants as obtained by regression analysis were 0.38 ± 0.012, 0.45 ± 0.017, 0.54 ± 0.035, 0.08 ± 0.003, and 0.37 ± 0.027 min^{-1} for SMX, SDM, SMT, ERM, and TYL, respectively. The results showed faster removal kinetics for sulphonamides containing the aromatic ring than for macrolides built of mostly saturated hydrocarbon structure. Degradation of contaminants containing unsaturated C-C bonds occurred faster at low pH, consistent with O_3 being the predominant oxidant and its bulk concentration higher at low pH. Degradation of erythromycin having a fully saturated structure was slower and more effective at higher pH or with added H_2O_2, both consistent with the enhanced production of $HO^•$ under such conditions that contributes to removal of the saturated compound. Low pH favours degradation via molecular O_3 while high pH via HO radical; the optimal pH thus depends on target compounds being treated, and buffered pH at 7 facilitates removal of all tested compounds. The addition of H_2O_2 to ozonation abets contaminant removal, and at mole ratio of H_2O_2:O_3 of 5:1 it attains the highest degradation speed for all contaminants. However, a large excess of added H_2O_2 results in the reduced or no benefits relative to O_3 alone. Thus, only a small dose of H_2O_2 is desirable when widely disparate compounds are treated by ozonation. The rate of degradation occurring under buffered pH conditions followed this decreasing order: pH 3 > 7 > 11, except for ERM that showed pH 7 11 > 3. The amount of dissolved O_3 in aqueous solution is a result of two opposing processes-the supply of O_3 by gas-to-liquid transfer and the depletion of O_3 via reactions of O_3 with OH^- (> 48 M^{-1} s^{-1}) and other reactions. When OH^- is scarce (*i.e.*, low pH), the depletion reaction is slow allowing for accumulation of dissolved O_3 to a high level (typically 4–8 mg/L with air feed to an O_3 generator), whereas when OH^- is abundant (*i.e.*, high pH), the depletion reaction is rapid thus prohibiting dissolved O_3 to accumulate (dissolved O_3 rarely exceeds 1 mg/L at pH 12 or higher). Thus, degradation of compounds (aromatic compounds) susceptible to electrophilic attack by O_3 is more rapid at low pH, where aqueous O_3 is higher. Contrarily, compounds resistant to O_3, including saturated compounds such as ERM, undergo faster degradation at increased pH. While the contribution of O_3 to ERM degradation is reduced at high pH due to reduced concentration, degradation rate is compensated by an increased concentration of $HO^•$ resulting from increased hydrolysis of O_3. The reactivity of ozone towards pharmaceuticals was affirmed by treatment of an authentic pharmaceutical wastewater that contained the test compounds at lower concentrations from ppb to 26 ppm, again demonstrating complete degradation (> 97–99 %) in 20 min.

5.6.2 The H_2O_2/UV-C process

In a study conducted by Rosario-Ortiz *et al.* (2010), advanced oxidation using low pressure UV light coupled with hydrogen peroxide (UV-C/H_2O_2) was evaluated for the oxidation of six pharmaceuticals (meprobamate, carbamazepine, dilantin, atenolol, primidone and trimethoprim) in three wastewater effluents. The compounds were selected based on analytical capabilities as well as occurrence at the sites evaluated and structural diversity. Effluents used in the study were collected from three tertiary wastewater treatment facilities located in Las Vegas, Nevada (LVNV), the Rocky Mountain Region of Colorado (RMCO), and Pinellas County, Florida (PCFL). Samples were collected prior to any disinfection process. Former studies have indicated that all of the selected pharmaceuticals were detected in these effluents given above at ng/L (trimethoprim, dilantin) and even µg/L levels (meprobamate, atenolol). A custom-made low pressure UV collimated beam system was used for all the irradiation experiments. The system included two G15T8 germicidal lamps (General Electric, Fairfield, CT, USA), housed inside a wood box. The petri factor for the photoreactor was determined to be greater than 0.9. UV fluences of 300, 500, and 700 mJ/cm^2 were evaluated for the three samples. Exposure times varied between 6 and 20 min to achieve the desired UV fluence rate. For each UV fluence rate, H_2O_2 concentrations of 0, 2, 5, 10, 15, 20 mg/L were applied 30 s prior to UV radiation. Special attention was paid to the role of water quality (*i.e.*, alkalinity, nitrite, and specifically effluent organic matter) on oxidation efficiency and this information was used to explain the differences in pharmaceuticals removal between the three wastewaters. The polarity, size distribution, fluorescence index and average molecular weight of effluent organic matter were evaluated by using different analytical and instrumental methods. The TOC of the effluent samples varied from 6.6 to

10.3 mg/L and the UV transmittance ranged from 55 to 72 %. The alkalinity and nitrite concentrations, which influence the effectiveness of the treatment for pharmaceutical removal, also varied among the samples. The SUVA values were between 2.5 and 1.7 L/mg × m, indicating a low aromatic character. The removal of these pharmaceuticals varied between insignificant to more than 90 %. When the H_2O_2 dose was 0 mg/L, dilantin was removed by 0–65 % indicating that this compound could be partially removed via direct UV-C photolysis in LVNV, RMCO, and PCFL. However, removal of the other five pharmaceuticals was generally less than 10 % indicating that these compounds were not degradable via UV photolysis. Since the experimental conditions were the same for all three waters, the fraction of $HO^{\bullet}$ available for pCBA decomposition was dependent upon the overall scavenging capacity of the water matrix and higher for the LVNV sample than for the RMCO or PCFL samples, as evidenced by the greater decrease in pCBA. For a UV fluence of 300 mJ/cm^2 and 20 mg/L of H_2O_2, the pCBA ($HO^{\bullet}$ probe compound) percent removals were approximately 70, 54, and 44 % for LVNV, RMCO, and PCFL, respectively. At 500 mJ/cm^2 and 20 mg/L of H_2O_2, the removals for pCBA were 80, 58, and 52 % of the original values for LVNV, RMCO, and PCFL, respectively. At a dose of 700 mJ/cm^2 with 20 mg/L of H_2O_2, the pCBA concentration decreased by 82 and 72 % for LVNV and RMCO, respectively (no data was obtained for PCFL). Results indicated that the efficiency of UV-C/H_2O_2 treatment for the removal of pharmaceuticals from wastewater was a function of not only the concentration of the organic matter content of the effluent matrix but also its inherent reactivity towards $HO^{\bullet}$. The removal of pharmaceuticals also correlated well with reductions in ultraviolet absorbance at 254 nm which offers utilities a surrogate to assess pharmaceutical removal efficiency during UV-C/H_2O_2 treatment.

5.6.3 Fenton's reagent and combinations

Several studies are available in the scientific literature dealing with the Fenton's treatment of pharmaceuticals and the effect of pretreatment on the biotreatability of advanced oxidation products. Tekin *et al.* (2006) investigated the applicability of Fenton's oxidation to improve the biodegradability of effluents originating from a medium scale drug manufacturing plant producing various antiseptics and disinfectant solutions. The total number of products is 10 including an antiseptic gargle and an antiseptic liquid soap formulated with the same active ingredient as well as several other antiseptic solutions with different active ingredients and disinfectant solutions for medical use. The intermittent cleaning of the tanks used in the production processes and domestic utilisation of the tap water make up the wastewater of the plant. Typically, wastewater originating from the washing of equipments amount to 2–3 m^3/day and that from the washing of bottles increases this volumetric flow rate to approximately 6-7 m^3/day. Domestic wastewater from the plant is around 5 m^3/day, making up a total wastewater flow rate of 11–12 m^3/day. Due to the fact that not all of the products were produced during the treatability studies, Fenton's oxidation was carried out with synthetically prepared wash waters (SPWWs) made up by diluting each product. The synthetically prepared wash waters were thereafter mixed with domestic wastewater at a rate of domestic/pharmaceutical = 3.5/1.5 to mimic daily average discharge rates. The COD and BOD_5 values of the mixed wastewaters were in the range of 900–7000 mg/L and 85–3600 mg/L, respectively. Accordingly, the BOD_5/COD ratios varied between 0.09–0.59. Communication with the plant and an analysis of the flow records, though scarce, indicated that the wash waters contained the product with an average of 1/100 dilution. In the preparation of SPWWs, three different dilution factors, 1/50, 1/100 and 1/500, were used to account for the variations in pharmaceutical wastewater flows. The effects of temperature, initial pH, coagulation pH, initial COD (at 3 different dilutions), H_2O_2 and Fe^{2+} dosages (at optimised pH and temperature; for H_2O_2 concentrations between 0.8 and 5.0 M and Fe^{2+} doses between 0.002 and 0.033 M) were studied. The Fe^{2+} concentrations were kept constant at 0.033 M while studying the effect of H_2O_2 dosage, and H_2O_2 concentration was fixed at 2.5 M in the tests for optimum Fe^{2+} dosage. The experiments were conducted at 50°C. The pH of the heated samples was adjusted and required amounts of $FeSO_4{\cdot}7H_2O$ and H_2O_2 were added to the samples. The reaction solutions were stirred for 30 min and then another 30 min were allocated for precipitation. The pH of the decanted supernatant was then adjusted to initiate coagulation. Two hours were allowed for precipitation. COD was measured in all treated samples whereas the BOD_5 test was conducted only for selected runs. Since no significant differences were observed in the treatment efficiency for room temperature and 50°C all further experiments were carried out at room temperature due to practical and economic reasons. The optimum pH was determined as 3.5 and 7.0 for the first (Fenton's oxidation) and second stage (based on iron removal from reaction solution by chemical precipitation) of the treatment process, respectively. The COD removal efficiency was relatively higher for lower initial COD values (dilution rate = 1/500); however, at the lower dilution ratios (1/50 to 1/100), average COD removals did not differ considerably. The optimal H_2O_2/Fe^{2+} ratio varied highly with type of waste to be oxidised. In addition, on the condition that the H_2O_2/Fe^{2+} ratio is optimum, the required H_2O_2 and $FeSO_4$ doses can be reduced or tuned up significantly. In the present study, COD removal efficiency was highest when the molar ratio of H_2O_2/Fe^{2+} was 150–250 for all pharmaceutical chemicals. At a H_2O_2/Fe^{2+} ratio of 155, *e.g.* 0.3 M H_2O_2 and 0.002 M Fe^{2+}, provided 45–65 % COD removal. Keeping the ratio constant at 155 thereby increasing H_2O_2 to 5 M and Fe^{2+} to 0.032 M, resulted in COD removal efficiencies ranging from 77 % to 95 %. In general, treatment with Fenton's oxidation was found to improve the biodegradability of the wash waters. The COD removal efficiency attained in this unit was between 45 % and 50 %. After Fenton's oxidation, the BOD_5/COD ratio of the wastewater increased by a factor of about 3–5. The wastewater treatment plant employing Fenton's oxidation followed by aerobic degradation in sequencing batch reactors (SBR), built after the treatability studies provided an overall COD removal efficiency of 98 %, and compliance with the discharge limits. The efficiency of the Fenton's oxidation was around 45–50 % and the efficiency in the SBR system which has two reactors each having a volume of 8 m^3 and operated with a total cycle time of one day, was around 98 %, regarding the COD removal. The BOD_5 values measured at the outlet of the treatment plant by the competent authority, during this 2-month period were in the order of 15–20 mg/L which corresponded to an overall BOD_5 removal efficiency of 98 %. This was well below the discharge limit of 50 mg/L set in the discharge standards. This was also reflected in the toxicity dilution factor (TDF) measurements by the Provincial Authority, which indicated that toxicity was greatly reduced. The TDF is defined as the minimum dilution factor required for the survival of all 10 specimens of *Lebistes reticulatus* after 48 h. The TDF values measured by the Provincial Authority consistently reduced to 2, after the new system was put to operation from its earlier value of 5–10.

Badawy *et al.* (2009) studied the treatability of pharmaceutical wastewater from El-Nasr Pharmaceutical and Chemical Company, located in South-East of Cairo. Chloramphenicol, diclofenac, salicylic acid and paracetamol were produced and obtained from El-Nasr Company in a pure form. During the study, a continuous monitoring program was carried out for almost one year. Six composite samples from end-of-pipe treatment were collected over the working hours. The company discharges both industrial (6000 m^3/d) and municipal wastewater (128 m^3/d) into a nearby evaporation pond without any pretreatment. The generated wastewater was characterised by high values of COD (4100–13,023 mg/L), TSS (20–330 mg/L) and oil&grease (17.4–600 mg/L). In addition, the presence of refractory compounds decreases the BOD_5/COD ratio (0.25–0.30), with an average value of 0.27. Analysis of raw wastewater confirmed that pretreatment is required prior to discharge into public sewers to comply with the Egyptian Environmental laws and regulations. Fenton's treatment process was carried out at room temperature by adding various doses of $FeSO_4 \cdot 7H_2O$. The pH was adjusted at 3.0 ± 0.2 using 1N H_2SO_4 and kept at that value during the reaction. The required amount of H_2O_2 was fed by a dosing pump during a period of 15 min, and then the coagulation experiments were conducted with the Jar Test apparatus that was preceded with rapid mixing of the Fenton-treated effluent at 100 rpm for 5 min, slow mixing at 40 rpm for

30 min, and then standstill for 30 min. After settling the supernatant was filtered through 0.45 μm, treated with enzyme Catalase to remove residual H_2O_2. The pH values for oxidation and coagulation experiments were controlled at 3.0 ± 0.2 and 8.5 ± 0.2, respectively with 0.1N sulphuric acid or sodium hydroxide. Raw and Fenton-treated wastewaters were subjected to biological treatment in two-litre plexiglass laboratory columns. Air was fed continuously at a rate of 120 mL/min to ensure sufficient dissolved oxygen. Biomass taken from an activated sludge process of a municipal wastewater treatment plant was acclimated to both raw wastewater conditions and Fenton-treated effluent for a period of 4 weeks. During this period, nutrients was added to raw wastewater and Fenton-treated effluent to keep the ratio 100:5:1 for the parameters COD:N:P, respectively. In order to study the effect of aeration time and the sludge loads on the treatment process, the system was-operated for 24 h at a MLSS of 3–4 g/L. Samples were withdrawn at regular intervals to follow COD and TOC. Diclofenac, chloramphenicol, paracetamol drugs and their by-products p-aminophenol, phenol, benzoic acid, nitrobenzene and salicylic acid were detected in all biologically treated effluents and their mean concentrations were 5.60, 38.84, 69.68, 62.94, 130.18, 2.51, 32.84 and 1.03 mg/L, respectively. Phenylacetic acid, a by-product of diclofenac, was not detected in any of treatment effluents. COD and TOC removal efficiencies obtained during biological treatment depended mainly on the characteristics of the wastewater, especially on the presence of non-biodegradable substances and the BOD/COD ratio (= 0.25). COD and TOC removal efficiencies were 51 % and 56 % as well as 48 % and 36 % for wastewater at an initial COD of 11,987 and 4100 mg/L, respectively. These low COD removal efficiencies were attributed to the presence of refractory organic compounds. A high removal efficiency of 76 % was obtained for initial COD of 13,023 mg/L due to the fact that for this wastewater sample the BOD_5/COD ratio increased to 0.30. The efficiency of Fenton process as a pretreatment step of wastewater was also investigated in terms of pH and hydrogen peroxide concentration. The highest COD and TOC removals were obtained as 64 % and 62 %, respectively, within1.5 h at a pH 3. The effect of hydrogen peroxide on oxidation efficiency was investigated under the following operating conditions: reaction time = 1.5 h, pH = 3, Fe^{2+}/H_2O_2 = 1:100, and a COD/H_2O_2 ratio varying from 1:1.1 to 1:4.4. It was found that TOC and COD removal efficiency increased with increasing peroxide concentration. Maximum removal efficiency was attained at a COD/H_2O_2 ratio of 1:2.2. In order to obtain the optimal Fe^{2+} dose, experiments were carried out at different Fe^{2+}/H_2O_2 molar ratios being equivalent to 1:10, 1:25, 1:50 and 1:100. The results indicated that the maximum TOC and COD removal were found at a Fe^{2+}/H_2O_2 molar ratio of 1:50. Further increase of this ratio decreased the extent of oxidation as a consequence of the recombination of $HO^\bullet$ radicals with Fe^{2+}. Fenton treatment of wastewater

collected from the company's end-of-pipe was carried for 1.5 h, with a COD/H_2O_2 ratio of 1:2.2, at pH 3 and a Fe^{2+}/H_2O_2 ratio of 1:50. Under these conditions, COD abatement ranged from 67 % to 87 %. These results revealed that the quality of treated effluent is satisfactory and parameter values complied with the Egyptian Law for discharge of industrial wastewater into the public sewage system. Fenton as a pretreatment process prior to biological activated sludge treatment increased the biodegradability of the wastewater by converting persistence compounds in the wastewater into more easily biodegradable ones. Results have indicated that complete removal of chloramphenicol, diclofenac, p-aminophenol, benzoic acid, nitrobenzene and salicylic acid was achieved in the final effluent when coupling Fenton's oxidation with a biological activated sludge processes. The obtained results have also showed that the refractory compounds and their by-products cannot be readily removed by biological treatment and always remain in the treated effluent or adsorbed on the sludge flocs. The application of Fenton oxidation process as a pretreatment improved the removal of pharmaceuticals from wastewater and appears to be an effective solution to achieve compliance with the law legislation with respect to discharge in a determined receptor medium.

Combined treatment of pharmaceutical effluent generally is more effective than single stage operations. Attempts were made in a study conducted by Xin and Sun (2009) to examine the combined coagulation, Fenton and sedimentation process for the treatment of non-degradable antibiotic fermentation wastewater. The antibiotic fermentation wastewater containing cefpirome, latomoxef, aztreonam, cefoperazone, cefatridine, ropylene glycol, and ceftazidime was obtained from a pharmaceutical company located in Harbin, China. The characterisation of the effluent was as follows; COD = 3279 mg/L; $BOD_5/COD < 0.1$; pH = 3–4; TOC = 1296 mg/L; TN = 221 mg/L; SS = 195 mg/L; NH_4^+–N = 54 mg/L. The coagulation experiments were conducted with a Jar Test apparatus and using polyferric sulphate (PFS) as the coagulant. During a coagulation experiment, pH was adjusted in the range of 2–10, thereafter PFS was added in the range 50–800 mg/L with rapid mixing at 300 rpm for 1 min, followed by slow mixing at 50 rpm for 15 min and then standstill for 30 min. In Fenton's oxidation reaction reagents were added and thereafter the water sample was mixed rapidly for 30–60 min, depending on the designated oxidation time. In the coagulation step, the above oxidised sample was mixed slowly for 15 min at a pH of 7 and then allowed to standstill for 30 min. The efficacy of combined wastewater treatment was mainly evaluated in terms of the parameters colour and COD removal. The experimental results indicated that 67 % colour and 72 % COD were removed under optimised conditions of coagulation (PFS dosage 200 mg/L and pH 4.0). In addition, optimal parameters of the Fenton process were determined to be 150 mg/L H_2O_2, 120 mg/L $FeSO_4$ and 1 h of reaction time at pH 4. In order to simplify the Fenton's oxidation kinetic model, it was assumed that the major oxidation action was attributed to $HO^{\bullet}$ and was provided instantaneously. The kinetics of COD removal for antibiotic fermentation wastewater in the Fenton's oxidation system showed pseudo-first order behaviour with respect to COD. It could be concluded that there was no marked effect of the initial COD (= 213, 599 and 905 mg/L) on the obtained abatement rate coefficients and the average reaction rate constant was found to be 0.0047 s^{-1}. The half life of pseudo-first order reaction was only 147 s, so it demonstrated that the degradation of organic pollutants in Fenton system was quite fast. When Fenton-treated effluent pH was adjusted to 7.0, the pollutants could be further removed by a coagulation and sedimentation process. The overall colour, COD and SS removals reached 97 %, 97 %, and 87 % under the selected optimum coagulation and Fenton's treatment (oxidation+sedimentation) conditions, respectively.

Among several advanced oxidation processes, Fenton's oxidation or Fenton-like reaction appeared to be the most promising one in terms of cost-effectiveness and ease of operation. In recent years, more and more reports have been appearing concerning the use of microwave radiation to promote the oxidative degradation of biorefractory wastes due to its advantages of swiftness, high-efficiency and environmentally friendliness (Kawala and Atamańczuk 1998; Abramovitch *et al.* 1999; Lai *et al.* 2006; Quan *et al.* 2007). These investigations suggested that microwave radiation was in favour of promoting the degradation efficiency of traditional treatment methods. Wang *et al.* (2009) explored microwave enhanced Fenton-like process to treat high concentration pharmaceutical wastewater with an initial COD of 49,913 mg/L. The pharmaceutical wastewater sample was collected from Harbin pharmaceutical manufacturing group in China. The effluent contained ferment ramification, remnant penicillin and several of remnant menstrum, acetone, amyl butyric ester as well as formaldehyde. The characteristics of the wastewater sample were as follows: TOC = 11,540 mg/L, pH = 4.42, UV_{254} absorbance (100 fold dilution) = 1.491 and a BOD_5/COD ratio of 0.165, respectively. The morphology of the sediment sludge was observed with a transmission electron micrograph with accelerated voltage of 80 kV. The liquid/solid ratio (L/S) was measured as the following steps: transportation of treated effluent to a 50 mL cylinder and allowing standstill for 1 h. Then the volume of supernatant and precipitated sludge was measured. The ratio between the two values was denoted as the L/S value. Settlement rate (SR) was used as an indicator for the settling quality of sludge and to optimise the treatment performance. The SR index was measured as follows: transportation of treated effluent to a 50 mL cylinder and determination of the increasing supernatant volume after a standstill period of 10 min. The ratio of the increasing volume versus time was denoted as the SR value. The parameters L/S, SR and BOD_5/COD were followed for treatment process optimisation. In addition ferric iron concentration was determined in treated samples by using ICP-OES (Optima 5300 DV, PerkinElmer). Operating parameters were investigated and the optimal conditions were established as follows: a microwave power of 300 W (studied range = 100–500 W), a radiation time of 6 min (studied range: 2–10 min), initial pH of 4.4 (range of study: 2–11), a H_2O_2 dosage of 1300 mg/L (range = 3200–19,000 mg/L) and a $Fe_2(SO_4)_3$ dosage of 4900 mg/L (range of study = 2000–8000 mg/L). Within the present experimental conditions employed, the COD and UV_{254} removals reached to 58 % and 55 %, respectively, and the BOD_5/COD ratio was enhanced from 0.165 to 0.470 under these reaction conditions. The variation of molecular weight distribution indicated that both macromolecular substances and micromolecular substances were eliminated quite well but mainly via different removal mechanisms, namely Fenton's oxidation for micromolecules and coagulation for macromolecules. The structure of flocs revealed that ferric hydroxide ions seemed to connect with ferric hydroxides and/or organic molecules to form large-size particles by means of Van der Waals force and/or hydrogen bonds. Subsequently, these particles aggregated to form flocs and settled down. Comparing with traditional Fenton-like reaction (55 % COD removal; final BOD_5/COD ratio = 0.38) and conventional heating assisted Fenton-like reaction (52 % COD removal, final BOD_5/COD ratio = 0.29), microwave-enhanced Fenton-like process displayed superior treatment efficiency in terms of biodegradability enhancement, L/S ratio and COD removal efficiency. Microwave was in favour of improving the degradation efficiency, the settling quality of sludge, as well as reducing the yield of sludge and enhancing the biodegradability of effluent. Microwave enhanced Fenton-like process is believed to be a promising treatment technology for high concentration and biorefractory wastewater.

5.6.4 Photo-Fenton processes

Antibiotics are among the pharmaceuticals most commonly detected in the aquatic environment, because, due to their antibacterial nature, they are not treatable in the biological step of sewage treatment plants (Alexy *et al.* 2004). A typical case is that of sulphamethoxazole (SMX), an antibiotic used worldwide, currently prescribed for treatment of urinary infection. SMX has been identified along with other pharmaceuticals in many municipal sewage treatment plant effluents at concentrations varying from 0.01 to 2 μg/L (Hirsch *et al.* 1999; Andreozzi *et al.* 2003; Bueno *et al.* 2007) and has also been found at concentrations varying from 0.03 to 0.48 μg/L in surface water in different European countries (Hirsch *et al.* 1999). Although no specific adverse effects on human health have been ascribed to SMX up to now, some studies have demonstrated a relatively high incidence of bacterial resistance to this antibiotic (Le *et al.* 2005). Trovó *et al.* (2009) evaluated the photocatalytic degradation of the antibiotic SMX by solar photo-Fenton at pilot-plant scale in distilled water (DW) and in seawater (SW) using sunlight as the radiation source. SW was selected as the alternative medium due to the implementation of this antibiotic in aquaculture practices. SW was collected directly from the Mediterranean Sea in Almería, and had the following composition; $NaCl = 28$ g/L; $KCl = 0.7$ g/L; $CaCl_2{\cdot}2H_2O = 1.6$ g/L; $MgCl_2{\cdot}6H_2O = 4.8$ g/L; $NaHCO_3 = 0.11$ g/L; $MgSO_4{\cdot}7H_2O = 3.5$ g/L. SW was filtered, acidified and autoclaved prior to use. Degradation and mineralisation of SMX (50 mg/L; DOC = 23.75 mg/L) were strongly hindered in SW compared to DW. The influence of H_2O_2 (studied range: 30 to 210 mg/L, with step-wise additions of 30 mg/L just after the former H_2O_2 was totally consumed) and Fe^{2+} (2.6, 5.2 and 10.4 mg/L as $FeSO_4{\cdot}7H_2O$) concentrations on the efficiency of the photocatalytic process was evaluated. For SMX kinetics evaluation in DW, 210 mg/L H_2O_2 were added at the beginning of the reaction with no further addition. The solar pilot plant reactor consisted of a parabolic collector with a 3.0 m^2 irradiated surface and a total volume of 39 L with an illuminated volume of 22 L (Malato *et al.* 2002). Water circulates under turbulent flow conditions in the reactor absorber tubes in a closed recirculating system. The typical solar UV power on a perfectly sunny day around noon was accepted as 30 W/m^2. A commercial bioassay, Biofix® Lumi-10, based on inhibition of the luminescence emitted by the marine bacteria *Vibrio fischeri,* was performed for toxicity evaluation after the same sample pretreatment described for intermediates identification. Alteration in the light output was measured after 30 min of SMX exposure. *Daphnia magna* acute immobilisation tests were performed with a commercial test kit (Daphtox kit F™ magna, Creasel, Belgium) after 48 h of exposure time. Several blank experiments were performed at the initial SMX concentration of 10 mg/L, to assure that the results found during the photocatalytic tests were consistent and not due to hydrolysis and/or photolysis. No hydrolysis was detected at the pH studied (2.5, 4.8 and 9.0) after 48 h. On the other hand, 40 % removal of the initial SMX concentration (10 mg/L) was observed during the first 45 min of irradiation only experiments conducted at an initial pH of 4.8. After 7 h of irradiation 76 % of the initial concentration of SMX was removed. On the other hand, no mineralisation was observed after the same irradiation time. This demonstrates the formation of intermediates, which are more resistant to photodegradation than SMX itself. In the SW experiment, only a slight reduction of SMX concentration (14 % after 7 h) was observed. A significantly higher H_2O_2 demand could be observed for the abatement of SMX and DOC in SW than in DW. This may be associated with the very high concentration of inorganic ions, like chloride and sulphate, in SW, which decreases photo-Fenton oxidation efficiency. With 210 mg/L H_2O_2

concentration, residual DOC was found as 2.9 mg/L corresponding mainly to carboxylates, such as acetate, formate, pyruvate and oxalate. At 210 mg/L H_2O_2, no intermediates other than carboxylates were found. Maximum concentrations of acetate, formate, pyruvate and oxalate detected were 1.02, 0.395, 0.686 and 2.6 mg/L, respectively. Released nitrogen was measured as a combination of ammonium and nitrate. Both ammonium and nitrate were detected at different relative concentrations. The nitrogen content of SMX was converted mainly into ammonium, but only 85 % of the initial nitrogen was detected as ammonium or nitrate with 210 mg/L of H_2O_2. This incomplete nitrogen mass balance, frequently observed in such processes indicates that other nitrogen-containing compounds must be present in the solution or evaporated during the process. Results of the solar photo-Fenton experiments have also indicated that an increase in Fe^{2+} concentration from 2.6 to 10.4 mg/L showed only a slight improvement in SMX degradation and mineralisation. However, an increase in the H_2O_2 concentration up to 120 mg/L during photo-Fenton in DW decreased SMX solution toxicity from 85 % (v/v) to 20 % (v/v), according to results of the *Daphnia magna* bioassays. The same behaviour was not observed after photo-Fenton treatment in SW. In spite of 45 % mineralisation in SW, toxicity increased from 16 % to 86 % as shown by *Vibrio fischeri* bioassays, which suggests that the intermediates generated in SW are different from those in DW. For the identification of oxidation intermediates, accurate mass measures by the use of liquid chromatography coupled to time-of-flight mass spectrometry (LC-TOF-MS) provides a higher grade of confidence in the assignation of possible structures of the intermediates generated. Two initial transformation pathways were proposed for photo-Fenton oxidation in DW, involving the attack of the hydroxyl radical on the benzene and isoxazole rings. However, based on the SMX photoproducts identified (C1, C2, C5, C6 and C8) the isoxazole ring is more susceptible to $HO^{\bullet}$ attack than the benzene ring, which persists in most of the intermediates identified.

There are only few studies that combine the information of chemical, toxicity and biodegradability analysis to investigate the viability of the combination of photo-Fenton and biological treatment on actual pharmaceutical wastewater, not only model wastewater prepared by dissolving parent compound in water. For instance, Sirtori *et al.* (2009) studied the characterisation and treatment of a real pharmaceutical wastewater with a COD of 3420 mg/L and a DOC of 775 mg/L by a solar photo-Fenton/biotreatment combination. There were also many inorganic compounds present in the effluent matrix (conductivity = 7 mS/cm; chloride = 2800 mg/L; sodium = 2000 mg/L). The most important organic chemical in this wastewater was nalidixic acid (45 mg/L), an antibiotic of the quinolone group frequently used in the treatment of urinary tract infections involving gram-negative bacteria, and the main DOC component in the wastewater, *i.e.* acetate (1900 mg/L). All solar photochemical experiments were performed in a pilot plant made up of compound parabolic collectors that were composed of two modules with 12 Pyrex glass tubes mounted on a fixed platform tilted 37° (local latitude). The total illuminated area was 3 m^2 and the volume is 40 L, 22 L of which are irradiated volume. Toxicity of the initial wastewater and selected pretreated samples was evaluated with Biofix® Lumi-10, a commercial bioassay. The test was based on the inhibition of the luminescence emitted by the marine bacteria *Vibrio fischeri*. Biodegradability of the photo-Fenton pretreated pharmaceutical wastewater at different stages was evaluated by a Zahn-Wellens test (an adaptation of the EC protocol, Directive 88/303/EEC). Activated sludge from the El Ejido wastewater treatment plant (Almería-Spain), nutrients and the test material as the sole carbon source were placed together in a 0.25 L glass vessel equipped with an agitator. Samples analysed were considered biodegradable when DOC abatement due to biodegradation percentage was over 70 % (US EPA 1997). The selected biological reactor was an immobilised biomass reactor (IBR). The IBR consisted of a 160 L flat-bottom tank filled with 90–95 L of polypropylene 15-mm Pall Ring supports colonised by activated sludge from the conventional aerobic wastewater treatment plant located at El Ejido (Almería). The system is also provided with a 100 L conditioner tank with pH control connected to the IBR through a recirculation pump. The operation flux was 500 L/h. Dissolved oxygen, pH and temperature were automatically measured and registered. Total volume of both tanks was 150 L. Start-up and adaptation of the biological reactor began with the immobilisation of the sludge on the ring supports, which took 2 days. After this, the system was maintained with controlled additions of glucose and ammonium chloride, keeping the carbon/nitrogen ratio at 100/20. The next step was to adapt the sludge to high NaCl content (approximately 5 g/L). During the whole adaptation process the analytical controls used to evaluate the IBR state were the total suspended solid, DOC, pH, ammonium and nitrate. Firstly, the photo-Fenton degradation of 30 mg/L nalidixic acid in saline water (5 g/L of NaCl) was investigated. These conditions were selected to simulate the ionic strength of the real wastewater. In these solar photo-Fenton experiments the pH was kept at 2.6–2.8, the temperature at 30–40°C and Fe^{2+} concentration was selected as 20 mg/L, the optimum iron concentration for solar photoreactors found in previous studies. Thereafter, two studies of photo-Fenton treatment of the real wastewater were performed; one with an excess of H_2O_2 (kinetic study) and another with controlled H_2O_2 dosing (biodegradability and toxicity studies). 90 % of the initial DOC was removed in 400 min of illumination time and the total H_2O_2 consumed was 180 mM. Nalidixic acid had completely disappeared at 190 min with 72 mM of H_2O_2 consumed. In view of these considerations, it is not recommended to mineralise 90 % of DOC, because of the very long time and huge amount of hydrogen peroxide needed (around 6 g/L) for such a high mineralisation level. In the other experiment with controlled H_2O_2, nalidixic acid degradation was complete in the presence of 66 mM H_2O_2. All samples were at least 90 % biodegradable at the end of the Zahn-Wellens test. In the test samples in which nalidixic acid concentrations were 20.7, 8.5 and 4.6 mg/L, 70 % biodegradability was attained only after 10 days of. On the other hand, samples with very low concentrations (<1 mg/L) or without nalidixic acid were biodegradable after 3 days. Results indicated that untreated samples need much longer adaptation periods than treated samples and that nalidixic acid (at concentration as low as 4.6 mg/L) is also detrimental in the sense that they reduce biodegradation efficiency. These results demonstrated that photo-Fenton should be performed until total degradation of nalidixic acid before coupling a biological treatment. During the bioassay, the nalidixic acid concentration remained constant, showing that it cannot be biodegraded. Hence, photo-Fenton process was applied first to enhance the biodegradability, followed by a biological treatment step in an Immobilised Biomass Reactor. Biodegradability and toxicity bioassays showed that photo-Fenton should be performed until total degradation of nalidixic acid before coupling a biological treatment. Analysis of the average oxidation state demonstrated the formation of more oxidised intermediates during photo-Fenton treatment. With this information, the photo-Fenton treatment time (190 min) and H_2O_2 dose (66 mM) necessary for adequate biodegradability of the wastewater could be determined. The IBR operated in batch mode was able to reduce the remaining DOC to less than 35 mg/L. Ammonium consumption and NO_3^- generation demonstrated that nitrification was also attained in the IBR. Overall DOC degradation efficiency of the combined photo-Fenton and biological treatment was over 95 %, of which 33 % correspond to the solar photochemical process and 62 % to the biological treatment.

Méndez-Arriaga *et al.* (2010) carried out the degradation of the worldwide non-steroidal anti-inflammatory drug ibuprofen (IBP) employing the photo-Fenton reaction by use of solar artificial irradiation. Non-photocatalytic experiments (complex formation, photolysis and UV/Vis-H_2O_2 oxidation) were executed to evaluate the isolated effects and additional differentiated degradation pathways of IBP. For the direct photolysis experiment, the IBP solution (initial concentration of 0.87 mM) was irradiated throughout the photoreactor during 2 h. On the other hand, several molar ratios of IBP-H_2O_2 (ranging from 1:0.001 to 1:10) were mixed in 30 mL vials at free pH conditions (pH 6.25 ± 0.25), constant controlled temperature (30°C) and continuous dark stirring for 24 h. Furthermore, H_2O_2 photolysis process was carried out irradiating a solution of 0.87 mM of IBP + 0.32 mM of H_2O_2 for 2 h. Direct UV photolysis did not change IBP and TOC concentrations. Similar results were observed for dark experiments with different ratios of IBP and H_2O_2. In contrast, UV-Vis/H_2O_2 process showed an evident decrease in the IBP concentration. When solution was irradiated in presence of H_2O_2, almost 40 % IBP degradation took place after 2 h, since in the UV spectrum of the used light source (Xe lamp and Duran-glass photoreactor) there is an overlap between the emitted light and the irradiation yet available to be absorbed for H_2O_2 at around 290–310 nm. Hence, the photolytic cleavage of H_2O_2 generates hydroxylated-IBP by-products without any mineralisation. The dark Fenton reaction (conditions: 0.15 mM, 0.29 mM, 0.60 mM and 1.2 mM Fe^{2+}; 0.87 mM IBP; T = 30°C and 0.32 mM H_2O_2) promoted hydroxylation accompanied with 60 % IBP degradation and 10 % mineralisation. In contrast, photo-Fenton in addition promotes the decarboxylation of IBP and its total depletion is observed. In absence of H_2O_2, 60 % IBP abatement was observed in the Fe(II)/UV-Vis process (Fe^{2+} = 1.2 mM; initial pH = 5) after 2 h irradiation due to the complex formation between iron and the IBP-carboxylic moiety. Complete IBP abatement was observed for the same treatment system in the presence of 5 mM Fe^{2+}. The degradation pathway can be described as an interconnected and successive principal decarboxylation and hydroxylation steps. TOC depletion of 40 % was observed in photo-Fenton degradation. The iron-IBP binding was the key-point of the decarboxylation pathway. Both decarboxylation and hydroxylation mechanisms, as individual or parallel process are responsible for IBP removal in Fenton and photo-Fenton systems. The degradation of IBP by photo-Fenton oxidation goes through the simultaneous cleavage of the carboxyl acid moiety to produce formic acid or acetaldehyde as scavengers of hydroxyl radicals. As described before, the formation of highly soluble products including isopropanol, formic acid and acetone promotes stable dissolution. An increase in the biodegradability of the final effluent after photo-Fenton treatment was observed. Final BOD_5 of 25 mg/L was reached in contrast to the initial BOD_5 shown by the untreated IBP solution (BOD_5 < 1 mg/L).

5.6.5 Heterogeneous photocatalytic processes

Rizzo *et al.* (2009) studied the degradation kinetics and mineralisation of an urban wastewater treatment plant effluent contaminated with a mixture of pharmaceutical compounds composed of amoxicillin (AMX; 10 mg/L), carbamazepine (CBZ; 5 mg/L) and diclofenac (DFC; 2.5 mg/L) by TiO_2-mediated photocatalysis were investigated in both spiked distilled water and actual wastewater solutions (TOC of wastewater + pharmaceuticals mixture = 14.1 mg/L). Photocatalytic experiments were carried in a 300 mL cylindrical Pyrex vessel that was filled with 100 mL of the aqueous pharmaceutical solutions or the spiked wastewater. TiO_2 (Aeroxide P 25, Degussa) was added to achieve the desirable catalyst loading in the range of 0.2–0.8 g/L. The light source was a 125 W black light fluorescent lamp (light intensity = 4.7×10^{-7} einstein/s) emitting radiation between 300 and 420 nm (with a maximum emission band at 350 nm) under continuous magnetic stirring. A set of bioassays (*Daphnia magna*, *Pseudokirchneriella subcapitata* and *Lepidium sativum*) was performed to evaluate the potential acute toxicity of the oxidation intermediates. No significant variation in terms of TOC and UV absorbance after 120 min exposure to UV-A light alone was observed and the experiments under dark conditions (adsorption on 0.8 g/L TiO_2 in the absence of UV-A light) resulted in less than 15 % removal after 120 min. A pseudo-first order kinetic model was found to fit well the experimental data. The TOC and UV absorbance parameters strongly depended on the target contaminant and the photocatalyst loading. As expected, the half-life time ($t_{1/2}$) decreased as the photocatalyst loading increased and it was shorter for the degradation of the single compound compared to its mineralisation rate. The mineralisation rate of the wastewater contaminated with the pharmaceuticals was slow ($t_{1/2}$ = 86.6 min) compared to that of the same pharmaceuticals spiked in distilled water ($t_{1/2}$ = 46.5 min) probably due to the carbonate/carbonate alkalinity of the wastewater. The initial concentrations of AMX and CBZ were found to be toxic to *D. magna* (100 % immobilisation for AMX, 25 and 45 % immobilisation for CBZ after 24 and 48 h exposure times respectively). The toxicity of AMX was reduced to 45 and 55 % of immobilisation (24 and 48 h exposure times, respectively) after 120 min of photocatalytic treatment using 0.8 g/L TiO_2 at which the highest TOC removals were obtained for both AMX and CBZ. The toxicity reduction to *D. magna* for CBZ was quite low compared to AMX: 10 and 20 % of immobilisation was observed for 24 and 48 h exposure times, respectively, after 120 min of irradiation at the same TiO_2 loading. A slightly higher toxicity to *P. subcapitata* compared to *D. magna* was observed for all the samples tested. The photocatalytic treatment of the mixture of AMX, CBZ and DCF resulted in a more severe toxicity to *D. magna* than the single pharmaceutical solutions because of the presence of AMX and/or its oxidation intermediates (100 % of immobilisation after 120 min of irradiation at 0.2 g/L-0.8 g/L TiO_2 loadings). *P. subcapitata* results were in accordance with those of *D. magna.* Photocatalytic treatment did not completely reduce the toxicity under the investigated conditions (maximum catalyst loading and irradiation time of 0.8 g/L TiO_2 and 120 min respectively). The toxicity of the uncontaminated wastewater sample increased from 34 to 66 % when it was spiked with 2.5, 5.0 and 10.0 mg/L of AMX, CBZ and DCF initial concentrations, respectively. After 5 and 10 min of photocatalytic oxidation the toxicity of the wastewater + pharmaceuticals mixture decreased drastically to 37 and 21 %, respectively. This decrease followed by an increase can be attributed to initial rapid oxidation of parent compounds and subsequent formation of hardly oxidisable organic intermediates which took a longer time and caused an increase in the toxicity to 50 % inhibition after 30 min of oxidation as. Afterwards, the toxicity decreased to 30 % and thereafter increased back to 41 % after 60 and 120 min oxidation, respectively, indicating that oxidation and consequently the formation of intermediates continued throughout the treatment process. Similar to algae toxicity, the treated wastewater-drug mixture samples displayed a multi-phase toxicity profile to *L. sativum* by means of the germination index (GI) was stimulated when it was spiked with the drug mixture. The increasing GI after 5 min photocatalysis (increasing GI indicates toxicity reduction, even stimulation) revealed that the formation of toxic oxidation intermediates was low. However, by increasing the irradiation time from 5 to 20 min, the formation of relatively toxic oxidation intermediates is expected to increase and therefore the stimulating synergistic effect due to the combination between the pharmaceuticals and the wastewater characteristics can no longer balance the increased toxicity. Ultimately, after 120 min photocatalytic treatment a drastic decrease in the GI was observed, indicating the formation of oxidation intermediates which were toxic to *L. sativum* seeds.

5.6.6 Wet peroxide oxidation

Fenton-like reactions have been efficiently used as oxidation processes for the treatment of real industrial wastewaters. Nevertheless, works described in the literature have used homogeneous Fenton reactions, maintaining the acidic pH medium and post-treatment steps to remove the iron species. A step forward is carried out in the work of Melero *et al.* (2009) using an Fe_2O_3/SBA-15 material as a Fenton-like heterogeneous catalyst for the treatment of a pharmaceutical wastewater by catalytic wet oxidation in the presence of hydrogen peroxide (CWPO), in a continuous up-flow fixed bed reactor (FBR). The wastewater (COD = 1901 mg/L; TOC = 860 mg/L; BOD_5 = 381 mg/L; pH = 5.6; nitrate = 500 mg/L; chloride = 3380 mg/L; SS = 40.6 mg/L; average oxidation state ((AOS) = 0.70) was obtained from a large scale drug and fine chemical manufacturing industrial plant located in the south of Madrid, Spain. This industry is focused on the synthesis of specific drugs for cardiovascular and venous diseases, as well as diabetes. Iron oxide supported into a SBA-15 meso-structured material was prepared by co-condensation of iron and silica (tetraethoxysilicate) sources under acidic conditions and templated with Pluronic 123. The procedure described in the literature was slightly modified, changing the ageing pH up to 3.5 by addition of an appropriate amount of ammonia aqueous solution, in order to promote the formation and precipitation of iron oxo-hydroxyde particles. The resultant solution was aged at 110°C for 24 h under static conditions. The solid product was recovered by filtration, and calcined in air at 550°C for 5 h, removing the template and promoting the crystallisation of iron oxides. For the preparation of the catalytic packed bed, Fe_2O_3/SBA-15 nanocomposite material was extruded by blending the fresh powder catalyst (75 % w/w) with sodium bentonite (25 % w/w) and synthetic methylcellulose polymer which acts as binder in the extrusion process. All the components were kneaded under high shear conditions until a homogeneous paste was achieved a adding controlled dosage of deionised water. The paste was then passed through a 5 mm circular die using a ram extruder. The shaped-rod product was dried for 3 days in a controlled temperature and humidity chamber with temperature and relative humidity ranges of 20–40°C and 70–10 %, respectively. After that, conformed catalyst was calcined at 650°C in air for 2 h, using a slow ramp of temperature to remove water gradually, burning off the organic content and enhancing the mechanical strength of the calcined materials by sintering the inorganic binder. The final Fe_2O_3/SBA-15 pellets were obtained by crushing and sieving the extruded material to particle sizes ranging from 1.0 to 1.6 mm. Some physicochemical characteristics of the pellets were; pore diameter: 8.1 nm; BET surface area: 264 m^2/g; iron content: 14 %; average particle size: 1000–1600 μm, whereas the powdered form of Fe_2O_3/SBA-15 had the following characteristics: pore diameter: 7.9 nm; BET surface area: 495 m^2/g; iron content: 19 %; average particle size: 14.4 μm. The powder catalyst can be described as a composite material that contains different iron oxides particles (mainly crystalline hematite) embedded on a mesostructured SBA-15 matrix, and a small amount of ionic Fe^{3+} species included in the framework by isomorphous substitution of Si ions. In order to elucidate the influence of temperature, initial oxidant concentration and initial pH of the reaction medium Preliminary catalytic wet peroxide oxidation experiments were performed in a 100 mL glass autoclave reactor, with a total reaction volume of 75 mL, and mechanically stirred at 350 rpm. An appropriate amount of catalyst powder suspended in the wastewater (0.6 g/L) was placed into the glass reactor. Then, the system was pressurised up to 7 bar with air and heated to the selected reaction temperature. When the required temperature was achieved, the corresponding amount of hydrogen peroxide was added to the reactor, and the reaction started. Total reaction time was fixed at 4 h. Samples were withdrawn every 30 min and filtered through a 0.22 μm nylon filter. In that case, a temperature of 80°C using an initial oxidant concentration corresponding to twice the theoretical stoichiometric amount for complete carbon depletion and initial pH of 3 resulted in 50 % TOC removal after 200 min treatment. Negligible TOC conversion was achieved in the absence of oxidant. The iron detected within the reaction medium after 240 min of reaction hardly reached 6 mg/L, which corresponds to a loss of 6 % of the iron initially present in the solid catalyst. Particularly, the catalytic run carried out at 80°C, initial pH of 3 and initial hydrogen peroxide concentration of 10,800 mg/L allowed a TOC conversion of around 58 % after 240 min, accompanied by an iron leaching degree of only 4.8 %, which corresponds to a concentration of iron in the reaction medium of 4.8 mg/L. Thereafter, the powder catalyst was extruded with bentonite to prepare pellets that could be used in an up-flow fixed bed reactor (FBR) working under atmospheric pressure. The FBR consisted of a jacketed glassware tube of 1.2 cm inner diameter and 15 cm length. A catalyst mass of 2.9 g was placed in the catalytic packed bed between two beds of spherical inert glass particles. An appropriate mixture of wastewater and hydrogen peroxide was pumped to the packed bed reactor by means of a Gilson 10SC HPLC pump operating at variable flow. The temperature of the system on the upper part of the bed was controlled at 80°C using an external recirculating heating bath with a stream of silicon oil through the jacket of the glass reactor. Time t = 0 min was considered to be when the feed flow reaches the top of the catalytic bed. An aqueous solution of the wastewater (860 mg/L of TOC content) and hydrogen peroxide (10,800 mg/L, equivalent to twice the stoichiometric amount for the theoretical complete mineralisation of the TOC content) acidified to pH 3 was used in these FBR experiments. Results in the up-flow FBR indicated that the catalyst shows high activity in terms of TOC mineralisation (*ca.* 60 % under steady-state conditions) and a high stability of the supported iron species. The oxidant consumption efficiency decreased gradually with increasing hydrogen peroxide loading in the feed. However, whereas this value hardly varies when the initial oxidant concentration is changed from 5400 mg/L to 10,800 mg/L (efficiencies = 0.33 and 0.26, respectively), there is an outstanding decrease when the hydrogen peroxide loading is increased to 16,400 mg/L (efficiency = 0.18). These catalytic results indicate that the use of an initial oxidant concentration twice the stoichiometric amount allows the obtaining of high organic degradation with a reasonable use of hydrogen peroxide. Upper oxidant loadings in the feed stream do not improve the TOC degradation, yielding to an inefficient use of hydrogen peroxide. Nevertheless, in all cases, the efficiency is far from the unit value (which means maximum efficiency in the use of the oxidant) which indicates that not all the oxidant is efficiently used for TOC mineralisation due to the formation of intermediate compounds more refractory to be oxidised or even scavenging reactions of the hydroxyl radicals. The activity of the catalyst was kept constant for at least 55 h of reaction in the range of 50–60 % TOC conversion, showing a high stability of iron species, and the iron concentration in the outlet effluent being almost negligible. Likewise, the catalyst particles did not grind during the process. As result of the oxidation process the COD removal attained in the system was about 81 %. The chloride amount increased after treatment; the pharmaceutical wastewater could contain chlorinated organic compounds that can be degraded during the Fenton-like oxidation, releasing chloride ions into the reaction medium.

Furthermore, the BOD_5/COD ratio increased from 0.20 to 0.30, whereas the AOS changed from 0.70 to 2.35.

5.6.7 Power ultrasound

In the aquatic environment, diclofenac (DCF), an anti-inflammatory non-steroidal agent, amoxicillin (AMX), a moderate-spectrum β-lactam antibiotic, and carbamazepine (CBZ), an anticonvulsant and mood-stabilising drug, are considered as the most frequently detected pharmaceuticals. Their presence in the effluents of sewage treatment works has often been documented at concentrations up to 2 μg/L (Zhang *et al.* 2008). The application of sonolysis (ultrasound; US) for wastewater treatment is an area of increasing interest. The aim of the study conducted by Naddeo *et al.* (2009) was to evaluate the effect of US on the degradation of the pharmaceuticals DCF ($C_{14}H_{10}Cl_2NNaO_2$), AMX ($C_{15}H_{12}N_2O$) and CBZ ($C_{16}H_{19}N_3O_5S$) in single solutions and mixtures spiked into urban wastewater (pH = 7.5; TOC = 4.4 mg/L; COD = 10.5 mg/L; UV_{254} = 9.97 m^{-1}) to study their combined/synergistic impact. Several operating conditions, such as power density (25–100 W/L), initial substrate concentrations (2.5–10 mg/L), initial solution pH (3–11), and air sparging was varied for the evaluation and understanding of the process. The degradation (as assessed by measuring UV absorbance), the generation of hydroxyl radicals (as assessed by measuring H_2O_2 concentration), the mineralisation efficiency (in terms of TOC and COD removals), and the aerobic biodegradability (assessed by the BOD_5/COD ratio) were monitored during sonication. A Sonics Vibracell™ VCX-750 (Sonics&Materials Inc., U.S.A.) ultrasound generator operating at a fixed frequency of 20 kHz and equipped with a titanium horn (tip diameter = 1.3 cm) was employed. Reactions were carried out in a 300 mL cylindrical Pyrex vessel (Schott Duran, Germany), which was filled with 200 mL of wastewater spiked at various initial concentrations of pharmaceuticals (2.5 mg/L DCF; 2.5 and 5 mg/L AMX; 2.5, 5 and 10 mg/L CBZ and their mixtures. The applied electrical power was varied between 25 and 100 W/L. The effect of sparging of pure air on the process efficiency was also evaluated. Ecotoxicity to *Daphnia magna*, *Pseudokirchneriella subcapitata* and *Lepidium sativum* before and after treatment was also examined. Results have indicated that the degradation of pharmaceuticals was enhanced at increased power densities, under acidic conditions and in the presence of dissolved air. As well known, pharmaceuticals predominantly exist in their ionic form at pH values greater than 7.0 while they are principally found in molecular/protonated form at lower pH values Decomposition increased under acidic pH due to the fact that the molecular state of drugs is more hydrophobic than the ionic one, and the solute is expected to accumulate in the vicinity of the cavitation bubbles where reactions between pharmaceuticals and the radicals occur at enhanced rates. On the contrary, their more hydrophilic ionic states prevail at pH values greater than 7.0. At the extreme alkaline state (pH = 11), the rate of H_2O_2 production becomes three times faster than when the pH is initially 3 due to the fact that at acidic pH, free radicals have a greater chance to attack pharmaceuticals and their degradation by-products rather than recombining and yielding H_2O_2. The reaction rate increases with increasing initial concentration of single pharmaceuticals but it remains constant in the mixtures, indicating different kinetic regimes (*i.e.* first and zero order respectively). Mineralisation is a slow process as reaction by-products are more stable than pharmaceuticals to total oxidation; however, they are also more readily biodegradable. The toxicity of the wastewater samples before and after contamination with pharmaceuticals both in mixtures and in single substance-containing solutions was observed more severely on *P. subcapitata*; a fact that raises concerns in regards to the discharge of such effluents. *D. magna* displayed less sensitivity compared to *P. subcapitata* because it belongs in a lower taxonomic species than *D. magna*. The germination index of *L. sativum* in the presence of the drugs was stimulated instead of inducing any toxic influence that could be attributed to the fact that low drug concentrations acted as a provider of additional nutrients to the wastewater.

Levodopa is the most frequently prescribed drug for the treatment of Parkinson disease. Levodopa effects on the cellular death by oxidative stress and its neurotoxicity on animals have been demonstrated (Mormont and Laloux 2002). Paracetamol, a non-steroidal anti-inflammatory recalcitrant drug, is one of the most frequently detected pharmaceutical products and its transformation into toxic compounds during chlorination in wastewater treatment plants is well described by Bedner and Maccrehan (2006). The aim of this paper authored by Isariebel *et al.* (2009) was to evaluate the influence of several parameters of the US process on the degradation of paracetamol and levodopa. Experiments were carried out at 574, 860 and 1134 kHz of ultrasonic frequency with horn-type sonicator and actual power values of 9, 17, 22 and 32 W at 20°C. Initial concentrations of 25, 50, 100 and 150 mg/L of both products were used. US reactions were carried out in a 500 m L cylindrical glass reaction vessel cooled with water circulation. 300 mL of the aqueous solution at initial drug concentrations of 25, 50, 100 and 150 mg/L were prepared and subjected to ultrasonic irradiation. Treatment efficiency was assessed following changes in pharmaceuticals concentration and COD. The sonochemical degradation of both products follows a pseudo-first order reaction kinetics. 4 h sonication of a reaction mixture initially containing 25, 50, 100 and 150 mg/L levodopa resulted in 91, 70, 54 and 46 % of degradation, respectively. In the case of paracetamol, the extent of degradation was 95, 82, 70 and 56 %, for the reaction mixture initially containing 25, 50, 100 and

150 mg/L, respectively, after 4 h of sonolysis. Although complete drug degradation was achieved within 8 h US treatment, some DOC always remained in solution showing that long lived intermediates were recalcitrant to ultrasound irradiation. Pollutant conversion and COD removal were found to decrease with increasing the initial solute concentration and decreasing power. The best results were obtained with 574 kHz frequency. Investigations using 1-butanol as a radical scavenger and H_2O_2 (590 mg/L) as a promoter revealed that pollutant degradation proceeds principally through radical reactions, although some differences were observed between both molecules. 5 % of degradation was reached for paracetamol and 17 % for levodopa in the presence of the free radical scavenger, suggesting that some minor degradation can also take place in the bulk solution. Because paracetamol has a higher log P_{OW} than levodopa (0.89 instead of –2.9, respectively), one could expect that its degradation at the interface will reach a higher extend than levodopa and thus that the influence of 1-butanol will be also higher. For levodopa, due to its higher hydrophilicity the mechanism in the bulk solution is more important than for paracetamol. Addition of H_2O_2 had a positive effect on degradation rate, but the optimum concentration of hydrogen peroxide depends on the pollutant.

5.6.8 Process combinations (hybrid systems)

Recent papers report integrated biological+chemical oxidation to be more effective in pharmaceutical effluent treatment (Gebhardt and Schröder 2007; Badawy *et al.* 2009; Xing and Sun 2010; Mandal *et al.* 2010). For problematic and recalcitrant effluents that still have a high COD (> 800–900 mg/L) even after biological treatment, chemical oxidation processes also have to be combined. The treated wastewater investigated by Boroski *et al.* (2009) consists of refractory materials and high organic content of hydrolyzed peptone residues from pharmaceutical factory. The effluent was obtained from a pharmaceutical and cosmetic factory located in north western Parana State, Brazil. It was collected from various baths in an equalisation tank of the last stage of the biological treatment. The biotreated effluent had the following characteristics; COD = 1753 mg/L; BOD_5 = 200 mg/L; sulphate = 894 mg/L; ammonia-N = 220 mg/L; organic-N = 344 mg/L; nitrite = 384 mg/L; phosphate = 17 mg/L.; pH = 7.3. The EC apparatus used in this investigation consisted of a pair of iron electrodes (dimensions: 12.50 cm × 2.50 cm × 0.10 cm of thickness) with an active surface area of 31.25 cm^2 distanced by 2.0 cm. Electrodes were connected to a DC power supply of 5 A and 12 V (Hayama 1-212), operating with current density of 763 A/m^2. Electrocoagulation (EC) was carried out in a batch-type system by using 500 mL of original samples (as collected from the factory) at the pH values of 4.0, 5.0, 6.0, 7.0 and 10.0. The effluent subjected to EC treatment under optimised reaction conditions was submitted to the photocatalysis unit. An UV artificial light source (3×250 W low-pressure mercury lamps without bulbs) was used to treat 300 mL effluent in a 500 mL of borosilicate glass Erlenmeyer photo-reactor (80 cm × 50 cm) fitted with three UV lamps at the top 15 cm away from the samples. The light power was measured by a light meter (Newport Optical Power Meter Model 1830-C) at 400 nm. Four fans set on the box side walls were used to reduce the heat caused by the UV lamps. pH, time, photocatalyst (Degussa P 25) and hydrogen peroxide concentration were selected as the process variables. For pH optimisation, experiments were conducted in 0.25 g/L of TiO_2 without H_2O_2. At pre-set pH, experiments were carried out in 0.25, 0.50 and 0.75 g/L of TiO_2 without H_2O_2. After pH and TiO_2 concentration optimisation, photocatalytic experiments were run in the presence of 2, 10 and 50 mM of H_2O_2. The necessary time of irradiation was investigated under the optimum conditions of pH, TiO_2 and H_2O_2. The combination of EC followed by heterogeneous photocatalysis (TiO_2) conditions was maximised. For EC optimum conditions were established as a current density of 763 A/m^2, a treatment time of 90 min and an initial pH of 6.0. During EC the majority of the dissolved organic and suspended material was removed (91 % and 86 % of the turbidity and COD, respectively). After EC, refractory residues still remained in the effluent. The subsequent photocatalysis conditions were optimised as follows; pH 3.0, 4 h irradiation, 0.25 g/L TiO_2 and 10 mM H_2O_2 exhibited the highest levels of inorganic (nitrite, sulphate, *etc.*) and organic pollutant (organic carbon and nitrogen from peptone liquor) elimination. The photocatalytic treatment of the investigated wastewater was favoured under acidic pH because effluents contain pollutants with a negative charge. These anions adsorb on the positively charged suspended photocatalyst; the point zero charge of the TiO_2 occurs near pH 6.5 (Chu and Wong 2004). The COD after EC was reduced to 160 mg/L and to 50 mg/L after photocatalytic treatment.

The compound 2-[3-(2-methylpropyl)phenyl] propanoic acid, commercially available as ibuprofen (IBP), is widely used as an anti-inflammatory and antipyretic drug especially prescribed for the treatment of fever, migraine, muscle aches, arthritis and tooth aches. Due to its widespread applications, several kilotons of ibuprofen have been synthesised worldwide. Industrial and domestic routes were identified as the major contamination pathways for IBP in aquatic environments. The sonolytic (at 213 kHz; power density range: 16, 35 and 55 W/L), photocatalytic and sonophotocatalytic degradations of IBP (0.02, 0.05, 0.07, 0.09 and 0.09 mM) in the presence of homogeneous (0.045 mM Fe^{3+} at pH 2.7) and heterogeneous photocatalysts (1 g/L Degussa P25 TiO_2) were studied by Madhavan *et al.* (2010). When compared with sonolysis and photocatalysis, a higher degradation rate was observed for sonophotocatalysis in the presence of TiO_2 or Fe^{3+} and also a slight synergistic enhancement was found with a synergy index of 1.3 and 1.6, respectively. Even though TiO_2 sonophotocatalysis showed an additive process effect in the mineralisation, a significant synergy effect was observed for the sonophotocatalysis in the presence of Fe^{3+}. This might be due to the formation of photoactive complexes between Fe^{3+} and IBP degradation products, such as carboxylic acids. High performance liquid chromatography (HPLC) and electrospray ionisation mass spectrometry (ESMS) techniques were employed for the identification of the degradation intermediates. The sonication of IBP led to the formation of its mono- and di-hydroxylated intermediates. Apart from the hydroxylated intermediates, products formed due to the oxidation of propanoic acid and isobutyl substituents of IBP were also observed.

5.6.9 Comparative studies

Arslan-Alaton and Dogruel (2004) comparatively examined a variety of advanced oxidation processes (AOP; O_3/OH^-, H_2O_2/UV, Fe^{2+}/H_2O_2, Fe^{3+}/H_2O_2, Fe^{2+}/H_2O_2/UV and Fe^{3+}/H_2O_2/UV) for the oxidative pretreatment of real penicillin (active substances: amoxicillin trihydrate; *e.g.* $C_{16}H_{19}N_3O_5S{\cdot}3H_2O$, molecular weight = 419.4 g/mole; assisted by the β-lactamase inhibitor clavulanate potassium; *e.g.* $C_8H_8KNO_5$, molecular weight = 237.3 g/mole) formulation effluent (average COD_0 = 1395 mg/L; TOC_0 = 920 mg/L; $BOD_{5,0} \approx 0$ mg/L; pH = 6.95; TSS = 145mg/L). The selected washwater corresponded to approximately 24 % of the total daily effluent (=150 m^3/day), of which 47 % was process water. Ozone was produced from dry air by a corona discharge PCI GL-1 model pilot scale ozone generator with a maximum capacity of 20 g/h. The mass transfer coefficient of ozone in the semi-batch reactor was determined for ozonation in acidic pure water by employing the indigo spectrophotometric method as 0.22 min^{-1}. The ozone dose applied after 1 h was 2750 mg/L wastewater. A 2000 mL capacity annular Plexiglass photoreactor equipped with a 21 W low pressure mercury arc lamp emitting UV-C light at 253.7 nm and being placed into a quartz sleeve housing was used throughout the photochemical experiments. The light intensity and effective pathlength of the UV-C lamp were determined as 3.65 W/L (= 1.73×10^{-4} einstein/L/s) and 1.72 cm, respectively. For the ozonation process the primary involvement of free radical species such as $HO^\bullet$ in the oxidative reaction could be demonstrated via inspection of ozone absorption rates. COD removal rates increased with increasing pH as a consequence of enhanced ozone decomposition rates at elevated pH values. Ozone decomposition to free radicals could also be confirmed by simple comparison of the absorbed ozone rates; only 38 % of the applied ozone dose was absorbed at pH = 3, whereas this rate increased to 67 and 72 % at pH = 7.0 and pH = 11.5, respectively. Alkaline ozonation and the photo-Fenton's reagent both appeared to be the most promising

AOP in terms of COD (49–66 %) and TOC (42–52 %) abatement rates, whereas the BOD_5 of the originally non-biodegradable effluent could only be improved to a value of 100 mg/L with O_3/pH = 3 treatment (BOD_5/COD = 0.08). During the ozonation experiments where the pH was kept constant with phosphate buffers COD removal rates did not slow down and were improved appreciably as compared to the unbuffered effluent samples. At pH = 7.0, for instance, a COD reduction similar to that at pH = 11.5 for the sample with no buffer could be achieved, and the highest COD removal rate (86 %) was obtained for the pH = 11.5-buffered sample. Again, ozone absorption rates of the control experiments were found relatively high (64 and 79 %) in comparison to ozonation at acidic pH, speaking for a shift in the reaction mechanism from a rather selective, molecular ozone oxidation to a faster, non-selective free radical reaction. The effluent was not degraded at all via direct UV-C photolysis. In addition to this observation, no increase in colour being measured at $\lambda = 436$ nm was obtained for mere UV-C photolysis speaking for the absence of any significant degradation, though the abrupt build-up of a pale pink colour (appearance of an absorption band at $\lambda = 525$ nm) instead delineates at least structural modifications, *i.e.* the possible formation of polymerization products, and hence probable changes in effluent composition. The addition of H_2O_2 only slightly improved the overall COD removal rates up to 22 % for 30 mM, which is relatively speaking to the ozonation experiments a rather poor oxidation rate. Increasing the H_2O_2 dose to 40 mM resulted in a significant inhibition of the COD abatement rate resulting in 11 % COD removal because H_2O_2 acts as a radical scavenger when overdosed depending upon the reaction conditions. Only 25 and 30 % of the introduced H_2O_2 could be used in the present study due to competitive UV light absorption of the effluent. The Fenton process was considerably enhanced in terms of COD removal rates upon the introduction of UV-C light. UV-C light introduction was more pronounced for the Fenton-like process involving Fe(III)-salt. Moreover, comparing the initial reaction rates $R_{COD,i}$ (in terms of COD abatement rates) of the Fenton and Fenton-like reactions (1 mM Fe(II)/(III) salt, 20 mM H_2O_2 at pH 3) in the absence and presence of UV-light it can be clearly seen that the reactions involving Fe^{2+} were approximately 3 times faster than those involving Fe^{3+}-iron ($R_{COD,i,Fe}{}^{2+}$ in the range of 108-130 mg/L COD/min). In terms of overall COD abatement rates, however, the photo-Fenton process initiated with Fe^{3+}-iron reaches the COD removal efficiency of the photo-Fenton process 20–30 min faster than during dark Fenton/Fenton-like reactions. Residual H_2O_2 measurements have demonstrated that during the Fenton and photo-Fenton reactions H_2O_2 was completely consumed at the end of the experiments. Evaluation on COD and TOC removal rates per applied active oxidant (AOx) and oxidant (Ox) on a molar basis revealed that alkaline ozonation and particularly the UV-light assisted Fenton processes enabling good oxidation yields (1–2 mole COD and TOC removal per AOx and Ox) by far outweighed the other studied AOP. Separate experimental studies conducted with the penicillin active substance amoxicillin trihydrate indicated that the aqueous antibiotic substance can be completely eliminated after 40 min advanced oxidation applying photo-Fenton's reagent (pH = 3; Fe^{2+}:H_2O_2 molar ratio = 1:20) and alkaline ozonation (at pH = 11.5), respectively. Only slight improvements of the biodegradability expressed as BOD_5 and BOD_5/COD were obtained after pre-oxidation with different, optimised AOP. The active ingredient amoxicillin could be completely eliminated after 40 min treatment with ozone at pH = 11.5 and photo-Fenton's reagent at pH = 3 under optimised conditions. Due to the fact that 400 mg/L is the practically highest expected concentration of active ingredient in the formulation effluent it can be inferred that its complete elimination should also be expected for real penicillin formulation wastewater. Active ingredient abatement was accompanied by 72 % COD-51 % TOC (O_3/pH 11.5) and 81 % COD-58 % TOC (photo-Fenton/pH 3) removals.

Klavarioti *et al.* (2009) recently reviewed the application of Advanced Oxidation Processes (AOP) for the treatment of pharmaceutical compounds. Herein it is emphasised that (i) several recent publications have been devoted to monitoring pharmaceuticals in various aqueous matrices (*i.e.* water and/or wastewater) (ii) regarding treatment efficiency, AOP are generally capable of completely destroying the specific pharmaceutical in question but this is not necessarily accompanied by total mineralisation. In several cases, degradation by-products are more biodegradable and less toxic than the original substrate, thus implying that a biological post-treatment may be feasible, (iii) regarding the type of AOP employed, heterogeneous photocatalysis with semiconductors, ozonation as well as Fenton and alike reactions are the most investigated types, (iv) regarding the water matrix, most studies deal with model aqueous solutions and surface waters (*i.e.* from rivers or lakes), while actual wastewaters from sewage treatment plants or effluents from pharmaceutical industrial units have received less attention, (v) the most common pharmaceuticals studied were diclofenac, carbamazepine, sulphamethoxazole, clofibric acid and 17β-estradiol, (vi) there is an increasing interest on the use of AOPs for the removal of pharmaceuticals from water/wastewater and this is reflected in the increasing number of journal articles published in the recent years (during the 1998–2008 period).

5.7 CONCLUDING REMARKS

The occurrence and fate of pharmaceuticals in the environment and particularly in aquatic media have received considerable attention by scientists during the last two decades. Although they appear at relatively low concentrations ranging between ng/L-μg/L levels, they may impose serious negative effects on the environment. Pharmaceutical processes use numerous raw materials and generate significant wastes and emissions of considerable variability and complexity. However, compared to other industries, production and process profiles are relatively well established and environmental management procedures have been standardised. In designing bulk manufacturing processes special consideration has been given to the availability of the starting materials and their toxicity as well as wastes (mother liquors, filter residues, by-products, cleaning and rinse water) and emissions generated. Due to purity concerns, solvents that are considered as the most toxic and/or hazardous compounds of pharmaceutical manufacturing activities are often not reused but sold for non-pharmaceutical uses, fuel blending, recycled or destroyed via incineration. Specific chemicals produced in pharmaceutical manufacturing processes enter aquatic and terrestrial environments mainly through non-point sources and different routes via discharge of industrial, agricultural and domestic wastewater. These chemicals are in some cases not only refractory but also seriously inhibit conventional biological treatment processes. Some of these compounds are human and veterinary antibiotics. Hence, their production is usually being isolated from other manufacturing activities. A considerable scientific literature exists in the treatability of different drugs in aqueous solutions and/or real pharmaceutical wastewater, however scientific reports have focused on the chemical oxidation of aqueous model pollutants (analgesics, antibiotics, hormones, *etc.*) being typical for this industry. Consequently it becomes very difficult to draw general information from the available data about the merits of chemical oxidation processes for the treatment of actual pharmaceutical wastewater. Recent treatability studies have demonstrated that most pharmaceutical formulations can be efficiently treated with emerging processes including AOP and ozonation. From the practical point of view, treatment-at-source may be a more realistic option at drinking water plants where ground and surface waters can be chemically oxidised to achieve destruction of pharmaceuticals together with other micropollutants as well as at pharmaceuticals manufacturing plants where formulation effluents are generated. Given the relatively high concentration of organics in such effluents, a process train comprising chemical and biological oxidation may be technically and economically feasible.

REFERENCES

Abellan, M. N., Bayarri, B., Gimenez J. and Costa, J. (2007) Photocatalytic degradation of sulfamethoxazole in aqueous suspension of TiO_2. *Appl. Catal. B* **74**, 233–241.

Abramovitch, R. A., Huang, B. Z., Abramovitch, D. A. and Song, J. G. (1999) In situ decomposition of PCBs in soil using microwave energy. *Chemosphere* **38**, 2227–2236.

Alexy, R., Kumpel, T. and Kümmerer, K. (2004) Assessment of degradation of 18 antibiotics in the closed bottle test. *Chemosphere* **57**, 505–512.

Andreozzi, A., Raffaele, M. and Nicklas, P. (2003) Pharmaceuticals in STP effluents and their solar photodegradation in aquatic environment. *Chemosphere* **50,** 1319–1330.

Arslan-Alaton, I. and Caglayan, A. E. (2006) Toxicity and biodegradability assessment of raw and ozonated procaine penicillin G formulation effluent. *Ecotoxicol. Environ. Safety* **63**, 131–140.

Arslan-Alaton, I. and Dogruel, S. (2004) Pre-treatment of penicillin formulation effluent by advanced oxidation processes. *J. Hazard. Mater*. **112**, 105–113.

Badawy, M. I., Wahaab, R. A. and El-Kalliny, A. S. (2009) Fenton-biological treatment processes for the removal of some pharmaceuticals from industrial wastewater. *J. Hazard. Mater.* **167**, 567–574.

Balcioglu, I.A. and Ötker, M. (2003) Treatment of pharmaceutical wastewater containing antibiotics by O_3 and O_3/H_2O_2 processes. *Chemosphere* **50,** 85–95.

Bedner, M. and Maccrehan, W. (2006) Transformation of acetaminophen by chlorination produces the toxicants 1,4-benzoquinone and N-acetyl-p-benzoquinone imine. *Environ. Sci. Technol.* **40**, 516–522.

Bendz, D., Paxeus, N. A., Ginn, T. R. and Loge, F. J. (2005) Occurrence and fate of pharmaceutically active compounds in the environment, a case study: Hoje River in Sweden. *J. Hazard. Mater.* **122**, 195–204.

Benitez, F. J., Acero, J. L., Real, F. J. and Roldán, G. (2009) Ozonation of pharmaceutical compounds: Rate constants and elimination in various water matrices. *Chemosphere* **77**, 53–59.

Benner, J., Salhi, E., Ternes, T. and von Gunten, U. (2008) Ozonation of reverse osmosis concentrate: kinetics and efficiency of beta blocker oxidation. *Water Res.* **42**, 3003–3012.

Boroski, M., Rodrigues, A. C., Garcia, J. C., Sampaio, L. C., Nozaki, J. and Noboru, H. (2009) Combined electrocoagulation and TiO_2 photoassisted treatment applied to wastewater effluents from pharmaceutical and cosmetic industries. *J. Hazard. Mater.* **162**, 448–454.

Bueno, M.J.M., Agüera, A., Gómez, M. J., Hernando, M. D., Garcia-Reyes, J. F. and Fernández-Alba, A. R. (2007) Application of liquid chromatography/quadrupole-linear ion trap mass spectrometry and time-of-flight mass spectrometry to the determination of pharmaceuticals and related contaminants in wastewater. *Anal. Chem.* **79,** 9372–9384.

Buxton, C.L. Greenstock, W.P. Helman and A.B. Ross (1988) Critical review of rate constants for oxidation of hydrated electrons, hydrogen atoms and hydroxyl radicals (AOH/AO^-) in aqueous solutions. *J. Phys. Chem. Ref. Data* **17**, 513–886.

Chu, W. and Wong, C. C. (2004) The photocatalytic degradation of dicamba in TiO_2 suspensions with the help of hydrogen peroxide by different near UV irradiation. *Water Res.* **38,** 1037–1043.

Gebhardt, W. and Schröder, H. F. (2007) Liquid chromatography-(tandem) mass spectrometry for the follow-up of the elimination of persistent pharmaceuticals during wastewater treatment applying biological wastewater treatment and advanced oxidation. *J. Chromatogr. A* **1160**, 34–43.

Hirsch, R., Ternes, T., Haberer, K. and Kratz, K.-L. (1999) Occurrence of antibiotics in the aquatic environment. *Sci. Total Environ.* **225,** 109–118.

Huber, M. M., Korhonen, S., Ternes T. A. and von Gunten, U. (2005) Oxidation of pharmaceuticals during water treatment with chlorine dioxide. *Water Res.* **39,** 3607–3617.

Isariebel, Q. P., Carine, J. L., Javier, J. H. U., Anne-Marie, W. and Henri, D. (2009) Sonolysis of levodopa and paracetamol in aqueous solutions. *Ultrason. Sonochem.* **16**, 610–616.

Kabdasli, I., Gürel, M. and Tünay, O. (1999) Pollution prevention and waste treatment in chemical synthesis processes for pharmaceutical industry. *Wat. Sci. Tech.* **39**, 265–271.

Kaniou, S., Pitarakis, K., Barlagianni, I. and Poulios, I. (2005) Photocatalytic oxidation of sulfamethazine. *Chemosphere* **60**, 372–380.

Kawala, Z. and Atamańczuk, T. (1998) Microwave-enhanced thermal decontamination of soil. *Environ. Sci. Technol.* **32,** 2602–2607.

Klavarioti, M., Mantzavinos, D. and Kassinos, D. (2009) Removal of residual pharmaceuticals from aqueous systems by advanced oxidation processes. *Environ. Int.* **35**, 402–417.

Lai, T. L., Lee, C. C., Wu, K. S., Shu, Y. Y. and Wang, C. B. (2006) Microwave-enhanced catalytic degradation of phenol over nickel oxide. *Appl. Catal. B* **68,** 147–153.

Le, T. X., Munekage, Y. and Kato, S. (2005) Antibiotic resistance in bacteria from shrimp farming in mangrove areas. *Sci. Tot. Environ.* **349,** 95–105.

Lin, A. Y.-C., Lin, C.-F., Chiou, J.-M. and Hong, P.-K.A. (2009) O_3 and O_3/H_2O_2 treatment of sulfonamide and macrolide antibiotics in wastewater. *J. Hazard. Mater.* **171,** 452–458.

Lishman, L., Smyth, S. A., Sarafin, K., Kleywegt, S., Toito, J., Peart, T., Lee, B., Servos, M., Beland, M. and Seto, P. (2006) Occurrence and reductions of pharmaceuticals and personal care products and estrogens by municipal wastewater treatment plants in Ontario, Canada. *Sci. Tot. Environ.* **367**, 544–558.

Madhavan, J., Grieser, F. and Ashokkumar, M. (2010) Combined advanced oxidation processes for the synergistic degradation of ibuprofen in aqueous environments. *J. Hazard. Mater.* **178**, 202–208.

Malato, S., Blanco, J., Vidal, A. and Richer, C. (2002) Photocatalysis with solar energy at a pilot-plant scale: an overview. *Appl. Catal. B* **37,** 1–15.

Mandal, T., Maity, S., Dasgupta, D. and Datta, S. (2010) Advanced oxidation process and biotreatment: their roles in combined industrial wastewater treatment. *Desalination* **250**, 87–94.

Melero, J. A., Martínez, F., Botas, J. A., Molina, R. and Pariente, M. I. (2009) Heterogeneous catalytic wet peroxide oxidation systems for the treatment of an industrial pharmaceutical wastewater. *Water Res.* **43**, 4010–4018.

Méndez-Arriaga, F., Esplugas, S. and Giménez, J. (2010) Degradation of the emerging contaminant ibuprofen in water by photo-Fenton. *Water Res.* **44**, 589–595.

Mormont, E. and Laloux, P. (2002) Therapeutic trends for the treatment of Parkinson's disease (Stratégie thérapeutique dans le traitement initial de la maladie de Parkinson). *Louvain Med.* **121**, 93–99.

Naddeo, V., Meric, S., Kassinos, D., Belgiorno, V. and Guida, M. (2009) Fate of pharmaceuticals in contaminated urban wastewater effluent under ultrasonic irradiation. *Water Res* **43**(16), 4019–4027.

Pereira, V. J., Linden, K. G. and Weinberg, H. S. (2007) Evaluation of UV irradiation for photolytic and oxidative degradation of pharmaceutical compounds in water. *Water Res.* **41**, 4413–4423.

PhRMA (1996) *Opportunities and Challenges for Pharmaceutical Innovation: PhRMA Industry Profile*, Pharmaceutical Research and Manufacturers of America, Washington, D.C.

PhRMA (1997) *Pharmaceutical Industry Waste Minimization Initiatives*, Pharmaceutical Research and Manufacturers of America, Washington, D.C.

Quan, X., Zhang, Y. B., Chen, S., Zhao, Y. Z. and Yang, F. L. (2007) Generation of hydroxyl radical in aqueous solution by microwave energy using activated carbon as catalyst and its potential in removal of persistent organic substances. *J. Mol. Catal. A: Chem.* **263,** 216–222.

Rizzo, L., Meric, S., Guida, M., Kassinos, D. and Belgiorno, V. (2009) Heterogenous photocatalytic degradation kinetics and detoxification of an urban wastewater treatment plant effluent contaminated with pharmaceuticals. *Water Res.* **43**, 4070–4078.

Rosario-Ortiz, F. L., Wert, E. C. and Snyder, S. A. (2010) Evaluation of UV/H_2O_2 treatment for the oxidation of pharmaceuticals in wastewater. *Water Res.* **44**, 1440–1448.

Sarmah, A. K., Meyer, M.T. and Boxall, A.B.A. (2006) A global perspective on the use, sales, exposure pathways, occurrence, fate and effects of veterinary antibiotics (VAs) in the environment. *Chemosphere* **65**, 725–759.

Sirtori, C., Zapata, A., Oller, I., Gernjak, W., Agüera, A. and Malato, S. (2009) Decontamination industrial pharmaceutical wastewater by combining solar photo-Fenton and biological treatment. *Water Res.* **43**, 661–668.

Standard Industrial Classification Manual (SIC 1987) Executive Office of the President, Office of Management and Budget, Washington D.C.

Tekin, H., Bilkay, O., Ataberk, S. S., Balta, T. H., Ceribasi, İ. H., Sanin, F. D., Dilek, F. B. and Yetis, Ü. (2006) Use of Fenton oxidation to improve the biodegradability of a pharmaceutical wastewater. *J. Hazard. Mater.* **136**, 258–265.

Trovó, A. G., Nogueira, R. F. P., Agüera, A., Fernandez-Alba, A.R., Sirtori, C. and Malato, S. (2009) Degradation of sulfamethoxazole in water by solar photo-Fenton. Chemical and toxicological evaluation. *Water Res.* **43**, 3922–3931.

Tünay, O., Samuk, B., Ölmez, T. and Kabdaşlı, I. (2004) Application of advanced oxidation to enhance biodegradability of pharmaceutical industry wastewaters. *Fres. Env. Bull.* **13**, 965–968.

United Sates Environmental Protection Agency (US EPA) (1997) *Office of Compliance Sector Notebook Project: Profile of the Pharmaceutical Manufacturing Industry*. Office of Enforcement and Compliance Assurance, Washington D.C.

United Sates Environmental Protection Agency (US EPA) (1998) *Pharmaceutical Manufacturing Category Effluent Limitations Guidelines*, Pretreatment Standards and New Source Performance Standards. Final Rule (40 CFR Parts 136 and 439, Washington, D. C.

United Sates Environmental Protection Agency (US EPA) (2004) *Pharmaceutical Manufacturing Point Source Category*, Federal Register Environmenal Documents dated May 6th, 2004, vol. 69, Nr. 88, Rules and Regulations, pp. 25324–25325, Washington, D. C.

U.S. Department of Commerce (1992 and 1995) *Census of Manufacturers for Drugs*, Industry Series, U.S. Department of Commerce, Bureau of the Census, Washington, D.C.

Van der Bruggen, B., Lejon, L. and Vandecasteele, C. (2003) Reuse, treatment, and discharge of the concentrate of pressure-driven membrane processes. *Environ. Sci. Technol.* **37**, 3733–3738.

Xing, Z.-P. and Sun, D.-Z. (2009) Treatment of antibiotic fermentation wastewater by combined polyferric sulphate coagulation, Fenton and sedimentation process. *J. Hazard. Mater.* **168**, 1264–1268.

Yang, Y., Wang, P., Shi, S. and Liu, Y. (2009) Microwave enhanced Fenton-like process for the treatment of high concentration pharmaceutical wastewater. *J. Hazard. Mater.* **168**, 1238–245.

Yao, D. C. C. and Haag, W. R. (1991) Rate constants for direct reactions of ozone with several drinking water contaminants. *Water Res.* **25,** 761–773.

Zhang, Y. J., Geissen, S. U. and Gal, C. (2008) Carbamazepine and diclofenac: removal in wastewater treatment plants and occurrence in water bodies. *Chemosphere* **73**, 1151–1161.

Zhou, J. L., Zhang, Z. L., Banks, E., Grover, D. and Jiang, J. Q. (2009) Pharmaceutical residues in wastewater treatment works effluents and their impact on receiving river water. *J. Hazard. Mater.* **166**, 655–661.

Chapter 6

Pulp and paper industry

6.1 PULP AND PAPER INDUSTRY

6.1.1 Overview

Pulp and paper industry is a major source of industrial pollution worldwide. One important characteristic of pulp and paper industry is the interconnection of operations between pulp mills and downstream processing of pulp into paper, paperboard and building paper. Another important characteristic of the industry is the range of processes, chemical inputs and outputs used in pulp manufacture (Shammas 2009). The pulp and paper industry comprises two types of facilities: (i) pulp and paper mills process raw wood fibre or recycled fibre to make pulp and/or paper; and (ii) converting facilities that use these primary materials to manufacture more specialised products such as paperboard boxes, writing paper, and sanitary paper (US EPA 2002). The specific industry components covered in this industry are the following (US EPA 2002):

- pulp mills;
- paper mills;
- paperboard mills;
- paperboard containers and boxes;
- miscellaneous converted paper products.

Pulp mills separate the fibres of wood or the other materials, such as rags, linters, wastepaper, and straw in order to produce pulp. The Mills can use chemical, semi-chemical, or mechanical processes, and can produce co-products such as turpentine and tall oil (US EPA 2002). Paper mills are primarily engaged in manufacturing paper from wood pulp and other fibre pulp, and can also manufacture converted paper products (US EPA 2002). Paperboard mills are primarily engaged in manufacturing paperboard, including paperboard coated on the paperboard machine, from wood pulp and other fibre pulp; and can also manufacture converted paperboard products (US EPA 2002). Paperboard containers and boxes are produced in the manufacture of corrugated and solid fibre boxes and containers from purchased paperboard. The principal commodities of this industry are boxes, pads, partitions, display items, pallets, corrugated sheets, food packaging, and non-food (*e.g.*, soaps, cosmetics, and medicinal products) packaging (US EPA 2002). Miscellaneous converted paper product facilities produce a range of paper, paperboard, and plastic products with the purchased material. Common products include paper and plastic film packaging, specialty paper, paper and plastic bags, manila folders, sanitary paper products, envelopes, stationery, and the other products (US EPA 2002).

Manufacturing processes and untreated wastewater characteristics (*i.e.*, pollutant loadings) are the principal factors used to develop the subcategorisation of the pulp and paper industry. In 1993, US EPA made several revisions after reviewing the subcategorisation proposed in 1982 (US EPA 1982; US EPA 1993a) by considering several factors such as process technologies and products manufactured, raw materials, wastewater characteristics, discharge rates and treatability, wastewater pollution control technologies, *etc.* Table 6.1 presents the revised (US EPA 1982) and original subcategory groupings (US EPA 1993a) of pulp and paper industry.

6.1.2 Pulp and paper manufacturing processes

The pulp and paper industry includes mills that manufacture market pulp, paper and/or paperboard from wood or non-wood pulp. Wood and non-wood pulp may be manufactured on-site, obtained from other mills by purchase or intra-company transfer or derived from pre- and/or post-consumer reclaimed fibre (secondary fibre) (US EPA 1993a).

In this subsection the manufacturing processes used in the industry, including fibre furnish preparation and handling, pulping, chemical recovery, pulp processing, bleaching, stock preparation, and pulp, paper and paperboard making will be described. The only processes common to all mills are pulp stock preparation and final product manufacture. The primary processes used at a typical bleached chemical pulp mill include (US EPA 1993a):

Table 6.1 Revised and original subcategory groupings of pulp and paper industry (US EPA 2002)

Subpart	Revised Subcategory	Previous Subcategory (Previous Subpart in Parenthesis)
A	Dissolving Kraft	Dissolving Kraft (F)
B	Bleached Papergrade Kraft and Soda	Market Bleached Kraft (G) BCT Bleached Kraft (H) Fine Bleached Kraft (I) Soda (P)
C	Unbleached Kraft	Unbleached Kraft (A) ➢ Linerboard ➢ Bag and Other Products Unbleached Kraft and Semi-Chemical (D, V)
D	Dissolving Sulphite	Dissolving Sulphite (K) ➢ Nitration ➢ Viscose ➢ Cellophane ➢ Acetate
E	Papergrade Sulphite	Papergrade Sulphite (J, U) ➢ Blow Pit Wash ➢ Drum Wash
F	Semi-Chemical	Semi-Chemical (B) ➢ Ammonia ➢ Sodium
G	Mechanical Pulp	Groundwood-Thermo-Mechanical (M) Groundwood-Coarse, Molded, News (N) Groundwood-Fine Papers (O) Groundwood-Chemi-Mechanical (L)
H	Non-Wood Chemical Pulp	Miscellaneous mills not covered by a specific subpart
I	Secondary Fibre Deink	Deink Secondary Fibre (Q) ➢ Fine Papers ➢ Tissue Papers ➢ Newsprint
J	Secondary Fibre Non-Deink	Tissue from Wastepaper (T) Paperboard from Wastepaper ➢ Corrugating Medium ➢ Non-Corrugating Medium Wastepaper-Molded Products (W) Builders' Paper and Roofing Felt
K	Fine and Lightweight Papers from Purchased Pulp	Non-Integrated Fine Papers (R) ➢ Wood Fibre Furnish ➢ Cotton Fibre Furnish Non-Integrated Lightweight Papers (X) ➢ Lightweight Papers ➢ Lightweight Electrical Papers
L	Tissue, Filter, Non-Woven and Paperboard from Purchased Pulp	Non-Integrated ➢ Tissue Papers (S) ➢ Filter and Non-Woven (Y) ➢ Paperboard (Z)

- fibre furnish preparation and handling (log slashing, debarking, chipping, and screening);
- pulping (chemical, semi-chemical and/or mechanical);
- chemical recovery;
- pulp processing (deknotting, brown stock washing, screening, cleaning, and thickening);
- bleaching;
- stock preparation (mixing, refining, and addition of wet additives).

If a mill manufactures market pulp, paper or paperboard, the pulp from bleach plant proceeds to a paper machine. Paper and paperboard making processes consist of wet end operations, dry end operations, pulp or paper drying and finishing. The processes mentioned above are discussed in the following subsections. A simplified flow diagram of an integrated mill including chemical pulping, bleaching, and paper production steps is delineated in Figure 6.1 (Smook 2002).

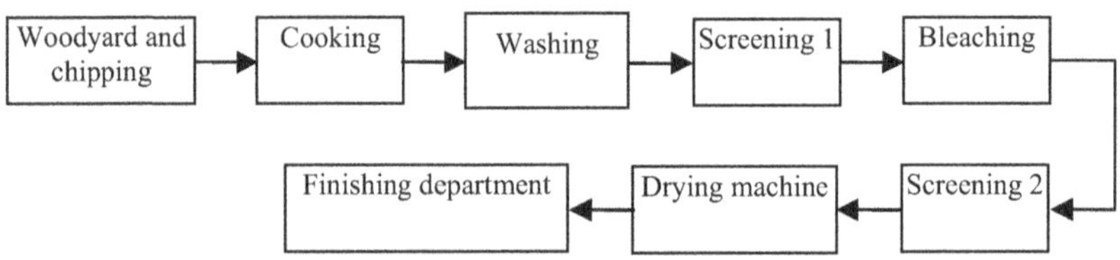

Figure 6.1 Simplified flow diagram of an integrated mill (chemical pulping, bleaching, and paper production) (Smook 2002).

6.1.2.1 Fibre furnish and fibre furnish preparation and handling

Furnish is the blend of fibrous materials used to make pulp. The major source of fibre for paper products are the vegetative tissues of vascular plants. Although almost any vascular plant could be used for paper production, the economics of scale require a high fibre yield for paper manufacture. A wide variety of fibre furnish types are used in the industry, but the most widely used by far is wood (US EPA 1993a). Overall, wood furnish averages approximately 50 percent of pulp content industry-wide (US EPA 1992). The fibrous particles used to make paper are made of cellulose, a primary component of the cell walls of vascular plant tissues. The cellulose fibres must be removed from a chemical matrix (*e.g.*, lignin, hemicelluloses, and resins) and result in a mixture of relatively pure fibres (US EPA 2002). Secondary fibres comprise of the next most common furnish constituent. Secondary fibres consist of pre-consumer fibres (*e.g.*, mill waste fibres) and post-consumer fibre. Post-consumer fibre sources are diverse, corrugated boxes, and news print are the two of the most common sources of secondary fibres. Secondary fibres must be processed to remove contaminants such as glues or bindings, but depending on the end product, may or may not be processed to remove ink contaminants or to brighten the pulp (US EPA 2002). Other fibre furnishes commonly used in the industry are non-wood fibres, including cotton linters, cotton rags, flax, hemp, bagasse, tobacco, and the other materials such as synthetic fibres (polypropylene) and inorganics (US EPA 1993a). These substances are not widely used, however, as they are typically for low volume, specialty grades of paper (US EPA 2002).

Furnish is prepared for pulp production by a process designed to supply a homogenous pulping feedstock. In the case of round wood furnish (logs), the logs are cut to manageable size and then debarked. The bark of those logs not fit for lumber is usually either stripped mechanically or hydraulically with high powered water jets in order to prevent contamination of pulping operations (US EPA 2002). Bark is typically incinerated in bark-burning furnaces which may produce energy depending on the moisture content of the bark (US EPA 1993a). Debarked logs are generally chipped using blades mounted on a rotating disk. After chipping, wood chips are passed over vibratory screens to remove oversized chips and fines. Oversized chips remain on the upper screen and are recycled to the rechipper. Fines are collected and usually burned with bark. After screening, chips are generally stored in large outside piles and are moved to subsequent operations via conveyors or augers (US EPA 1993a). Secondary fibre preparation and handling involves sorting and classifying the materials into quality grades. The bales of secondary fibres are charged directly to the continuous pulper where pulping, and removal of contaminants is performed (US EPA 1993a). Non-wood fibres are handled in ways specific to their composition. Steps are always taken to maintain fibre composition and thus pulp yields (US EPA 2002).

6.1.2.2 Pulping process

Pulping processes convert raw materials into fibres that can be formed into a sheet. Pulping may be performed chemically or mechanically using a variety of processes. The selection of pulping method depends upon the final product to be produced and the raw material used. Although, secondary fibres have previously undergone pulping, they require additional processing to make them amenable to forming into a sheet (US EPA 1993a). Three basic types of wood pulping processes and secondary fibre pulping techniques are discussed below.

Chemical pulping: Chemical pulping involves mixing the raw materials with cooking chemicals under controlled temperature and pressure conditions to yield a variety of pulps with unique properties. Chemical pulping separates the fibres of wood by dissolving the lignin bond holding the wood together (Shammas 2009). The three types of chemical pulping differ primarily in the types of chemical solutions and chemical recovery processes used (US EPA 1993a). The specialty paper products rayon, viscose, acetate, and cellophane are made from dissolving pulp, a variant of standard kraft or sulphite chemical pulping processes (US EPA 2002).

Kraft pulping (or sulphate) process uses a sodium-based alkaline pulping solution (liquor) consisting of sodium sulphide (Na_2S) and sodium hydroxide (NaOH) in 10 percent solution. This liquor (white liquor) is mixed with the wood chips (in pulp) and a liquid that contains the dissolved lignin solids in a solution of reacted and unreacted pulping chemicals (black liquor). The black liquor undergoes a chemical recovery process to regenerate white liquor for the first pulping step. Overall, the kraft process converts approximately 50 percent of input furnish into pulp (US EPA 2002).

Sulphite pulping process uses acid solutions of sulphurous acid (H_2SO_3) and bisulphite ion (HSO_3^-) to degrade the lignin bonds between wood fibres. The bisulphite may be an ionic salt of calcium, magnesium, sodium or ammonium. Sulphite pulps have less colour than kraft pulps and can be bleached more easily. The efficiency and effectiveness of the sulphite process are also dependent on the type of wood furnish and the absence of bark. For these reasons, the use of sulphite pulping has declined in comparison to kraft pulping over time (US EPA 2002).

The soda process, from which the kraft process evolved, uses an alkaline solution of only NaOH. The kraft process has virtually replaced with the soda process due to the economic benefits of chemical recovery and improved reaction rates (US EPA 2002).

Semi-chemical pulping: Semi-chemical pulping involves processing wood chips in a relatively mild chemical solution prior to mechanical refining for fibre separation. The chemical solution is usually neutral sodium sulphite/sodium carbonate liquor which softens and, to a limited extent, dissolves the lignin to promote fibre separation. Other semi-chemical pulping processes include using green liquor for pulping, the Permachem process, and the two-stage vapour process (US EPA 1993a). The yield of semi-chemical pulping ranges from 55 to 90 percent, depending on the process used, but pulp residual lignin content is also high since bleaching is more difficult (US EPA 2002).

Mechanical pulping: Mechanically produced pulp is of low strength and quality. Such pulps are used principally for newsprint and other nonpermanent paper goods. Mechanical pulping uses physical pressures instead of chemicals to separate furnish fibres. The processes are stone groundwood pulping, refiner mechanical pulping, thermo-mechanical pulping (TMP), chemi-mechanical pulping

(CMP), and chemi-thermo-mechanical pulping (CTMP). The stone groundwood process simply involves mechanical grinding of wood in several high-energy refining systems. The refiner mechanical process involves refining wood chips at atmospheric pressure. The thermo-mechanical process uses steam and pressure to soften the chips before mechanical refining. In the chemi-mechanical process, chemicals can be added throughout the process to aid the mechanical refining. The chemi-thermo-mechanical process involves the treatment of chips with chemicals for softening followed by mechanical pulping under heat and pressure. Mechanical pulping typically results in high pulp yields, up to 95 % compared to chemical pulping yields of 45–50 %, but energy usage is also high. To offset its structural weakness, mechanical pulp is often blended with chemical pulp.

Secondary fibre pulping: Secondary fibre pulping includes all operations performed to repulp secondary fibre and to remove contaminants. Common contaminants include ink, adhesives and coatings, polystyrene foam, dense plastic chips, plastic films (polyethylene), wet strength resins, latex and synthetic fibre (US EPA 1993a). Contaminant removal processes depend on the type, and source of secondary fibre to be pulped. Mill paper waste can be easily repulped with minimal contaminant removal. Recycled post-consumer newspaper, on the other hand, may require extensive contaminant removal, including deinking, prior to reuse (US EPA 2002).

6.1.2.3 Chemical recovery processes

Chemical recovery is a crucial component of the chemical pulping process, it recovers process chemicals from the spent cooking liquor for reuse. The chemical recovery process has important financial and environmental benefits for pulp and paper mills. The kraft, sulphite, and semi-chemical pulping processes all use chemical recovery systems. The kraft chemical recovery process consists of black liquor concentration, recovery boiler, recausticizating and calcining (US EPA 2002). There are a variety of sulphite chemical pulping recovery systems. Heat and sulphur can be recovered from all liquors generated, however the base chemical can only be recovered from magnesium and sodium base processes (US EPA 2002).

6.1.2.4 Pulp processing

Pulp processing is performed after pulping and prior to bleaching (if performed) or stock preparation. Pulp processing removes impurities such as uncooked chips, and recycles any residual cooking liquor via the washing process (US EPA 2002). The primary pulp processing operations include defibering, deknotting, brown stock washing, pulp screening and centrifugal cleaning (US EPA 1993a). If pulp is to be stored for long periods of time, drying steps are necessary to prevent fungal or bacterial growth (US EPA 2002). Defibering is performed immediately after chemical and semi-chemical pulping to completely separate the fibre. Deknotting removes knots and uncooked chips (rejects) prior to brown stock washing (US EPA 1993a). Residual spent cooking liquor from chemical pulping is washed from the pulp using brown stock washers. Efficient washing is critical to maximise return of cooking liquor to chemical recovery and to minimise carryover of cooking liquor (known as brown stock washing loss) into the bleach plant, because excess cooking liquor increases consumption of bleaching chemicals. Specifically, the dissolved organic compounds (lignins and hemicelluloses) contained in the liquor will bind to bleaching chemicals and thus increase bleach chemical consumption. In addition, these organic compounds function as precursors to chlorinated organic compounds (*e.g.*, dioxins, furans), increasing the probability of their formations. The most common washing technology is rotary vacuum washing, carried out sequentially in two or four washing units. Other washing technologies include diffusion washers, rotary pressure washers, horizontal belt filters, wash presses, and dilution/extraction washers (US EPA 2002). Pulp screening removes remaining oversized particles such as bark fragments, oversized chips, and uncooked chips. Centrifugal cleaning (also known as liquid cyclone, hydrocyclone, or centricleaning) is used after screening to remove relatively dense contaminants such as sand and dirt (US EPA 2002).

6.1.2.5 Bleaching

Bleaching is defined as any process that chemically alters pulp to increase its brightness. Bleached pulps create papers that are whiter, brighter, softer, and more absorbent than unbleached pulps (US EPA 2002). The type(s) of fibre furnish and pulping processes used, as well as the desired qualities and end use of the final product, greatly affect the type and degree of pulp bleaching required (US EPA 1993a). Typical bleaching processes for each pulp type are discussed below.

Chemical pulp is bleached in traditional bleach plants where the pulp is processed through three to five stages of chemical bleaching and water washing. The number of cycles is dependent on the whiteness desired, the brightness of initial stock pulp, and plant design (US EPA 2002).

Typically bleaching stages alternate acid and alkaline conditions. Chemical reactions with lignin during the acid stage of the bleaching process increase the whiteness of the pulp. The alkaline extraction stages dissolve the lignin/acid reaction products. At the washing stage, both solutions and reaction products are removed. Chemicals used to perform the bleaching process must have high lignin reactivity and selectivity to be efficient (US EPA 2002). The most common bleaching chemicals used are elemental chlorine (Cl_2), chlorine dioxide (ClO_2) and hypochlorites (HOCl, NaOCl, and $Ca(OCl)_2$), hydrogen peroxide (H_2O_2), oxygen (O_2), ozone (O_3), sodium hydroxide (NaOH), sulphur dioxide (SO_2), and sulphuric acid (H_2SO_4). Because of environmental and health concerns about dioxins, pulp mills now use elemental chlorine free (ECF) and total chlorine free (TCF) bleaching technologies. The difference between ECF and TCF is that ECF may include chlorine dioxide (ClO_2) and hypochlorites (HOCl, NaOCl, and $Ca(OCl)_2$) based technologies (US EPA 2002).

Semi-chemical pulps are typically bleached with H_2O_2 in a bleach tower (US EPA 1993a).

Mechanical pulps are bleached with H_2O_2 and/or sodium hydrosulphite ($Na_2S_2O_4$). Bleaching chemicals are applied either without using separate equipment during the pulp processing stage (*i.e.*, in-line bleaching), or in bleaching towers. Full bleaching of mechanical pulps is generally not practical due to bleaching chemical cost and the negative impact on pulp yield (US EPA 2002).

Deinked secondary fibres are usually bleached in a bleach tower, but may be bleached during the repulping process. Bleach chemicals may be added directly into the pulper. The following are examples of chemicals used to bleach deinked secondary fibres: hypochlorite (HOCl, NaOCl, $Ca(OCl)_2$), H_2O_2 and $Na_2S_2O_4$ (US EPA 2002).

6.1.2.6 Stock preparation

At this final stage, the pulp is processed into the stock used for paper manufacture. Market pulp, which is to be shipped off-site to paper or paperboard mills, is simply dried and baled during this step. Processing of pulp in integrated mills includes pulp blending specific to the desired paper product, dispersion in water, beating and refining to add density and strength, and addition of any necessary wet additives. Wet additives are used to create paper products with special properties or to facilitate the papermaking process. Wet additives include resins and waxes for water repellency, fillers such as clays, silicas, talc, inorganic/organic dyes for colouring, and certain inorganic chemicals (calcium sulphate, zinc sulphide, and titanium dioxide) for the improved texture, print quality, opacity, and brightness (US EPA 2002).

6.1.2.7 Paper and paperboard making processes

The paper and paperboard making process consists of wet and dry end operations. In wet end operations, the processed pulp is converted into a paper product via a paper production machine. In this production system, the pulp slurry is deposited on a moving wire belt that carries it through the first stages of the process. Water is removed by gravity, vacuum chambers, and vacuum rolls. This wastewater is recycled to the slurry deposition step of the process due to its high fibre content. The continuous sheet is then pressed between a series of rollers to remove more water and compress the fibres (US EPA 2002). In dry end operations, after pressing, the sheet enters a drying section, where the paper fibres begin to bond together as steam heated rollers compress the sheets. In the calender process the sheet is pressed between heavy rolls to reduce paper thickness and produce a smooth surface. Coatings can be applied to the paper at this point to improve gloss, colour, printing detail, and brilliance. Lighter coatings are applied on-machine, while heavy coatings are performed off-machine. The paper product is then spooled for storage (US EPA 2002).

6.1.3 Wastewater sources

All pulp and papermaking processes require the use of water, however, water use, wastewater sources and wastewater characteristics for any mill are dependent upon the combination of raw materials, processes used and products manufactured (US EPA 1993a). Pulp mills that use logs as raw material may consume water for three purposes to prepare wood for pulping: log conveyance, log washing and wet debarking. The major source of wastewater in wood preparation is, the debarking plant. Wastewater sources for pulping vary according to the specific type of pulping process performed. Wastewater sources from chemical pulping include digester relief and blow condensates, wastewater from screen room, cleaners, deckers and spills from the digester area where inadequate spill control and containment are practiced (US EPA 1993a). The mechanical pulping processes use water as a coolant, a carrier to sluice pulp from the body of the grinder and diluents for subsequent pulp screening and cleaning steps. In secondary fibre pulping, water is used to solubilise the furnish (waste paper, old newspapers, cardboards *etc.*). When deinking is not necessary, the contaminants are removed physically (via sedimentation, flotation and filtration) (US EPA 1993a). These processes are the main wastewater sources of secondary fibre processing.

The wastewater sources for chemical recovery vary according to the specific type of chemical recovery operations performed. During the recovery of kraft pulping chemicals, the weak black liquor is concentrated to typically more than 60 percent liquor solids in multi-stage evaporators. The condensates from the evaporators comprise the excess water resulting from liquor concentration (US EPA 1993a). In kraft pulping chemical recovery processes, wastewater is also produced from the washing step of the solids precipitates from the recovery cycle. Although, recovery of pulping chemicals is not practiced at sulphite mills as extensively as at kraft mills, sulphite pulp wash water (weak red liquor) is evaporated generating an evaporator condensate wastewater (US EPA 1993a).

Pulp bleaching is a multi-stage process that uses different chemicals and conditions in each stage, with washing performed between stages. Primary water use and wastewater generation are from inter-stage pulp washing. The high chloride content of bleaching wastewaters makes them incompatible with pulping chemical recovery processes. In paper making processes, water that drains from the wet end of the paper machine is known as white water and is largely captured and reused in stock preparation or on the machine, after removal of entrained solid by savealls. White water may also be recycled to the woodyard, mill utilities, brown stock washers, and bleach line washers. However, excess white water is also a wastewater source of these processes. Table 6.2 summarises the typical wastewater generation rates based on processes used in the pulp and paper industry for specific subcategories determined for the industry (US EPA 1993a).

6.1.4 Wastewater characterisation

The characteristics of the wastewater generated from various processes of the pulp and paper industry depend upon the types of process and wood materials, process technology applied, management practices, internal recirculation of the effluent for recovery, and the amount of water to be used in the particular process (Pokhrel and Viraraghavan 2004). The conventional pollutant parameters as present in pulp and paper industry wastewaters are characterised as biochemical oxygen demand (BOD), total suspended solids (TSS), and pH (US EPA 1993a).

The priority and nonconventional pollutants that have been determined for the pulp and paper industry are chlorinated dioxins and furans, chlorinated phenolic compounds, volatile organic compounds (VOCs), adsorbable organic halides (AOX), chemical oxygen demand (COD), colour and other compounds such as resin and fatty acids, metals, semi-volatile compounds, and pesticides/herbicides (US EPA 1993a). The major sources of pollutant released in pulp and paper industry occur at the pulping and bleaching stages.

Pulp, paper and paperboard mills that use chemical pulp and bleached wood with chlorine and chlorine derivatives generate significant discharges of toxic pollutants from the pulping and bleaching processes. Such toxic pollutants include chlorinated dioxins and furans, particularly 2,3,7,8-tetrachlorodibenzofuran (2,3,7,8-TCDF) and 2,3,7,8-tetrachlorodibenzo-p-dioxin (2,3,7,8-TCDD) and the other chlorinated dibenzo-p-dioxins (CDDs), and chlorinated dibenzofurans (CDFs) (US EPA 1993a). VOCs (such as terpenes, alcohols, phenols, methanol, acetone, chloroform, methyl ethyl ketone (MEK), chloromethane, trichloroethane) and chlorinated phenolic compounds are also formed from the same processes (US EPA 1993a; US EPA 2002). Chlorinated phenolic compounds found in pulp and paper wastewaters cover chlorinated phenols, chlorinated guaiacols, chlorinated

Table 6.2 Typical wastewater generation rates for pulp and paper industry (US EPA 1993a)

Process	Subcategory	Wastewater generation (m^3/OMMT* of final product)
Wood Preparation	Dissolving Kraft	7.2
	Bleached Papergrade Kraft and Soda	1.8
	Unbleached Kraft	0.5
	Dissolving Sulphite	19.2
	Papergrade Sulphite	3.1
	Semi-Chemical	1.3
	Mechanical Pulping	5.2
Chemical Pulping	Dissolving Kraft	28
	Bleached Papergrade Kraft and Soda	16.4
	Unbleached Kraft	7.5
	Dissolving Sulphite	29.8
	Papergrade Sulphite	34.3
	Non-Wood Chemical	31.9
Chemical Recovery	Dissolving Kraft	12.4
	Bleached Papergrade Kraft and Soda	8.8
	Unbleached Kraft	4.2
	Dissolving Sulphite	18.1
	Papergrade Sulphite	10.4
	Semi-Chemical	2.5
	Non-Wood Chemical	12.8
Bleaching	Dissolving Kraft	68.7
	Bleached Papergrade Kraft and Soda	27.3
	Dissolving Sulphite	113.4
	Papergrade Sulphite	65.8
	Non-Wood Chemical	13.7
Pulp Handling and Paper-making	Dissolving Kraft	11.9
	Bleached Papergrade Kraft and Soda	23.2
	Unbleached Kraft	17.5
	Dissolving Sulphite	8.5
	Papergrade Sulphite	53.7
	Semi-Chemical	12.9
	Mechanical Pulping	30.9
	Non-Wood Chemical	329.2
	Secondary Fibre Deink	30.0
	Secondary Fibre Non-Deink	21.7
	Fine and Lightweight Paper from Purchased Pulp	64.2
	Tissue, Filter, Nonwoven and Paperboard from Purchased Pulp	60.2

* OMMT = off-machine tons.

catechols and chlorinated vanillins. AOX is a measure of the halogenated organic compounds that adsorb onto granular activated carbon. For pulp and paper industry wastewaters, essentially all of the halogenated organic substances measured as AOX are chlorinated compounds which result from the bleaching of pulps with chlorine and chlorine derivatives (US EPA 1993a).

Table 6.3 represents the process steps, material inputs, process outputs and major wastewater pollutants of a pulp and paper mill using kraft chemical pulping and chlorine-based (*e.g.*, hypochlorite or chlorine dioxide) bleaching which are both commonly used processes in the pulp and paper industry. The general characteristics of the final effluent produced from dissolving kraft, bleached papergrade kraft and soda, dissolving sulphite subcategories in terms of pollution loads per air dried metric ton (ADMT) of pulp and concentration (mg/L) are given in Table 6.4 (US EPA 1993a). As aforementioned the characteristics of wastewaters generated from various processes as well as final effluent of the pulp and paper industry highly depend upon the type of process, type of wood materials, process technology applied *etc.*. However, in general the final effluents of pulp and paper industry wastewaters are high in BOD, at 10–40 kg/ADMT; TSS, 10–50 kg/ADMT; COD, 20–200 kg/ADMT; and chlorinated organic compounds, which may include dioxins, furans, and other adsorbable organic halides, at 0–4 kg/ADMT (World Bank Group 1998). Wastewater from chemical pulping contains 12–20 kg of BOD/ADMT. The corresponding values for mechanical pulping wastewater are 15–25 kg BOD/ADMT. For chemi-mechanical pulping, BOD discharges are 3 to 10 times higher than those for mechanical pulping. Pollution loads for some processes, such as those using non-wood raw materials, could be significantly different (World Bank Group 1998).

6.1.5 Pollution prevention and control

The chemical recovery systems used in chemical pulping processes are an example of pollution prevention technologies that have evolved alongside process technologies. An efficient chemical recovery system is a crucial component of chemical pulping mill operation because of the chemical recovery process regenerating process chemicals, in reducing natural resource usage and associated costs, as well as discharges to the environment and producing energy (US EPA 2002). The most significant environmental issues are the discharge of chlorine-based organic compounds (from bleaching) and of other toxic organics. Efforts to prevent chlorinated releases have, therefore, focused on source reduction and material substitution techniques such as defoamers, bleaching chemical, or wood chip substitution. Such source reduction efforts and material substitutions usually require substantial changes in the production process (US EPA 2002; Shammas

Table 6.3 Major wastewater pollutants of a kraft chemical pulped bleached paper production (US EPA 2002)

Process step	Material inputs	Process outputs	Major wastewater pollutants[1]
Wood Preparation	Wood logs, Chips, Sawdust	Furnish chips	➢ TSS ➢ BOD
Chemical Pulping Kraft process	Furnish chips	Black liquor Pulp	➢ Colour ➢ BOD ➢ COD ➢ AOX ➢ VOCs (terpenes, alcohols, phenols, methanol, acetone, chloroform, MEK)
Bleaching[2]	Chemical pulp Hypochlorite Chlorine dioxide	Bleached pulp	➢ Dissolved lignin ➢ Carbohydrates ➢ Colour ➢ COD ➢ AOX ➢ VOCs (acetone, methylene chloride, chloroform, MEK, chloromethane, trichloroethane) ➢ Chlorinated dioxins and furans ➢ Chlorinated phenolic compounds ➢ Inorganic chlorine compounds
Paper making	Additives Bleached Pulp Unbleached Pulp	Paper/ paperboard products	➢ Particulate wastes ➢ Organic compounds ➢ Inorganic dyes ➢ COD ➢ Acetone
Chemical recovery system			
Evaporators	Black liquor	Strong black liquor	Evaporator condensates ➢ BOD ➢ TSS
Recovery furnace	Strong black liquor	Smelt, energy	–
Recausticising	Smelt	Regenerated white liquor Lime mud Slaker grits	➢ Waste mud solids
Calcining (lime kiln)	Lime mud	Lime	–

[1]Pollutant outputs may differ significantly based on mill processes and material inputs (e.g., wood chip resin content).

[2]Pollutant list based on elemental chlorine-free (ECF) bleaching technologies.

Table 6.4 Wastewater characteristics of pulp and paper industry (US EPA 1993a)

Parameters	Subcategory Dissolving Kraft	 Bleached Papergrade Kraft and Soda	 Dissolving Sulphite
2,3,7,8-TCDD (pg/L)	10–21	8–13	10–11
2,3,7,8-TCDF (pg/L)	10–320	9.3–14	10–55
AOX (mg/L)	4.26–13.60	3.33–18	2.16–9.27
COD (mg/L)	380–690	160–740	–
Colour (mg/L)[1]	965–1865	300–1210	250–850
		Subcategory	
Parameters	Dissolving Kraft	Bleached Papergrade Kraft and Soda	Dissolving Sulphite
2,3,7,8-TCDD (kg/ADMT)	1.30×10^{-9}–3.20×10^{-9}	6.07×10^{-10}–1.45×10^{-9}	8.99×10^{-10}–2.76×10^{-9}
2,3,7,8-TCDF (kg/ADMT)	1.48×10^{-9}–4.15×10^{-8}	6.07×10^{-10}–9.74×10^{-10}	2.01×10^{-9}–1.22×10^{-8}
AOX (kg/ADMT)	0.582–2.21	0.211–1.48	0.25–1.79
COD (kg/ADMT)	55.3–93.4	11.1–77.3	–
Colour (kg/ADMT)	162–268	19.1–129	41.3–192

ADMT: air dried metric ton; [1]Colour units reported as mg/L relative to the platinum standard.

2009). In addition to the major process changes aimed at reducing toxics releases, the industry is implementation of a number of pollution prevention techniques to reduce water use and pollutant releases (BOD, COD, and TSS) such as: dry debarking, recycling of log flume water, improved spill control, bleach filtrate recycle, closed screen rooms, and improved storm water management (US EPA 2002). Pollution prevention techniques and process recommendations found to be effective at pulp and paper industry are provided below (US EPA 1993a, b; 2002; World Bank Group 1998; Sumathi and Hung 2006)

New developed pulping technologies such as organic solvent pulping, acid pulping, and biopulping can be used instead of conventional pulping processes. For pollution prevention and control cleaner bleaching methods based on elemental chlorine free (ECF), total chlorine free (TCF), microbial systems (bio-bleaching), extended delignification can considerably reduce the chlorinated organics and other toxic organics. Extended delignification may be achieved through prolonged cooking, oxygenation, ozonation and addition of chemical catalyst. Installation of on-line monitoring systems at appropriate locations and controlled dosing of bleach chemicals can aid the reduction of chlorinated organics in effluents. Detailed explanations on the pollution prevention options and descriptions of additional alternative pulping and bleaching processes can be found at the Pollution Prevention and Toxics' 1993 report, Pollution Prevention Technologies for the Bleached Kraft Segment of the U.S. Pulp and Paper Industry (US EPA 1993b).

Pollution from the pulp and paper industry can be minimised by various internal process changes and management measures such as best available technology (BAT). The technology selected by the US EPA to define BAT (BAT-US) performance for the pulp and paper industry, includes process changes and internal controls. US EPA developed a number of process technology combinations capable of reducing the discharge of priority and nonconventional pollutants (chlorinated phenolics, AOX and COD) of concern in the pulp and paper industry, and configured these combinations into BAT-US options. The BAT-US options considered for each subcategory have been numbered to indicate the degree of pollutant control achieved by each option, with Option 1 being the lowest degree of control and each increasing number providing increased control (US EPA 1993a). For bleached papergrade kraft and soda subcategory, five basic options have been determined as BAT-US for the control of chlorinated phenolics and AOX. These options are (US EPA 1993a);

- Option 1-Split Addition of Chlorine
- Option 2-Substitution of Chlorine Dioxide for Chlorine
- Option 3-Oxygen Delignification or Extended Cooking with Substitution of Chlorine Dioxide for Chlorine
- Option 4-Oxygen Delignification or Extended Cooking with Complete Substitution of Chlorine Dioxide for Chlorine
- Option 5-Oxygen Delignification and Extended Cooking with Complete Substitution of Chlorine Dioxide for Chlorine

Totally chlorine-free bleaching of papergrade kraft pulp is also considered as BAT-US by US EPA. US EPA developed two BAT-US (Option A and B) options for the control of COD discharges from the bleached papergrade kraft and soda subcategory (US EPA 1993a). Option A includes;

- effective brown stock washing,
- pulping liquor spill prevention and control,
- end-of-pipe wastewater treatment at a level of performance to best practicable technology (BPT-US) Option 2.

Option B included all of the elements of Option A with the addition of;

- closing the brown stock pulp screen room.

US EPA did not developed separate BAT-US options for the control of colour, but analysed the performance of options developed for the priority and nonconventional pollutants. The BAT-US options have been developed by US EPA for the other subcategories including dissolving kraft, unbleached kraft, dissolving sulphite, papergrade sulphite and semi-chemical are given in Table 6.5.

The issue of the integrated pollution prevention and control (IPPC) for the pulp and paper industry has been also defined in "*Reference Document on Best*

Table 6.5 BAT-US options for dissolving kraft, unbleached kraft, dissolving sulphite, papergrade sulphite and semi-chemical subcategories (US EPA 1993a)

Subcategory	BAT-US Option
Dissolving Kraft	1-Substitution of chlorine dioxide for chlorine 2-Oxygen delignification with substitution of chlorine dioxide for chlorine 3-Oxygen delignification with complete substitution of chlorine dioxide for chlorine *For Control of COD* Similar to Option B defined for bleached papergrade kraft and soda subcategory (Subpart B)
Unbleached Kraft	*For Control of COD* Similar to Option B defined for bleached papergrade kraft and soda subcategory (Subpart B)
Dissolving Sulphite	1-Oxygen delignification with complete substitution of chlorine dioxide for chlorine (Retaining hypochlorite) 2-Totally chlorine-free bleaching: Oxygen delignification, ozone and peroxide bleaching *For Control of COD* Similar to Option B defined for bleached papergrade kraft and soda subcategory (Subpart B)
Papergrade Sulphite	1-Oxygen delignification with complete substitution of chlorine dioxide for chlorine (Eliminating hypochlorite) 2-Totally chlorine-free bleaching: Oxygen delignification and peroxide bleaching *For Control of COD* Similar to Option B defined for bleached papergrade kraft and soda subcategory (Subpart B)
Semi-Chemical	*For Control of COD* Similar to Option B defined for bleached papergrade kraft and soda subcategory (Subpart B)

Available Techniques in the Pulp and Paper Industry" by European Commission (EC 2001). Similar to those of US EPA, the effective management systems, efficient raw material, energy and water usage, the substitution by less harmful substance, as well as minimisation, recovery and recycling of wastewater and wastes have been described in this document (EC 2001).

6.1.6 Treatment technologies

Two major approaches can be used to control the discharge of pollutants originating from pulp and paper mills. These approaches are: (i) in-process technology changes to prevent or reduce pollutant formation; (ii) end-of-pipe wastewater treatment technologies to remove pollutants from process wastewaters prior to discharge (US EPA 1993a). Plant process modifications and cleaner technologies have the potential to reduce the pollution load in pulp and paper mill effluents. However, this approach cannot eliminate waste generation. End-of-pipe treatment technologies are essential for meeting the prescribed limits for discharged pollutant concentrations such as colour, AOX, BOD, COD, and priority pollutants (Sumathi and Hung 2006). Conventional end-of-pipe wastewater treatment systems used in the pulp and paper industry typically comprise of primary, secondary, and tertiary (optional) stages of effluent treatment which are delineated in Figure 6.2.

Process technologies that are currently applied can be broadly classified as the physico-chemical and biological treatment methods. The application of these treatment methods is nearly universal across most segments of the industry (US EPA 1993a). The function of primary treatment is to remove suspended solids from the wastewater. Biological/secondary treatment is used to reduce the organic content in the wastewater and to destroy toxic organics. Because pulp mill wastewater is deficient in nitrogen and phosphorus relative to its high

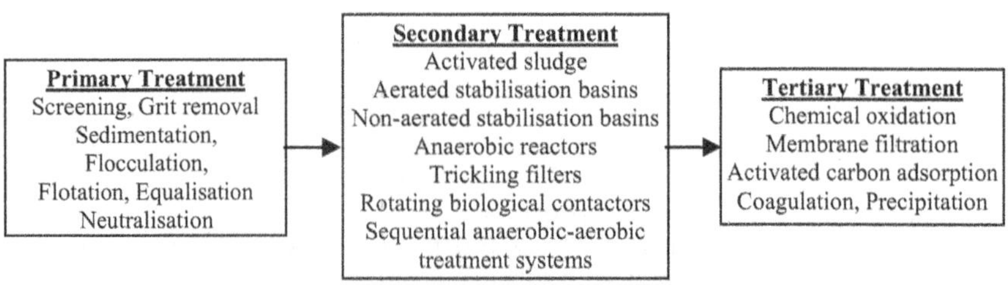

Figure 6.2 Conventional end-of-pipe wastewater treatment methods at pulp and paper industry (US EPA 1993a; Sumathi and Hung 2006).

carbon load, these nutrients are usually added to the process to enhance microbial activity (Shammas 2009). Several physico-chemical methods such as chemical coagulation, membrane separation and advanced oxidation processes are also used for the treatment of pulping and pulp bleaching effluents in order to remove toxic pollutants.

6.2 CHEMICAL OXIDATION PROCESSES

In pulp and paper industry wastewaters, refractory contaminants such as wood resin, dioxins, furans, catechols, guaiacol, lignin, and other chlorinated phenolic compounds, and colour are so recalcitrant that they cannot be completely degraded by conventional biological treatment systems (Kurniawan *et al.* 2006; Matta *et al.* 2007). The wastewater components causing colour (lignin and its derivatives) are difficult to degrade biologically because of strong linkages in their molecular structure (especially the biphenyl type of carbon-carbon bond) (Soloman *et al.* 2009). Besides an important group of compounds present in pulp and paper industry wastewaters is responsible for the toxicity of the wastewater and may adversely affect the microbial systems in the biological treatment systems (Delos Santos Ramos *et al.* 2009). Therefore, these non-biodegradable toxic contaminants have to be eliminated before biological treatment. As described before, in certain cases, efficient treatment technologies such as chemical oxidation for the treatment of pulp and paper industry wastewaters are necessary. In recent years, various chemical oxidation technologies have been applied as a pretreatment or post-treatment option to facilitate the removal of chlorine based organic compounds, colour and toxicity from pulp and paper industry wastewaters. Implementation of chemical oxidation technologies for the treatment of pulp and paper industry wastewaters can be summarised as follows:

- removal of toxic and bio-recalcitrant organic compounds before biological treatment in order to reduce the toxic effects and increase the biodegradability;
- removal of residual organic compounds after biological treatment;
- removal of highly resistant colour.

A brief summary of the application of chemical oxidation technologies as pretreatment or post-treatment options for pulp and paper industry wastewaters are given in the following subsections.

6.2.1 Pretreatment applications

Eskelinen *et al.* (2010) studied the treatability of bleaching effluents from pulp and paper mills using individual physico-chemical treatment methods like chemical precipitation and electrochemical processes. In their study, AOPs such as ultrasonic irradiation in combination with Fenton-like oxidation (Fe^{3+}/H_2O_2) or photo-Fenton degradation ($Fe^{2+}/H_2O_2/UV$) were also separately tested. Effects of operating parameters such as pH and oxidant's dose on the removal of organic compounds were investigated in terms of COD removal. Among the treatment methods tested, at the same initial COD concentrations ranging from 1150 to 1500 mg/L, the highest COD removal efficiency (90 %) was obtained by the application of chemical precipitation using 5 g/L of CaO at pH 12. An integrated ultrasonic irradiation with Fenton-like oxidation (12 %), photo-Fenton's oxidation (20 %) and electrooxidation treatment (28 %) yielded significantly lower COD removal efficiencies than that of chemical precipitation at the operating conditions of Fe^{3+} = 1 g/L; H_2O_2 = 3 g/L, pH = 6.9, agitation speed = 200 rpm, and contact time = 60 min. The UV irradiation during photo-Fenton's oxidation slightly improved the removal of COD from 12 % to 20 %. It has been concluded that the subsequent biological treatment application such as conventional activated sludge system needs to be undertaken to ensure the degradation of target compounds in the bleaching effluents prior to their discharges.

Effluent from a kraft mill was exposed to power ultrasound at 357 kHz at powers of 2.4 W/cm^2 or 4.0 W/cm^2 with the aim of reducing colour, turbidity and COD (Shaw and Lee 2009). The absorbance measurements showed a decolourisation of the effluent at wavelengths above 250 nm, indicating loss of aromatic chromophores. Effluent turbidity also decreased. However, there was no observable decrease in COD which was attributed to the presence of bicarbonate and sulphate ions in the final effluent, which reacted with $HO^{\bullet}$ and effectively blocked the oxidation of organics in the effluent.

Wet oxidation (WO) of debarking wastewater, which was highly contaminated by fatty and resin acids, tannins, lignin and their derivates, from the pulp and paper industry was studied by Kindsigo and Kallas (2009). The WO experiments were performed at various temperatures, partial oxygen pressures, and pH values. The best results were achieved at 190°C, 1 MPa of oxygen partial pressure, pH 5 and at an initial dilution rate 1:50 (4.2 g COD/L). Under these conditions, 98 % of tannins/lignin, 61 % of COD and 93 % of colour were removed and a biodegradability ratio (BOD/COD) of 73 % was achieved, after 2 hours of WO treatment. WO was recommended as a pretreatment to eliminate toxic components and to reduce the high level of organics produced by the pulp and paper industry. It was concluded that the combination of WO with a biological treatment could be an effective option for the treatment of debarking wastewaters. Laari *et al.* (1999) evaluated the efficiency of wet oxidation for the treatment of thermo mechanical pulp processing waters. The major objective of this research was to reduce the concentration of lipophilic wood extractives (LWE) and to treat concentrated residues from evaporation and membrane filtration by low-pressure catalytic wet oxidation. The wet oxidation of membrane and evaporation concentrates was effective in reducing 50 % of COD at 150°C and enhancing the biodegradability of wastewater. The removal of organics from paper and pulp industry wastewater was investigated employing WO and catalytic wet oxidation (CWO) techniques using a high-pressure reaction system by Akolekar *et al.* (2002). The pulp and paper wastewater treatments were conducted in selected different environments such as (i) wastewaters with different organic concentrations, (ii) wastewaters with different pH (11–14), (iii) a range of reaction temperatures (413–463 K), (iv) different catalysts (Cu, Mn, Pd, Cu/Mn, Mn/Pd, Cu/Pd), and (v) catalyst loading (1–8 g). The single transition (Cu, Mn) or noble metal (Pd) catalysts showed appreciable total organic removal, but the bi-metal catalysts (Cu/Mn, Cu/Pd, Mn/Pd) exhibited even higher activity for organics removal. Much higher TOC removal (>84 %) was achieved by using the CWO process. The TOC removal was the highest by the Cu/Pd catalyst, followed by the Mn/Pd and Cu/Mn catalytic system.

The study of Uğurlu and Karaoğlu (2009) aimed to remove AOX, total nitrogen, and lignin degrading products from bleached kraft mill effluents (BKME) which had initial AOX, COD, lignin, and total nitrogen concentrations of 2.82 mg/L, 426 mg/L, 50 mg/L and 23 mg/L, respectively. The photocatalytic degradation of these pollutants in BKME was investigated considering the different process parameters, such as reaction time, H_2O_2 and TiO_2 concentrations. When the effect of H_2O_2 and reaction time were explored, it was observed that the AOX concentration increased from 3.0 to 11.0 mg/L by only UV irradiation (mercury lamp, 30W). However, when H_2O_2 was added, AOX concentration decreased from 2.82 to 0.0 mg/L. The optimum reaction conditions for the removal of AOX found to be an initial H_2O_2 concentration of 9 mg/L and reaction time of 50 min. The results of this study clearly demonstrated that the UV/TiO_2/H_2O_2 (pH = 8.5; H_2O_2 = 4.5 mg/L; TiO_2 = 25 mg/L) photocatalytic process could greatly enhance the removal of AOX, COD, lignin and total nitrogen from the BKME.

El-Bestawy *et al.* (2008) studied the integrated chemical-biological treatment of wastewaters generated from two different paper mills. The wastewater (WW1) having initial COD and tannin/lignin concentrations of 761–855 mg/L and 31.4–21.3 mg/L, respectively obtained from a chemical pulping mill using rice straw, sugarcane bagasse, and wastepaper as raw materials for the production of writing paper, bag paper, linerboard, and duplex board. The other paper mill was mechanical pulping mill using rice straw and scrap papers as raw materials to produce printing and writing paper, machine glazed paper, linerboard, corrugating medium, and low-grade board. The initial COD and tannin/lignin concentrations of wastewater (WW2) obtained from this mill were 504–613 mg/L and 11.2–13.3 mg/L, respectively. Based on their experimental results, optimised operational conditions for the batch chemical and the continuous biological treatment sequences were proposed for both wastewater samples. For WW1 sample, the sequence was as follows: (i) coagulation with 375 mg/L $FeCl_3$, (ii) oxidation with 50 mg/L H_2O_2 and (iii) biological treatment using activated sludge with 2000 mg/L initial biomass concentration and 90 min hydraulic retention time (HRT), while for WW2 sample, the sequence was (i) coagulation with 250 mg/L $FeCl_3$, (ii) oxidation with 45 mg/L H_2O_2 and (iii) biological treatment using activated sludge with 2000 mg/l initial biomass concentration and 60 min HRT. The authors concluded that the application of the proposed sequential treatment system removed almost all COD, BOD_5, high molecular weight compounds, and silica from WW1 and WW2 samples and produced high-quality effluents.

Rodrigues *et al.* (2008) investigated the combined treatment (coagulation-flocculation followed by heterogeneous photocatalysis) of post-bleaching effluent from a cellulose and paper industry having a COD value of 1303 mg/L. In their study, the optimised coagulation experimental conditions were found as: pH 6.0, 80 mg/L of $FeCl_3$, and 50 mg/L of chitosan and the optimum conditions for photocatalysis were given as: pH 3.0 in 0.50 g/L of TiO_2 and 10 mM of H_2O_2. The final effluent COD concentration after the combined treatment was reported as 246 mg/L. The combined treatment increased the biodegradability index (BOD_5/COD) of raw effluent from 0.11 to 0.71. The authors concluded that the biodegradability index for effluent fed into to coagulation-flocculation followed by photocatalysis was suitable for complete biological degradation.

In the study of Mounteer *et al.* (2007) a poorly biodegradable (BOD_5/COD = 0.3) industrial alkaline ECF bleaching filtrate was treated using different advanced oxidation processes (O_3/H_2O_2, H_2O_2/UV and TiO_2/UV) to evaluate their use in combined chemical-biological treatment aimed at increasing recalcitrant COD removal and improving final effluent quality. The initial COD, BOD_5, lignin, carbohydrate, and colour concentrations of the wastewater sample were 1867 mg/L, 556 mg/L, 14 mg phenol/L, 111 mg glucose/L, and 605 mg Pt/L, respectively. The O_3/H_2O_2 process increased effluent biodegradability up to 68 % as a result of increasing BOD and decreasing COD. Increasing the O_3 dose had a greater effect on improvement of biodegradability, and removal of lignin and colour efficiencies than increasing the H_2O_2 dose. A combined oxidant dose of 5 mM O_3 and 2 mM H_2O_2 resulted in 75 % lignin removal, 40 % colour removal and 6 % carbohydrate loss without mineralising the organic carbon. The photocatalytic processes (H_2O_2/UV and TiO_2/UV) led to a decrease in effluent biodegradability through combined decrease in BOD and increase in COD and did not result in efficient lignin or colour removal. The authors concluded that the photocatalytic oxidation was apparently inhibited by the high chloride and COD levels in the alkaline filtrate, and might be more efficient in recalcitrant COD removal if performed after biological treatment.

Amat *et al.* (2005) studied the ozone and/or UV systems for the treatment of effluents from a paperboard industry. Three different batches of wastewaters, with COD values of around 4500 mg/L, 7000mg/L and 11000 mg/L were used for the investigation of the effect of ozone and O_3/UV combination at pH 6 and 9. They reported the following trends for the ozone and O_3/UV systems: (i) better results were achieved at basic pH than acidic pH values for both processes, (i) the synergetic effect of O_3/UV was not very significant, most probably due to the high turbidity of the wastewaters and (iii) higher degradation yields were obtained in more concentrated wastewater samples.

The treatment of a pulp and paper wastewater having initial COD of 964 mg/L and colour of 4236 mg Pt/L by Fenton-like AOP was investigated by Tambosi *et al.* (2006). The main parameters that govern the Fenton-like process (pH and Fe(III) and H_2O_2 concentrations) were studied and maximum COD, colour and aromatic compound removals were obtained as 75, 98 and 95 %, respectively, under optimum operating conditions of Fe(III) = 400 mg/L; H_2O_2 = 500–1000 mg/L; pH = 2.5; followed by coagulation at pH 5.0. The biodegradability of the wastewater increased from 0.4 to 0.7 under optimum conditions and no residual H_2O_2 was found after treatment. However, partially or non-oxidised compounds existing in the treated wastewater presented higher acute toxicity to *Artemia salina* than the untreated wastewater.

In the study of Xu *et al.* (2007), a bleaching wastewater effluent taken from a pulp and paper mill was treated with solar photo-Fenton process. The mill used wastepaper as raw material and its effluent contained 332 mg/L of TOC and 1286 mg/L of COD. The highest percentage of TOC removal in the first 15 min was about 60 % at initial pH values of 2.8 (H_2O_2 = 1883 mg/L, 10:1 of H_2O_2/Fe(II), temperature = 30°C) and 3.5 (H_2O_2 = 3766 mg/L, 10:1 of H_2O_2/Fe(II), temperature = 30°C). Under the extended reaction times, the TOC removal efficiency reached 82 % at 120 min and was 94 % at 180 min. The result of the study indicated that the solar photo-Fenton process had a potential to effectively remove TOC from the wastepaper pulp effluent on actual scale. Rodriguez *et al.* (1999) evaluated Fenton-type reactions facilitated by catecholic compounds such as 2,3-dihydroxybenzoic acid, 3,4-dihydroxybenzoic acid, and 1,2-dihydroxybenzene in the treatment of pulp bleaching effluent. Their research indicated that 2,3-dihydrobenzoic acid was the most effective compound in enhancing hydroxyl radical formation in the iron-hydrogen peroxide reaction system at pH 4.0 with the concomitant reduction in the AOX concentration and toxicity of the bleach effluents.

In another study, TiO_2-photocatalytic degradation of a cellulose elemental chlorine free bleaching effluent was evaluated using multivariate experimental design (Pérez *et al.* 2001). The effluent was characterised by general parameters such as AOX (58 mg organic chlorine/L), TOC (487 mg/L), COD (1400 mg/L), colour (2100 mg Pt/L), lignins and tannins (40 mg phenol/L), acute toxicity (EC_{50} = 36.5) and by the analysis of chlorinated low molecular weight compounds using gas chromatograph/mass spectroscopy (GC/MS). Under optimised conditions (TiO_2 = 1.04 g/L; pH = 7.2 and H_2O_2 = 0.1 mM) after 30 min of reaction more than 60 % of the toxicity was removed and after 420 min of reaction none of the initial chlorinated low molecular weight compounds were detected, suggesting an extensive mineralisation which was corroborated by 95 and 50 % AOX and TOC removals, respectively. In another study conducted by Pérez *et al.* (2002) the degradation of the organic content of a bleaching kraft mill effluent (BKME) was carried out by using Fenton reagent and irradiation providing the conditions needed for the simultaneous occurrence of Fenton and photo-Fenton reactions. Concentrations of Fe(II) between 0 and 800 mg/L, and H_2O_2 between 0 and 10,000 mg/L were chosen for the investigation. For the simultaneous concurrence of all reactions, pH 3 was found to be the optimal value. The authors reported that the initial concentrations of Fe(II) in solution above 400 mg/L were found to be detrimental, while up to the values tested, the reactivity was larger the higher the concentration of H_2O_2. The authors also reported that the solar light irradiation turned out to be particularly efficient. Temperature was found to be a key parameter that markedly increased reaction rates, although a residual TOC persisted and could not be eliminated even at high temperatures (25 to 70°C).

Bedoui *et al.* (2009) investigated the H_2O_2/UV treatment of an actual pulp and paper mill wastewater. To determine optimal conditions, the effects of some experimental parameters (initial pH, initial H_2O_2 concentration, temperature, and organic load) on the H_2O_2/UV process efficiency in terms of absorbance at 330 and 281 nm, COD, and TOC removals were followed. The results showed that the H_2O_2/UV process led successfully to the almost-complete elimination of colour, COD, and TOC from the wastewater initially containing 270 mg/L TOC and 680 mg/L COD under original pH (pH 11.25) using 2.1 g/L H_2O_2 at 28 °C. Taking into account of the results obtained in the study it has been suggested that several steps were involved in the photochemical oxidation of organics, starting with degradation of lignin derivatives and tannins to obtain aromatic intermediates which further underwent oxidative ring opening, leading to aliphatic carboxylic acids. In the final steps, carboxylic acids were mineralise into CO_2 and H_2O.

In the study of Medeiros *et al.* (2008), the effect of ozone oxidation in removing high molecular weight (HMW) organics in order to improve the biodegradability of an alkaline bleach plant effluent (COD = 4782 mg/L; TOC = 1970 mg/L; BOD_5 = 338 mg/L, and colour = 4031 C.U.) was investigated using a semi-batch reactor under initial pH values of 12 and 7. Ozonation using 0.6 mgO_3/mL_{ww} was capable of improving biodegradability of the alkaline bleach plant effluent for both initial pH values 12 and 7. The colour removal was lower at pH 7 than that of pH 12 indicating selective reaction of ozone with organic compounds particularly those with higher molecular weight ranges. Ozonation resulted in a significant reduction of HMW fractions of the effluent. In particular, greater than 80 % of the organics with molecular weight higher than 5 KDa were removed during the ozonation process.

Kreetachat *et al.* (2007) studied the effects of ozonation process on lignin-derived compounds found in a raw pulp and paper mill wastewater (BOD_5 = 550 mg/L; COD = 2000 mg/ L; TOC = 650 mg/L). The characteristic compounds of the pulp and paper mill wastewater were identified as the lignin-derived compounds and aliphatic compounds used in the pulp and paper production process (i.e. *n*-alkanes, fatty alcohols, fatty acid and esters). The phenolic/aliphatic compound ratio as calculated by GC/MS chromatogram was 0.62/0.38. The BOD_5/COD ratios were also used for evaluation of biodegradability of ozonated effluents. The biodegradability increased with increasing ozonation time (the BOD_5/COD value increased from 0.10 to 0.32 with ozone flow rate of 4.0 L/min). After 45 min, the ozonation of effluents yielded almost colourless effluent with over 90 % decolourisation efficiency. Hostachy *et al.* (1997) reported detoxification and an increase in the biodegradability of bleaching effluents by ozonation at low dosages [0.5–1 kg/ADMT (air-dried metric ton)] of pulp. The researchers observed significant elimination of the residual COD by catalysed ozone treatment of hardwood and softwood pulp and paper mill wastewaters. Such a treatment method may provide reutilisation of treated process waters and reduces consumption of freshwater during pulping steps.

Yeber *et al.* (2007) investigated the colour removal from a kraft bleaching effluent by TiO_2/UV/O_2 and TiO_2/UV/Cu (II) processes. In their study TiO_2 and Cu (II) concentrations and the reaction times were optimised by using a wastewater sample having initial COD, BOD and colour concentrations of 1600 mg/L, 449 mg/L and 1500 mg Pt/L, respectively. Experimental results indicated that colour removal was 94 % at acidic pH (3.0) at 0.75 g/L TiO_2 and in the presence of Cu (II) (1 mg/L) as an electron acceptor. Under these conditions, the biodegradation of the effluent, represented as the ratio of BOD/ COD, increased from 0.3 to 0.6. 70 % of COD was removed, and the ecotoxicity, measured by *Daphnia magna*, was reduced. The authors concluded that the photocatalytic oxidation to remove the colour of the kraft mill bleaching effluent was effective under the following conditions: short reaction time, acidic pH values, and without the presence of oxygen due to the existence of Cu (II) in the effluent. Moreover, residual Cu (II) was a minimum (0.05 mg/ L) and was not toxic to the following biological treatment stage. Balcioglu *et al.* (1999) observed enhanced biodegradability (increase in BOD_5/COD ratio) of raw kraft pulp bleaching effluents and improved quality of the biologically pretreated effluents following TiO_2 photocatalytic oxidation. Yeber *et al.* (1999) described the photocatalytic (TiO_2 and ZnO) treatment of bleaching effluents from two pulp mills. Photocatalysis resulted in the enhanced biodegradability of effluents with concomitant reduction in the toxicity.

The treatability of a pulp mill effluent by different AOPs consisting of UV, H_2O_2/UV, TiO_2/UV and H_2O_2/TiO_2/UV in lab-scale reactors for TOC and toxicity removals were performed by Catalkaya and Kargi (2007). Initial COD, TOC, BOD_5 and AOX concentrations of the pulp mill effluent were 592 mg/L, 171 mg/L, 240 mg/L and 1.936 mg/L, respectively. Each treatment method resulted in some degree of TOC and toxicity removals from the pulp mill effluent. Either TiO_2/UV or H_2O_2/TiO_2/UV treatment could be used to obtain high TOC removals. However, TiO_2/UV treatment seemed to be more advantageous yielding higher toxicity removal as compared with H_2O_2/TiO_2/UV. Approximately, 79.6 % TOC and 94 % toxicity removals were obtained by the TiO_2/UV with a TiO_2 concentration of 0.75 g/L at pH 11 within 60 min. In another study of Catalkaya and Kargi (2008) H_2O_2, Fenton, UV photolysis, H_2O_2/UV, photo-Fenton, ozonation and O_3/H_2O_2 processes were applied for the treatment of a pulp mill effluent having initial COD of 400 mg/L, TOC of 110 mg/L, BOD_5 of 240 mg/L and AOX of 1.94 mg/L. H_2O_2 (50 mM) treatment alone resulted in very low TOC (5.1 %), AOX (34 %) and colour (24 %) removals at pH 11. UV treatment alone found also as an ineffective method yielding unacceptable low colour (6.6 %), TOC (4.1 %), and AOX (17 %) removals. 41 %, 11.1 % and 18.5 % colour, TOC and AOX removals were obtained by H_2O_2/UV treatment. Fenton oxidation yielded very high colour (85 %), TOC (88 %) and AOX (89 %) removals within 30 min at a pH of 5, which made this treatment more favourable among the others tested. Removals obtained with the photo-Fenton treatment were comparable with those obtained with the Fenton's reagent. Photo-Fenton treatment yielded 85 % TOC, 82 % colour and 93 % AOX removals within 5 min indicating the effectiveness of this treatment. Ozonation alone was not as effective as the treatment by the Fenton oxidation yielding low TOC (29 %) and AOX (62.4 %) removals. O_3/H_2O_2 improved the AOX removal, however yielded reduced colour and TOC removals as compared to the ozone treatment alone. On the basis of the experimental results the authors concluded that either Fenton or photo-Fenton treatment could be used for effective removal of colour, TOC or AOX from the paper mill wastewater. However, photo-Fenton treatment seemed to be more advantageous requiring much less reaction times and therefore smaller reactor volumes as compared to the Fenton treatment.

Boroski *et al.* (2008) investigated the electrocoagulation (EC) method followed by photocatalysis to treat effluents from cellulose and paper factories. The initial

COD value of the wastewater was 1310 mg/L and the biodegradability index (BOD_5/COD ratio) was 0.12. For EC, the experiment with aluminium (Al) and iron (Fe) electrodes yielded similar COD removal efficiencies. EC/Al treatment at fixed pH 5.0 and current density of 153 A/m^2 was carried out at several electrolysis times. Its efficiency was analysed by COD results: after 30 min an average reduction of 55 % was obtained whereas after 60 min the reduction was 56 %. For the EC/Fe experiments, the pH range of 4.0–7.0 resulted in COD values were 52 %. By applying 30 min of EC/Fe, 153 A/m^2 and pH 6.0, the COD values, UV-Vis absorbance and turbidity underwent an intense decrease. For the subsequent UV photocatalysis (mercury lamps) TiO_2 was employed and the favourable operational conditions were 0.25 g/L of the catalyst and solution pH 3.0. The addition of H_2O_2 (50 mM) highly increased the photo-process performance. By employing the H_2O_2/TiO_2/UV process, the COD reduction was 88 % compared to pretreated effluents and photo-bleaching was achieved. The biological tests using *A. salina* confirmed that the investigated treatments produced effluent containing lower levels of toxic substances.

Khansorthong and Hunsom (2009) studied colour and total COD reduction in a wastewater from a pulp and paper mill industry by using electrocoagulation techniques in batch mode to find the optimum conditions. In their study the effect of key parameters including the type of polyelectrolyte, current density, initial pH of the wastewater, and the circulating flow rate of wastewater in the reactor were investigated. A low-cost iron plate with a total surface area of 0.0161m^2 was used for the sacrificial electrodes, arranged in monopolar configurations. The wastewater was obtained from a pulp and paper mill plant using wood pulp as raw material and the final product was paper pulp for making writing paper and packaging paper. The initial COD, BOD_5 and colour were 1114 mg/L, 1095 mg/L and 1984 Pt-Co unit, respectively. The optimum conditions for the treatment of the pulp and paper industry wastewater in a batch electrochemical reactor was found to be at a current density of 20.7 A/m^2, an initial pH of wastewater of 7.58, a circulating flow rate of wastewater of 2.6 l/min and a 45 min operating time. Under these conditions, colour, COD and BOD_5 were reduced by 97.0, 87.8 and 91.2 %, respectively. Low energy consumption and operating costs were obtained at 1.2 kWh/m^3 and € 0.23/m^3, respectively at the optimum conditions.

El-Ashtoukhy *et al.* (2009) reported that the use of electrochemical technique for the treatment of paper mill effluents where rice straw was used as a raw material, reduced the COD from an average value of 5500 mg/L to 160 mg/L. The electrochemical oxidation experiments were carried out in a cylindrical agitated vessel lined with lead sheet as anode while a concentric cylindrical stainless steel sheet screen was placed as a cathode. The colour removal ranged from 53 % to 100 % depending on the operating conditions. Their calculation indicated that the energy consumption ranges from 4 to 29 kWh/m^3 of the effluent depending on the operating conditions. The authors concluded that the electrochemical oxidation was a powerful tool for treating paper mill effluents where rice straw was used as a raw material. Patel and Suresh (2008) investigated the removal of pentachlorophenol (PCP) as a model AOX compound in water as well as in a pulp bleaching effluent of a bamboo based mill by electrochemical treatment in batch mode. The authors reported that the PCP removal in neutralised raw pulp bleach effluent spiked with 10 mg/L PCP (pH = 6.7, chloride = 1830 mg/L, COD = 1410 mg/L) was very slow and incomplete even after 2 h of electrochemical treatment at a current density of 15 mA/cm^2. Electrochemical treatment of potash alum pretreated (dosage = 2000 mg/L, pH = 6.7) effluent spiked with 10 mg/L PCP could achieve >90 % removal of initial colour, COD and PCP in less than 1 h. The estimated cost of combined treatment (potash alum coagulation + electrochemical) was reported as € 0.57–0.73 per cubic meter of the raw pulp bleach effluent. Wang *et al.* (2007a) studied electrochemical treatment of a paper mill wastewater using three-dimensional electrodes with Ti/Co/SnO_2-Sb_2O_5 anode. The residual COD concentration was measured as 186 mg/L providing a COD removal of 86 %, and colour removal was obtained as 75 % by applying 167 mA/cm^2 current density at pH 11 and 15 g/L NaCl. In another study conducted by Wang *et al.* (2007b) the electrochemical oxidation of a pulp and paper making wastewater assisted by transition metal (Co, Cu) modified kaolin in a batch reactor with a graphite plate as electrodes was investigated. The optimum operating conditions were found as pH 3, a current density of 30 mA/cm^2, and a catalyst dose of 30 g/L. At these conditions removal was obtained as 96.8 % in 73 min resulting 53 mg/L effluent COD.

In the study of Soloman *et al.* (2009), electrochemical treatment of a wastewater (with an initial COD of 1669 ± 23 mg/L) obtained from a small-scale, agro-based pulp and paper industry was investigated using a RuO_2 coated titanium electrode in various types of reactor configurations, *i.e.* batch, batch recirculation, recycle, and single pass systems. The results obtained from batch studies showed that the operating parameters of current density = 2.5 A/dm^2, electrolysis duration = 5 h, supporting electrolyte concentration = 3 g/L, as optimal for the overall system performance. Batch recirculation mode of operation was superior for COD removal (73.3 *vs.* 64 %) when compared with the batch system with the same specific electrode surface, but with little reduction in energy figures (power consumption = 24.32 *vs.* 23.98 kWh/kg COD removed). The pollutant removal performance of the batch recirculation system improved considerably with an increase in the circulation flow rate. An enhancement in voltage (3.9–3.6 V) and in COD removal (56–73.3 %) resulted in considerable improvement in current efficiency (37.89–49.59 %) and power consumption (34.5–24.3 kWh/

kg COD removed) as the circulation flow rate increased from 20 to 100 L/h. At the best operating conditions in the recycle system, 59 % of COD was removed, corresponding to a current efficiency of 68.9 % and specific energy consumption of 18.46 kWh/kg COD removed. At these conditions, the biodegradability index (the ratio of BOD to COD) of the wastewater was improved from 0.18 ± 0.01 to 0.36 ± 0.01 which made the remaining pollutants more easily biodegradable.

6.2.2 Post-treatment applications

Pereira *et al.* (2009) evaluated the toxicity of a secondary bleached kraft pulp mill effluent, after tertiary treatment with photo-Fenton oxidation, using a battery of freshwater species. In their experimental study two samples which had COD concentrations of 392 and 232 mg/L were used. In the photo-Fenton experiments 2 mL of iron sulphate heptahydrate (0.5 M) and 70 mL of H_2O_2 (30 % v/v) were added to 500 mL of wastewater samples and UV lamp was turned on. The final pH was 4.0. The photo-Fenton/UV treatment has proved to be the most effective in reducing the colour and the COD of the effluent. Nevertheless, extremely low EC_{50} values were reported in this study for almost all species, after tertiary treatment with photo-Fenton process.

Gommers *et al.* (2007) studied the recalcitrant COD degradation by an integrated system of ozonation and membrane bioreactor. Screening tests with different types of oxidation technologies (H_2O_2/UV, photo-Fenton, O_3, O_3/H_2O_2 and O_3/UV) showed that ozonation of biologically treated effluent could reduce the initial COD (465 mg/L) 40 % with an ozone dose of 0.4–0.8 gO_3/g COD and BOD/COD ratio could be increased from 0.02 up to 0.19. Neither combination of O_3 with UV and/or H_2O_2 nor the process H_2O_2/UV or photo-Fenton provided any improvement in COD reduction or BOD increase, unless applied very high doses. Their experimental results showed that for the removal of recalcitrant COD down to 125 mg/L, the requirement was 1.2 g O_3/L which made the process expensive, with running costs for energy and oxygen consumption of € 3.0/m^3.

Tünay *et al.* (2008) evaluated the applicability of chemical oxidation processes (ozonation, high pH ozonation, O_3/H_2O_2 and Fenton oxidation) to polish biologically treated effluent of a plant manufacturing hardboard from waste paper to comply with the discharge limit of 120 mg/L COD. In their study a conventional treatment scheme, comprised of chemically assisted settling preceding an activated sludge treatment, was employed to produce secondary effluent from raw wastewater samples. A combination of alum (200 mg/L), lime and polyelectrolyte produced 24 % COD removal reducing the wastewater strength from 2500 to 1900 mg/L COD. The activated sludge process at optimum organic loading of 0.6 g COD/gVSS.d removed a further 80 % COD, yielding a secondary effluent having COD concentrations of 260–400 mg/L. Oxidation by using molecular ozone at pH 8.5 (ozone dose = 40 mg O_3/min) of biologically treated effluents yielded around 60 % COD removal in one hour which was not sufficient to reach the discharge standard. However, increasing the oxidation time up to two hours provided a COD removal of 80 %. O_3/H_2O_2 oxidation and high pH ozonation conducted at pH 13 provided COD removals comparable with those of ozonation while their ozone requirements were significantly greater than for ozonation. Fenton oxidation provided comparable COD removals with the use of reasonable doses of chemicals (0.5:1 mol/mol Fe^{2+}/H_2O_2 and 2:1 mol/mol H_2O_2/COD) as compared to other oxidation processes. The authors concluded that the application of chemical oxidation methods following biological treatment provided an efficient means to reduce the COD of waste paper processing effluents to comply with the direct discharge standards.

Sevimli (2005) investigated the effectiveness of chemical oxidation by applying ozonation, combination of O_3 and H_2O_2 and Fenton processes for decolourisation and residual COD removal of biologically pretreated pulp and paper industry effluents. The characteristics of the investigated biologically pretreated effluent were given as COD = 470 mg/L; and colour = 680 Pt-Co; ultraviolet absorbance at 254 nm (UV_{254}) = 2.48 1/m and pH = 7.1. The author stated that the combinations of O_3/H_2O_2 were only slightly more effective than ozonation for COD and colour removals. In the Fenton oxidation studies, the removal efficiencies of COD, colour and UV_{254} for biologically treated pulp and paper industry effluents were found to be about 83, 95, and 89 %, respectively. Experimental studies indicated that Fenton oxidation was a more effective process for the reduction of COD, colour, and UV_{254} as compared to ozonation and O_3/H_2O_2. In the same study the operating costs of investigated chemical oxidation technologies were also evaluated. The operating costs were found as € 0.27 /m^3 for ozonation (ozone dose = 175 mg/L), € 0.20 /m^3 for O_3/H_2O_2 process (ozone dose = 100 mg/L; H_2O_2 = 50 mg/L) and € 0.15 /m^3 for Fenton process (H_2O_2 = 50 mg/L; $FeSO_4.7H_2O$ = 375 mg/L; pH = 4.0–4.5).

Helbe *et al.* (1999) described a tertiary treatment process involving ozonation in combination with a fixed-bed biofilm reactor for the reuse of a biologically treated effluent of a pulp and paper industry. Sequential ozonation and bioreactor treatment yielded maximum elimination of COD, colour and AOX from biologically treated effluent with minimum dosage of ozone. Further, the authors suggested that two-stage ozonation with intermediate biodegradation was more effective in terms of achieving higher removal of persistent COD. In an attempt to achieve water quality suitable for the recycle of a thorn paper-mill effluent, photo-Fenton's advanced oxidation and chemical coagulation were investigated as treatment options to remove both non-biodegradable COD and colour from prebioremediated black liquor effluent (Helmy *et al.* 2003). At the primary treatment (bioremediation) stage, COD and TSS removals ranged from 35–98 % and 12–89 %, respectively. Photo-Fenton's UV advanced oxidation and lime coagulation combination treatment ($FeSO_4$ = 6 g/L, H_2O_2 = 4 g/L, $Ca(OH)_2$ = 5 g/L) resulted in complete removal of COD, TSS and colour.

The oxidation of lignin in a biologically treated pulp and paper mill wastewater was studied (initial COD = 48 mg/L, initial lignin = 313 mg/L) using various treatment methods such as H_2O_2/UV photolysis, homogeneous, heterogeneous and UV-assisted heterogeneous Fenton reactions, catalysed by FeZSM-5 zeolite (Makhotkina *et al.* 2008). The most effective lignin oxidation (85 %), and mineralisation of (80 %) were observed in the wastewater treatment with H_2O_2 at its 0.1 M concentration under UV-radiation at pH 8.3. Rather poor performance of heterogeneous Fenton system in delignification of the wastewater was explained by the selective oxidation of other wastewater constituents, which was seen in the decreased COD, colour and turbidity. The significant improvement of the heterogeneous Fenton's reagent performance was observed in UV-assisted reaction, although, the presence of FeZSM-5 zeolite in FeZSM-5/H_2O_2/UV process resulted in a drastic decrease in treatment efficiency compared to oxidation with H_2O_2/UV.

6.3 CONCLUDING REMARKS

Pulp and paper industry is a globally growing industry that consumes a significant amount of natural resources, chemicals and energy. Due to this fact, minimising the environmental impact of this sector is important. Besides, very stringent regulations have forced the pulp and paper industry to integrate and evolve novel technologies in terms of internal process changes and treatment technologies to reduce the environmental impacts. Stringent regulations brought to pulp and paper mill discharges by provincial and federal authorities necessitate large capital expenditures by the industry. This may mean the early closure of old mills, and economic hardship for many communities. At the same time, however, these regulations will ensure environmental quality improvement and a reduction of the risk to human health that is posed by this important industrial activity. Many pulp and paper mills use chlorine as a bleaching agent to produce high-quality white pulp. The high organic content of pulping wastewater, coupled with the presence of chlorine, results in the production of many highly toxic chlorinated organic compounds. Of prime concern are chlorinated phenols, guaiacols, catechols, furans, dioxins, aliphatic hydrocarbons. Some members of this family known to be toxic, mutagenic, persistent, and bio-accumulating are thought to cause numerous harmful disturbances in biological systems, although, these compounds pose a human health risk through long-term exposure via drinking water and through bioaccumulation along the food chain. The pollutants emitted from the pulp and paper industry can be minimised by various internal process changes and management measures such as best available technology (BAT) application. In-plant reduction of toxic emissions is affected through the installation of new oxygen delignification processes, and replacement or partial replacement of chlorine with hydrogen peroxide or chlorine dioxide for pulp bleaching. End-of-pipe treatment of wastewaters can be accomplished by the integration of traditional biological treatment processes with chemical/advanced oxidation applications.

The review of chemical oxidation technologies, in this chapter, has revealed that the applied processes are effective and promising applications for the treatment of pulp and paper industry wastewaters. Ozonation is efficient in removing COD, TOC and colour as well as increasing the biodegradability of the wastewater in many cases. However, it is rather an expensive process. Wet oxidation in its modifications (catalytic wet oxidation) is an effective technology in reducing the COD content of the wastewater and enhancing the biodegradability. AOPs such as Fenton, photo-Fenton, H_2O_2/UV-C proved to be effective when they are applied as tertiary treatment process. Electrochemical methods proved to be an efficient treatment option for the treatment of pulp and paper industry wastewaters. Combinations of two or more physicochemical processes can be used for the enhancement of process removal efficiencies. Currently the most important tasks seem to be the assessment of the conditions of chemical oxidation processes depanding on the aim of chemical oxidation and its way of coupling with biological treatment.

REFERENCES

Akolekar, D.B., Bhargava, S.K., Shirgoankar, I. and Prasad J. (2002) Catalytic wet oxidation: an environmental solution for organic pollutant removal from paper and pulp industrial waste liquor. *Appl. Catal. A-Gen.* **236**, 255–262.

Amat, A.M., Arques, A., Miranda, M.A. and López, F. (2005) Use of ozone and/or UV in the treatment of effluents from board paper industry. *Chemosphere* **60**, 1111–1117.

Balcioglu, A.I. and Cecen, F. (1999) Treatability of Kraft pulp bleaching wastewater by biochemical and photocatalytic oxidation. *Water Sci. Technol.* **40**, 281–288.

Bedoui, A., Hasni, M., Elaloui L., and Bensalah, N. (2009) Degradation and mineralization of organic pollutants contained in actual pulp and paper mill wastewaters by a UV/ H_2O_2 process. *Ind. Eng. Chem. Res.* **48**, 3370–3379.

Boroski, M., Rodrigues, A.C., Garcia, J.C., Gerola, A.P., Nozaki, J., and Hioka, N. (2008) The effect of operational parameters on electrocoagulation–flotation process followed by photocatalysis applied to the decontamination of water effluents from cellulose and paper factories. *J. Hazard. Mater.* **160**, 135–141.

Catalkaya, E.C. and Kargi, F. (2007) Color, TOC and AOX removals from pulp mill effluent by advanced oxidation processes: A comparative study. *J. Hazard. Mater.* **139**, 244–253.

Catalkaya, E.C. and Kargi, F. (2008) Advanced oxidation treatment of pulp mill effluent for TOC and toxicity removals. *J. Environ. Manage.* **87**, 396–404.

De los Santos Ramos, W., Poznyak, T., Chairez, I. and Córdova, I.R. (2009) Remediation of lignin and its derivatives from pulp and paper industry wastewater by the combination of chemical precipitation and ozonation. *J. Hazard. Mater.* **169**, 428–434.

EC (2001) Integrated Pollution Prevention and Control (IPPC) *Reference Document on Best Available Techniques in the Pulp and Paper Industry*, European Commission, Edificio Expo-WTC, Seville, Spain.

El-Ashtoukhy, E.-S.Z., Amin, N.K. and Abdelwahab, O. (2009) Treatment of paper mill effluents in a batch-stirred electrochemical tank reactor. *Chem. Eng. J.* **146**, 205–210.

El-Bestawy, E., El-Sokkary, I., Hussein, H., and Abu Keela, A.F. (2008) Pollution control in pulp and paper industrial effluents using integrated chemical-biological treatment sequences. *J. Ind. Microbiol. Biotechnol.* **35**, 1517–1529.

Eskelinen, K., Särkkä, H., Kurniawan, T.A., and Sillanpä, M.E.T. (2010) Removal of recalcitrant contaminants from bleaching effluents in pulp and paper mills using ultrasonic irradiation and Fenton-like oxidation, electrochemical treatment, and/or chemical precipitation: A comparative study. *Desalination* **255**, 179–187.

Gommers, K., De Wever, H., Brauns, E., and Peys, K. (2007) Recalcitrant COD degradation by an integrated system of ozonation and membrane bioreactor. *Water Sci. Technol.* **55**, 245–251.

Helbe, A., Schlayer, W., Liechti, P.-A., Jenny, R. and Mobius, C.H. (1999) Advanced effluent treatment in the pulp and paper industries with a combined process of ozonation and fixed bed biofilm reactors. *Water Sci. Technol.* **40**, 345–350.

Helmy, S.M., El Rafie, S. and Ghaly, M.Y. (2003) Bioremediation post-photo-oxidation and coagulation for black liquor effluent treatment. *Desalination* **158**, 331–339.

Hostachy, J.-C., Lenon, G., Pisicchio, J.-L., Coste, C. and Legay, C. (1997) Reduction of pulp and paper mill pollution by ozone treatment. *Water Sci. Technol.* **35**, 261–268.

Khansorthong, S. and Hunsom, M. (2009) Remediation of wastewater from pulp and paper mill industry by the electrochemical technique. *Chem. Eng. J.* **151**, 228–234.

Kindsigo, M. and Kallas, J. (2009) Wet oxidation of debarking water: changes in lignin content and biodegradability. *Environ. Chem. Lett.* **7**, 121–126.

Kreetachat, T., Damrongsri, M., Punsuwon, V., Vaithanomsat, P., Chiemchaisri, C. and Chomsurin, C. (2007) Effects of ozonation process on lignin-derived compounds in pulp and paper mill effluents. *J. Hazard. Mater.* **142**, 250–257.

Kurniawan, T.A., Lo, W.H. and Chan, G.Y.S. (2006) Degradation of recalcitrant compounds from stabilized landfill leachate using a combination of ozone-GAC adsorption treatment. *J. Hazard. Mater.* **137**, 443–455.

Laari, A., Korhonen, S., Tuhkanen, T., Verenich, S. and Kallas, J. (1999) Ozonation and wet oxidation in the treatment of thermomechanical pulp (TMP) circulation waters. *Water Sci. Technol.* **40**, 51–58.

Makhotkina, O.A., Preis, S.V., and Parkhomchuk, E.V. (2008) Water delignification by advanced oxidation processes: Homogeneous and heterogeneous Fenton and H_2O_2 photo-assisted reactions. *Appl. Catal. B-Environ.* **84**, 821–826.

Matta, R., Hanna, K., and Chiron, S. (2007) Fenton-like oxidation of TNT using different iron minerals. *Sci. Total Environ.* **385**, 242–251.

Medeiros, D.R., Pires, E.C., and Mohseni, M. (2008) Ozone oxidation of pulp and paper wastewater and its impact on molecular weight distribution of organic matter. *Ozone Sci. Eng.* **30**, 105–110.

Mounteer, A.H., Pereira, R.O., Morais, A.A., Ruas, D.B., Silveira, D.S.A., Viana, D.B. and Medeiros, R.C. (2007) Advanced oxidation of bleached eucalypt kraft pulp mill effluent. *Water Sci. Technol.* **55**, 109–116.

Patel, U.D. and Suresh, S. (2008) Electrochemical teratment of pentachlorophenol in water and pulp bleaching effluent. *Sep. Purif. Technol.* **61**, 115–122.

Pereira, R., Antunes, S.C., Gonçalves, A.M.M., Marques, S.M., Gonçalves, F., Ferreira, F., Freitas, A.C., Rocha-Santos, T.A.P., Diniz, M.S., Castro, L., Peres, I., and Duarte, A.C. (2009) The effectiveness of a biological treatment with *Rhizopus oryzae* and of a photo-Fenton oxidation in the mitigation of toxicity of a bleached kraft pulp mill effluent. *Water Res.* **43**, 2471–2480.

Pérez, M., Torrades, F., Peral, J., Lizama, C., Bravo, C., Casas, S., Freer, J., and Mansilla, H.D. (2001) Multivariate approach to photocatalytic degradation of a cellulose bleaching effluent. *Appl. Catal. B-Environ.* **33**, 89–96.

Pérez, M., Torrades, F. García-Hortal, J.A., Domènech, X. and Peral, J. (2002) Removal of organic contaminants in paper pulp treatment effluents under Fenton and photo-Fenton conditions. *Appl. Catal. B-Environ.* **36**, 63–74.

Pokhrel, D. and Viraraghavan, T. (2004) Treatment of pulp and paper mill wastewater-a review. *Sci. Total Environ.* **333**, 37–58.

Rodrigues, A.C., Boroski, M., Shimada, N.S., Garcia, J.C., Nozaki, J. and Hioka, N. (2008) Treatment of paper pulp and paper mill wastewater by coagulation-flocculation followed by heterogeneous photocatalysis. *J. Photoc. Photobiol. A* **194**, 1–10.

Rodriguez, J., Contreras, D., Parra, C., Freer, J., Baeza, J. and Duran, N. (1999) Pulp mill effluent treatment by Fenton-type reaction catalyzed by iron complexes. *Water Sci. Technol.* **40**, 351–355.

Shammas, N. K. (2009) Waste management in the pulp and paper industry. In Handbook of Advanced Industrial and Hazardous Wastes Treatment, (eds L.K. Wang, Y.T. Hung and N.K. Shammas), pp. 461–515, CRC Press, Taylor & Francis Group, (ISBN 978-1-4200722-2-8).

Shaw, L.E. and Lee, D. (2009) Sonication of pulp and paper effluent. *Ultrason. Sonochem.* **16**, 321–324.

Sevimli, M.F. (2005) Post-Treatment of pulp and paper industry wastewater by advanced oxidation processes. *Ozone Sci. Eng.* **27**, 37–43.

Smook, G.A. (2002) *Handbook for Pulp & Paper Technologists*, 3rd edn, Angus Wilde Publications, Vancouver, Canada.

Soloman, P.A., Basha, C.A., Velana M., and Balasubramanian, N. (2009) Electrochemical degradation of pulp and paper industry waste-water. *J. Chem. Technol. Biotechnol.* **84**, 1303–1313.

Sumathi, S. and Hung Y.T. (2006) Treatment of pulp and paper mill wastes. In *Handbook of Industrial and Hazardous Wastes Treatment*, 2nd edn, (eds. L.K. Wang, Y.T. Hung, H.H. Lo and C. Yapijakis), pp. 495–543, Marcel Dekker Inc., (ISBN 0-8247-4114-5).

Tambosi, J.L., Di Domenico, M., Schirmer, W.N., José, H.J. and de FPM Moreira, R. (2006) Treatment of paper and pulp wastewater and removal of odorous compounds by a Fenton-like process at the pilot scale. *J. Chem. Technol. Biotechnol.* **81**, 1426–1432.

Tünay, O., Erdemli, E., Kabdaşlı, I. and Ölmez, T. (2008) Advanced treatment by chemical oxidation of pulp and paper effluent from a plant manufacturing hardboard from waste paper. *Environ. Technol.* **29**, 1045–1051.

Uğurlu, M. and Karaoğlu, M.H. (2009) Removal of AOX, total nitrogen and chlorinated lignin from bleached Kraft mill effluents by UV oxidation in the presence of hydrogen peroxide utilizing TiO_2 as photocatalyst. *Environ. Sci. Pollut. Res.* **16**, 265–273.

US EPA (1982) *Development Document for Effluent Limitations Guidelines New Source Performance Standards and Pretreatment Standards for the Pulp, Paper and Paperboard and the Builder' Paper and Board Mills Point Source Categories*, EPA 440/1-82/025, Effluent Guideline Division, Office of Water, U.S. Environmental Protection Agency, Washington, D.C., United States.

US EPA (1992) *1990 National Census of Pulp, Paper, and Paperboard Manufacturing Facilities. Response to 308 questionnaire. Part A: Technical Information*, Office of Water, U.S. Environmental Protection Agency, Washington, D.C., United States.

US EPA (1993a) *Technical Development Document for the Pesticide Formulating, Packaging, and Repackaging Effluent Limitation Guidelines, Pretreatment Standards and New Source Performance Standards*, EPA 821/R-93/019, Engineering and Analysis Division, Office of Science and Technology, U.S. Environmental Protection Agency, Washington, D.C., United States.

US EPA (1993b) *Handbook on Pollution Prevention Opportunities for Bleached Kraft Pulp and Paper Mills*, EPA 600-R-93-098, Office of Research and Development, U.S. Environmental Protection Agency, Washington, D.C., United States.

US EPA (2002) *EPA Office of Compliance Sector Notebook Project Profile of the Pulp and Paper Industry*, 2nd edn, EPA 310-R-02-002, Office of Compliance, Office of Enforcement and Compliance Assurance, U.S. Environmental Protection Agency, Washington, D.C., United States.

Wang, B., Kong, W. and Ma, H. (2007a) Electrochemical treatment of paper mill wastewater using three-dimensional electrodes with $Ti/Co/SnO_2$-Sb_2O_5 anode. *J. Hazard. Mater.* **146**, 295–301.

Wang, B., Gu, L. and Ma, H. (2007b) Electrochemical oxidation of pulp and paper making wastewater assisted by transition metal modified kaolin. *J. Hazard. Mater.* **143**, 198–205.

World Bank Group (1998) Pesticide Formulation and Pesticide Manufacturing. In *Pollution Prevention and Abatement Handbook 1998: Toward Cleaner Production*, pp. 395–3400, World Bank Publications, (ISBN 0-8213-3638-0).

Xu, M., Wang Q., and Hao, Y. (2007) Removal of organic carbon from wastepaper pulp effluent by lab-scale solar photo-Fenton process. *J. Hazard. Mater.* **148**, 103–109.

Yeber, M.C., Oñate, K. and Vidal, G. (2007) Decolorization of kraft bleaching effluent by advanced oxidation processes using copper (II) as electron acceptor. *Environ. Sci. Technol.* **41**, 2510–2514.

Yeber, M.C., Rodriguez, J., Baeza, J., Freer, J., Zaror, C., Duran, N. and Mansilla, H.D. (1999) Toxicity abatement and biodegradability enhancement of pulp mill bleaching effluent by advanced chemical oxidation. *Water Sci. Technol.* **40**, 337–342.

Chapter 7

Pesticide industry

7.1 PESTICIDE INDUSTRY

7.1.1 Overview

Hundreds of active ingredients and tens of thousands of formulations are used to control agricultural pests and disease-carrying vectors (Meister 1999). 1.5 million tonnes of pesticides are manufactured every year, producing a business worth US$30 billion and continues to grow (McKenzie 2001). Wide varieties of pesticides or classes of pesticides are produced by the pesticide industry and these can be categorised by the point of use. They control into the following nine types of pests (US EPA 1993):

- herbicides: used for weed control;
- insecticides: used for control of insects;
- rodenticides: used for control of rodents;
- fungicides: used for control of fungi;
- nematocides: used for control of particular class of worms, which are often parasites of animals and plants;
- miticides: used for control of mites, which are tiny arachnids that often infest prepared food or act as parasites on animals, plants or insects;

- disinfectants: used for control of bacteria and viruses;
- defoliants: used to remove leaves from growing plants;
- synergists: used in conjunctions with other substances to enhance the effect of each.

The pesticides or classes of pesticides may also be grouped based on their chemical structure. Similarities in molecular weight, polarity, and solubility may be a common point in pesticides with the same structure. These similarities may result in similar amounts and types of pollutants in the wastewater generated during manufacture of the pesticides. The grouping of pesticides based on their chemical structure was also found to be a valid method of grouping for the purpose of evaluating the control and/or removal of pesticides from wastewater (US EPA 1985). The interim final effluent limitations guidelines promulgated November 1976 for the pesticide point source category established six pesticides product groups (US EPA 1976). The product groups or families proposed by EPA are presented in Table 7.1 (US EPA 1976). Division of a point source category into groupings entitled "*subcategories*" provides a mechanism for addressing variations between products, raw materials, processes, and other parameters which result in distinctly different effluent characteristics (US EPA 1993). The primary bases for subcategorising pesticides plants in this industry were found to be product type and raw materials used. The pesticide point source category has been divided into four subcategories (US EPA 1993; 1996):

Subcategory A: Organic Pesticides Chemicals Manufacturing,
Subcategory B: Metallo-Organic Pesticides Chemicals Manufacturing,
Subcategory C: Pesticide Formulating, Packaging and Repackaging (PFPR), Including Pesticide Formulating, Packaging and Repackaging Occurring at Pesticide Manufacturing Facilities (PFPR/Manufacturers), and at stand-alone PFPR facilities,
Subcategory E: Repackaging of Agricultural Pesticides Products at Refilling Establishments.

Organic pesticide chemicals manufacturing subcategory covers the production of carbon-containing pesticides and organo-tin pesticides. Metallo-organic pesticides chemicals manufacturing subcategory includes the manufacture of metallo-organic pesticides that contain mercury, cadmium, arsenic or copper. Subcategory C and E involve the processes mixing, blending/diluting one or more pesticides with one or more other active/inert ingredients to obtain a product used for additional processing or an end-use product (US EPA 1996).

Table 7.1 Pesticide groups and families (US EPA 1976)

Pesticide classification	
HALOGENATED ORGANICS	➢ DDT and relatives
	➢ Chlorinated arylalkanoic acids
	➢ Aldrin-toxaphene group
	➢ Halogenated aliphatic hydrocarbons
	➢ Halogenated aromatic-type compounds*
	➢ Other chlorinated compounds
PHOSPHORUS-CONTAINING PESTICIDES	➢ Phosphates and phosphonates
	➢ Phosphorothioates and phosphorodithioates
	➢ Phosphorus-nitrogen compounds
	➢ Other phosphorus compounds
NITROGEN-CONTAINING PESTICIDES	➢ Aryl and alkyl carbamates and related compounds
	➢ Thiocarbamates
	➢ Anilides
	➢ Amides and amines (without sulphur)
	➢ Ureas and uracils
	➢ Triazines
	➢ Amines, heterocyclic (sulphur-containing)
	➢ Nitro compounds
	➢ Other nitrogen-containing compounds
METALLO-ORGANIC PESTICIDES	
BOTANICAL AND MICROBIOLOGICAL ORGANIC PESTICIDES*	

* not elsewhere classified.

7.1.2 Pesticide manufacturing processes

There are two stages in the production of pesticides; the manufacture of pesticides, followed by the formulation and packaging of the pesticides. A pesticide is manufactured by the chemical reaction of two or more raw materials often in the presence of solvents, catalysts, and acidic and basic reagents. The raw materials may include any of a large number of organic and inorganic compounds (US EPA 1993). Pesticide active ingredients may also be used as raw materials in manufacturing derivative pesticides typically through the formation of various salts and esters (US EPA 1993). In the pesticide industry, the types of processes used (batch or continuous), the process chemistry (reaction mechanism), and the intermediate/by-product manufacture are highly dependent upon the type of pesticide manufactured. A pesticide intermediate is defined as "*specific precursor compound formed in the process of manufacturing an active ingredient*" whereas by-product is identified as a stream from the reaction process other than intermediates or active ingredients, which can be sold (US EPA 1993). A simplified flow diagram for pesticide manufacture is presented in Figure 7.1 (US EPA 1976; 1993).

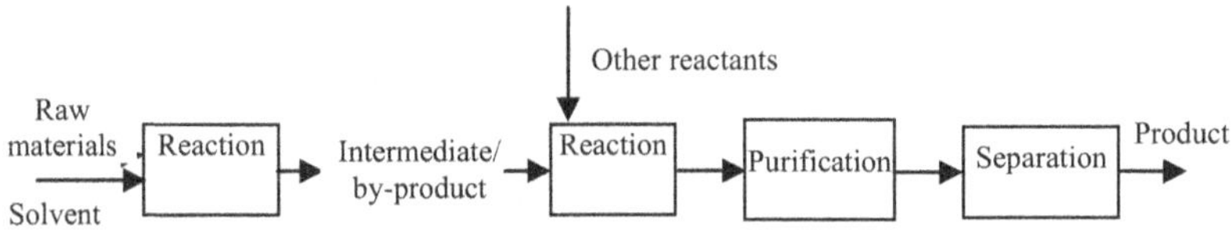

Figure 7.1 Pesticide manufacturing process (US EPA 1976; 1993).

Reaction mechanisms can include chlorination, alkylation, nitration, and many other substitution reactions. Purification is used to reclaim solvents or excess reactants as well as to purify intermediates and final products. Evaporation, extraction, and distillation are common purification processes. Separation processes include filtration, decantation, extraction, and centrifugation (US EPA 1976). The following sections describe examples of pesticide manufacturing processes for different group of pesticides and are not directly applicable to every pesticide manufactured in each structural category.

7.1.2.1 Halogenated organic pesticides

In this group of pesticides, in most cases the halogen component is chlorine. The chlorine groups generally are added via direct chlorination or via substitution from another chlorinated organic (US EPA 1976). Chlorinated aryloxyalkanoic acids are classified in this group of pesticides. Figure 7.2 shows a simplified process flow diagram for the manufacture of the chlorinated aryloxyalkanoic acids. The method of synthesis for 2,4-dichlorophenoxyacetic acid (2,4-D) is generally applicable to the majority of the class. The chlorinated aryloxyalkanoic acids can be prepared by charging equimolar quantities of a chlorophenol and a monochloroalkyl acid to a steam-heated kettle in the presence of dilute caustic. The reaction is carried on for several hours under reflux conditions, and then the reaction mass is acidified to approximately pH 1.0 with hydrochloric acid. The acidified liquor is sent to a crystalliser which is followed by centrifuge (US EPA 1976). Ester and amine salts are prepared by the reaction of the phenoxy alkyl acid with an alcohol or amine, respectively.

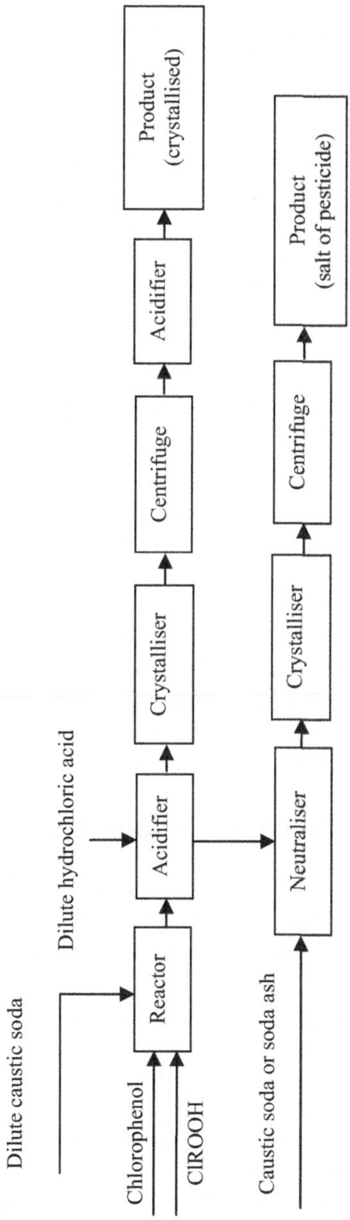

Figure 7.2 General process flow diagram of aryloxyalkanoic acid production facilities (US EPA 1976).

7.1.2.2 Nitrogen-containing pesticides

The nitrogen-containing pesticides include the greatest number of chemical types, the broadest raw material base and the most diverse process schemes. The process flow diagram to be described here is the amides and amines (without sulphur). These herbicides include two major groups; herbicides based on substituted anilide structures and chloroacetamide derivatives (US EPA 1976). This family of nitrogen-containing pesticides includes propachlor, alachlor, propanil, naptalam, and diphenamid. Figure 7.3 shows a typical process flow diagram of amide and amine (without sulphur) type pesticide. Briefly, the process is based on the reaction of an acetyl chloride with an amine. As an example alachlor is produced by the reaction of diethyl(chloro-methyl-chloroacetyl)anilide, anhydrous ammonia, and methanol. The process involves a reaction step prior to production formulation (US EPA 1977).

7.1.2.3 Metallo-organic pesticides

Metallo-organic group of pesticides includes the organic arsenicals and the dithiocarbamate metal (cadmium, copper, and mercury) complexes. A typical process flow diagram for the production of ethylene bisdithiocarbamate metal complex pesticide is depicted in Figure 7.4. The raw materials (carbon disulphide, ethylene diamine, and sodium hydroxide) are first reacted in a stainless steel, cooled vessel. The resulting concentrated intermediate solution is reacted (within 24 hours) with a metal sulphate and desired metal organic complex is precipitated. The slurry is washed with water to remove sodium sulphate, and then dried to less than 1 percent water content. Process by-products include sodium sulphate and small amounts of carbon disulphide and sodium hydroxide (US EPA 1976).

7.1.3 Pesticide formulating, packaging & repackaging processes

Pesticide formulation involves the processes of mixing, blending or diluting one or more pesticides with one or more other active or inert ingredients to obtain a product used for additional processing or an end-use (retail) product (US EPA 1996). Formulation does not involve an intended chemical reaction. The packaging of the formulated pesticide product depends on the type or formulation (liquid, dry, aerosol or pressurised gases). The formulating, packaging and repackaging can be performed in a variety of ways, ranging from very sophisticated and automated formulating and packaging lines to completely manual lines (US EPA 1996). Typical liquid formulating lines consist of storage tanks or containers to hold active and inert raw materials and a mixing tank for formulating the pesticide product.

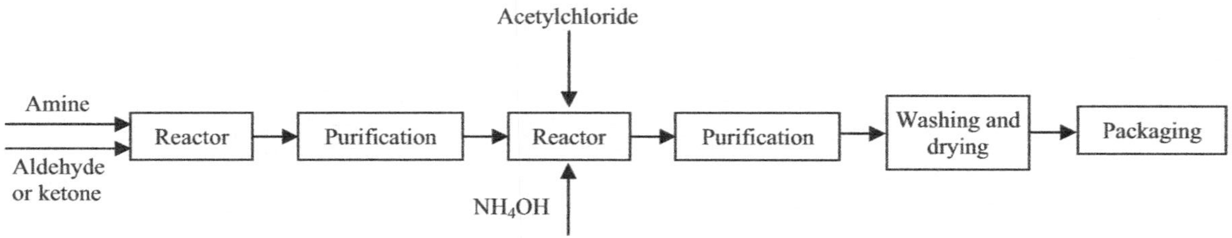

Figure 7.3 General process flow diagram for amide and amine production facilities (US EPA 1976).

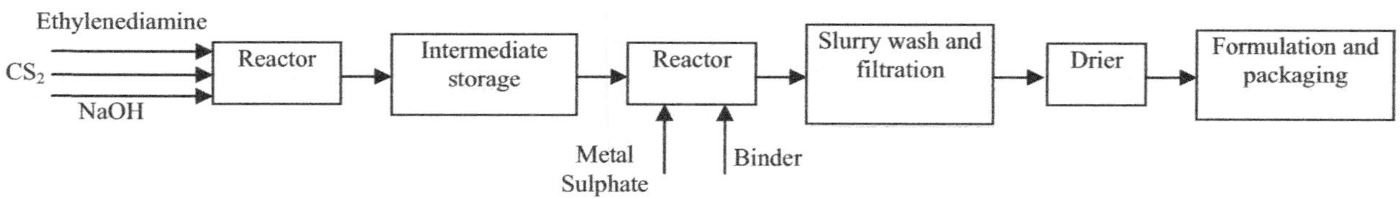

Figure 7.4 General process flow diagram for ethylene bisdithiocarbamate metal complex pesticide production facilities (US EPA 1976).

A storage tank may also be used on formulating line to hold the formulated pesticide product prior to packaging step (US EPA 1996). Because of the many types of dry pesticide products, dry pesticide formulating line can vary considerably. In general, similar to liquid formulating lines dry formulating lines have tanks or containers to hold the active ingredients and inert raw materials and may include mixing tanks, ribbon blenders, extruding equipments, high pressure, and temperature tanks for impregnating solids with active ingredients, vacuum or other type of drying equipment, tanks or bins for storage of formulated pesticide products, pelletisers, presses, milling equipment, sieves, and sifters (US EPA 1996). Liquid and dry formulating lines may also include piping and pumps to move raw materials from storage tanks to formulation equipment and to move formulated pesticide product to packaging equipment. Other items that may be part of the formulating line of liquid and dry formulation are pre-mixing tanks, stirrers, heaters, bottle washers (US EPA 1996). Some pesticides products are formulated and packaged as pressurised gases. Formulating and packaging operations for these products typically occur in one step in a closed-loop system (US EPA 1996). Liquid, dry, and aerosol formulations are packaged by simply transferring the final products into containers. Repackaging operations, except the "*raw material*" is an already formulated product that is packaged for sale. Repackagers often purchase formulated pesticide products, transfer the product to new containers with customer-specific labelling and sell them to distributors. A separate type of repackaging, called refilling, is usually performed by agrichemical facilities that transfer pesticide products from bulk storage tanks into mini-bulks (US EPA 1996).

7.1.4 Wastewater sources

The manufacture of a pesticide requires several types of input streams. These include raw materials, solvents, other reactants, and water. Water or steam may be added to the reaction medium to act as a solvent or carrier or water may be added during subsequent separation and purification steps (see Figure 7.1) (US EPA 1993). The stream leaving the process include the products, intermediates/by-products, liquid, and solid wastes. Liquid and solid wastes cover hazardous and non-hazardous organic and inorganic wastes as well as wastewaters. In addition to these, some chemical compounds may leave the manufacturing process in the form of air emissions (US EPA 1993). To provide ease of evaluation, wastewater sources can be classified into three main groups as follows (US EPA 1993):

- pesticide process wastewaters (water leaving the manufacturing process);
- other pesticide wastewaters (pesticide containing wastewater generated from sources not directly associated with the manufacturing process);
- the other facility wastewaters (wastewater from other manufacturing operations such as organic chemicals production or other facilities such as sanitary wastewater which is typically commingled and treated with pesticide containing wastewater).

Table 7.2 summarises the wastewaters generated during the manufacture of pesticides together with the definition of sources (US EPA 1993). Wastewaters generated in the pesticide formulating, packaging and repackaging (PFPR) industry is typically due to cleaning of equipment and related process areas and can be grouped as "*interior*" and "*exterior*" wastewater sources (US EPA 1996). The interior wastewater sources are; interior equipment cleaning, bulk tank rinsate, shipping container rinsate, and contact cooling water. The exterior wastewater sources are; floor, wall or exterior equipment wash water, leaks, and spills cleanup water, air or odour pollution control scrubber water, department of transportation (DOT) leak test water, safety equipment wash water and laboratory equipment water. Shower water, laundry water, fire protection test water, and contaminated precipitation runoff are excluded from the above wastewater sources and classified as other PFPR wastewater sources.

7.1.5 Wastewater characterisation

The pesticide manufacturing industry generates process wastewaters containing a variety of pollutants as a result of the wide variety of raw materials, processes used and products manufactured. Main pollutant parameters of pesticide manufacturing industry wastewaters are conventional pollutants (pH, biochemical oxygen demand (BOD_5), total suspended solids (TSS)), a variety of toxic pollutants and a large number of non-conventional pollutants (*i.e.* chemical oxygen demand (COD) and pesticides). The pollutants or pollutant groups likely to be present in process wastewater of pesticide manufacturing industry can be classified as volatile aromatics, halomethanes, cyanides, haloethers, phenols, polynuclear aromatics, heavy metals, chlorinated ethanes, and ethylenes, nitrosamines, phthalates, dichloropropane and dichloropropene, priority pollutant pesticides, dienes, 2,3,7,8-tetrachlorodibenzo-p-dioxin (TCDD) (US EPA 1985). According to the US EPA surveys (1985), COD concentrations range from 14 mg/L to 1220 g/L; total organic carbon (TOC) concentrations

Table 7.2 Wastewater sources of pesticide manufacturing process (US EPA 1993)

	Source	Description
PESTICIDE PROCESS WASTEWATERS	Water of reaction	Water which is formed during the reaction
	Process solvent	Water used to transport or support the chemical involved in reaction which is then removed from the process through a separation step
	Process stream washes	Water added to the carrier, spent acid or base which has been separated from the reaction mixture
	Product washes	Water added to the reaction medium in order to purify an intermediate product or active ingredient; or water which is used to wash crude product
	Spent acid/caustic	Spent acid and caustic streams are discharged from the process during the separation steps
	Product/process laboratory quality control wastewater	Water from laboratories
	Safety shower water	When used, any wastewater generated is process wastewater
	Steam jets or vacuum pumps*	Water which contacts the reaction mixtures or solvents or water stripped from the reaction mixture through the operation of a venture or vacuum pump
	Air pollution control scrubber blowdown*	Water used in air emission control scrubbers
	Equipment and floor washes*	Water used to clean process equipment and floors
	Pump seal water*	Water used to cool packing and lubricate pumps
	Shipping container cleanout*	Water used to clean out shipping containers
	Contact cooling water*	Water used to cool steam and other emissions from evaporating water from products
OTHER PESTICIDE WASTEWATERS	Showers used by employees	
	Laundries used to wash clothing of employees	
	Cleaning safety equipment used in pesticide production	
	Contaminated storm water	

* These water uses could result in the water becoming contaminated with pesticide or other compound used in the manufacturing process. These sources maybe intermittent or absent entirely.

change within a great wide range from 53 mg/L to 80 g/L; BOD_5 concentrations range from non-detected to 60 g/L and TSS varies from 2 mg/L to 4 g/L.

The pesticide formulating, packaging and repackaging facilities generates process wastewaters containing a variety of pollutants similar to pesticide manufacturing industry. Pesticides and solvents are the principal pollutants of concern. Although, their volumes are small, the wastewaters from pesticide formulating, packaging and repackaging facilities can be highly contaminated and toxic (Wong 2006). The process chemistry evaluation employed to predict the pollutant groups in pesticide manufacturing process wastewaters is briefly summarised as follows.

Volatile organics: Benzene, toluene, ethylbenzene, and chlorobenzene are the major volatile aromatics known to be present in the pesticide manufacturing process wastewaters. Benzene and its derivatives are widely used throughout the chemical industry as solvents and raw materials. Toluene and chlorobenzene are used as raw materials in the synthesis of pesticides, although they mainly function as a carrier solvent. Additionally, priority pollutants, aromatics, and chlorinated aromatics exist as impurities or as reaction by-products owing to the reactions of the basic raw materials and solvent compounds (US EPA 1985).

Halomethanes: Halomethanes, including methylene chloride, chloroform, and carbon tetrachloride (di-, tri-, and tetrachloromethane, respectively) are primarily used as raw material and solvents. Bromomethanes can be present in pesticides as raw material, by-products, or impurities, and in the case of methyl bromide, can function as a fumigant (US EPA 1985).

Cyanide: The primary raw materials which favour the generation of cyanides as either by-products or impurities are cyanamides, cyanates, thiocyanates, and cyanuric chloride. Cyanuric chloride is exclusively used in the production of triazine pesticides (US EPA 1985).

Haloethers: Haloethers contain an ether moiety and halogen atoms attached to the aryl and alkyl groups. Bis(2-chloroethyl) ether (BCEE) is used as a raw material then it is suspected to be present in the process wastewater. In the remainder of the pesticides the ethers are shown to be suspected raw material impurities (US EPA 1985).

Phenols: Phenols commonly present in the pesticide manufacturing wastewaters include chlorophenols, nitrophenols, and methylphenols (cresols). These compounds may be found throughout the pesticide process wastewaters as raw materials, impurities in raw materials or as by-products of reactions utilising the related compounds such as chlorobenzenes (US EPA 1985).

Polynuclear aromatics: Polynuclear aromatics consist of two or more benzene rings which share a pair of carbon atoms. Naphthalene derivatives such as alpha-naphthylamine and alpha-naphthol are used in pesticide processes, therefore, naphthalene is by far the most prevalent polynuclear aromatic pollutant in the industry. Acenaphthylene, anthracene, fluoranthene, and phenanthrene are present only as raw material impurities (US EPA 1985).

Metals: In the pesticide manufacturing industry metals are principally used as catalysts or raw materials which are incorporated into the active ingredients, e.g. metallo-organic pesticides. Metals commonly incorporated into metallo-organic pesticides include arsenic, cadmium, copper, and mercury. Metal such as manganese and tin are also used in pesticide industry (US EPA 1985).

Chlorinated ethanes and ethylenes: The chlorinated ethanes and ethylenes are used as solvents, cleaning agents, and intermediates. The main pollutants are 1,2-dichloroethane and tetrachloroethylene being used as solvents (US EPA 1985).

Nitrosamines: N-nitrosamines are a group of compounds characterised by a nitroso group attached to the nitrogen of an aromatic or aliphatic secondary amine. N-nitrosodi-N-propylamine is a suspected reaction by-product from the nitrosation of di-N-propylamine (US EPA 1985).

Phthalates: Dimethyl phthalate is known to be a raw material used in pesticide manufacturing processes (US EPA 1985).

***Dichloropropane and dichloropropene*:** 1,3-dichloropropene and combination pollutants 1,2 dichloropropane-1,3-dichloropropene are pesticide products as well as pollutants (US EPA 1985).

Dienes: The pollutant hexachlorobutadiene is suspected to be present in the pesticide process wastewater because it is a by-product of hexachlorocyclopentadiene synthesis and it is used as a solvent in the production of Mirex (US EPA 1985).

2,3,7,8-tetrachlorodibenzo-p-dioxin (TCDD): TCDD is believed to be a by-product in chemical processing generated by a halophenol or chlorobenzene starting material. TCDD is suspected to exist in wastewaters from pesticide manufacture that uses such raw materials as 2,4,5-trichlorophenol and 1,2,4,5-tetra-chlorobenzene, which are characteristic of TCDD precursors. A TCDD level as high as 111 mg/L may be found in drums of waste from the production of pesticide 2,4,5-trichlorophenol (US EPA 1985).

Priority pollutant pesticides: Priority pollutant pesticides are important group of pollutants originating from the pesticide industry. Pesticides identified as priority pollutants and/or persistent organic pollutants (POPs) by national regulations or by international agreements are as follows (US EPA 1985; SC 2001; EC 2004; 2006):

- Alachlor
- Aldrin
- Atrazine
- Alpha hexachlorocyclohexane
- Endrin
- Heptachlor
- Hexachlorobenzene
- Isoproturon

- Beta hexachlorocyclohexane
- Chlordane
- Chlordecone
- Chlorfenvinphos
- Chlorpyrifos
- DDT
- Dieldrin
- Diuron
- Endosulfan
- Lindane
- Mirex
- Pentachlorobenzene
- Pentachlorophenol
- Simazine
- Toxaphene
- Tributyltin
- Trifluralin

7.1.6 Pollution prevention and control

Source control applications can be extremely effective in reducing the cost for in-plant controls and end-of-pipe treatment and in some cases can eliminate the need for some treatment units entirely. The first and most cost-effective step which can be taken to reduce wastewater pollutant discharge is to control them at the source (US EPA 1985). The preparation of an inventory for the waste sources and the development of in-plant processes and equipment changes to reduce the amount of wastes are the main steps of the source control applications (Wong 2006). The following paragraphs address some of the source control applications available for pesticide industry.

Waste segregation is an important step in waste reduction. In the pesticide manufacturing industry processes wastewaters containing specific pollutants can often be isolated and disposed of or treated separately in a technically efficient and more economical manner (US EPA 1985). Highly acidic and caustic wastewaters are usually more effectively neutralised prior to being mixed with other wastes. Separate equalisation can be used for streams of highly variable characteristics (US EPA 1985).

Specific pollutants can be eliminated by requesting specification changes from raw material suppliers in cases where impurities are present and known to be discharged in process wastewaters. Every effort should be made to replace highly toxic and persistent ingredients with degradable and less toxic ones (World Bank Group 1998). Raw material recovery can be achieved through solvent extraction, steam stripping, and distillation operations. Additionally, dilute streams can be concentrated in evaporators and then recovered (US EPA 1985). Reverse osmosis and ultrafiltration process can also be used to recover and concentrate raw materials (World Bank Group 1998). By-products generated from the processes can be reused as raw materials or as raw material substitutes in other processes. Minimisation of material and product inventory can prevent the wastage that could lead to the formation of inactive but toxic isomers or by-products.

Good housekeeping practices and wastewater monitoring programs can affect considerably water consumption and can prevent permit violations due to spills and leaks. Dry clean-up of spills can be used instead of washing spilled wastes into floor drains (US EPA 1985). High-pressure hoses can be used for equipment cleaning to reduce wastewater. Equipment washdown waters and other process waters (such as leakage from pump seals) can be reused as makeup solutions for subsequent batches (World Bank Group 1998).

7.1.7 Treatment technologies

This section identifies the in-plant and end-of-pipe control and treatment technologies used for the removal of conventional, non-conventional and priority pollutants generated in the pesticide industry. In-plant control/treatment reduces or eliminates a particular pollutant before it is diluted in the main wastewater stream and may provide an opportunity for material removal (Wong 2006). End-of-pipe treatment is the final stage for meeting regulatory discharge requirements and protection of receiving water quality.

There are six primary in-plant recovery/treatment methods for removal of priority pollutants and pesticides in the pesticide manufacturing plants. These methods include steam stripping, activated carbon adsorption, chemical oxidation, resin adsorption, hydrolysis and metals separation (US EPA 1985). Volatile organics can be purged by steam stripping; semi-volatile organics and many pesticides can be removed by activated carbon adsorption; resin adsorption, chemical oxidation, and hydrolysis methods. Metal bearing wastes originating from the pesticide industry can be treated by proper separation methods such as chemical precipitation (US EPA 1985). End-of-pipe treatment methods used in the pesticide industry include equalisation, neutralisation, biological treatment, and filtration. Treatment technologies used for the removal of pollutant groups defined in Subsection 7.1.5, are summarised in Table 7.3 (US EPA 1985). The primary unit treatment recommended for each pollutant group is designated with "**1**". After treatment by the recommended primary unit, further removal is accomplished by follow-on treatment, which is designated with "**2**". There are other technologies also used for the treatment of pesticides industry wastewaters such as wet air oxidation, membrane processes, solvent extraction, and powdered activated carbon adsorption. However these processes are not considered as primary or secondary treatment unit (US EPA 1985).

Waste management in the pesticide formulating, packaging and repackaging facilities is simpler than in manufacturing plants because the waste volumes are much smaller (Wong 2006). The most common technologies currently used or applicable to facilities to treat pollutants in wastewaters from the

Table 7.3 Treatment technologies used for various pollutant groups in pesticide industry (US EPA 1985)

Treatment technology → Pollutant group ↓	Activated Carbon Adsorption	Resin Adsorption	Steam Stripping	Hydrolysis	Biological Treatment	Metal Precipitation	Chemical Oxidation	Filtration	Tertiary Activated Carbon Adsorption
Volatile organics	×	×	1	×	2	×	×	×	2
Halomethanes	×	×	1	×	2	×	×	×	2
Cyanide	×	×	×	×	2	×	1	×	2
Haloethers	×	×	1	×	2	×	×	×	2
Phenols	1	1	×	×	2	×	×	×	2
Polynuclear aromatics	×	×	×	×	2	×	×	×	2
Metals	×	×	×	×	2	1	×	2	2
Chlorinated ethanes-ethylenes	×	×	1	×	2	×	×	×	2
Nitrosamines	1	×	×	×	2	×	×	×	2
Phthalates	×	×	×	×	2	×	×	×	2
Dichloropropane-propene	×	×	1	×	2	×	×	×	2
Dienes	1	1	×	×	2	×	×	×	2
TCDD	1	×	×	×	×	×	×	×	2
Priority pollutant pesticides	1	1	×	1	2	×	1	×	2

1 = Primary removal technology
2 = Secondary removal technology
× = Indicated technology is considered as a treatment option but is not a primary or secondary recommendation

pesticide formulating, packaging and repackaging operations are: activated carbon adsorption, hydrolysis, reverse osmosis, chemical oxidation (by alkaline chlorination or ozone/UV) and chemical precipitation (for removal of metals) (US EPA 1996). Biological treatment (acclimated to specific pesticide-containing wastewaters) and steam stripping have also been shown to be effective in the treatment of wastewaters but these technologies are cost-prohibitive for the pesticide formulating, packaging and repackaging facilities (US EPA 1996).

US EPA proposed *"best available technologies economically achievable"* (BAT-US) for the treatment and control of pesticide industry wastewaters. In each case, technology options are considered and one option selected as the basis for regulations. In BAT-US definition described by US EPA, treatment systems are defined under four options. "*Option 1*" encompasses in-plant activated carbon adsorption, in-plant hydrolysis, and biological treatment. "*Option 2*" includes all of "*Option 1*" together with steam stripping, chemical oxidation, and metal precipitation. "*Option 3*" covers "*Option 2*" plus end-of-pipe multimedia filtration whereas "*Option 4*" encompasses "*Option 3*" plus end-of-pipe activated carbon adsorption (US EPA 1985).

7.2 CHEMICAL OXIDATION PROCESSES

Chemical oxidation processes have a high potential to reduce or completely eliminate the pesticide contamination during water and/or wastewater treatment. Several chemical oxidation treatment processes (as seen in Table 7.4) have been examined for their possible implementation to reduce pesticides concentrations in water and wastewater. Applications of chemical oxidation processes to actual pesticide manufacturing wastewaters are rather rare. Most of the studies have been conducted in reduced concentration of pesticides to represent or simulate the quality of contaminated water bodies (Felsot 1996; Kolpin *et al.* 1998; Ikehata and Gamal El-Din 2005a; 2005b). These studies of course are helpful since the target substrates are the same as those existing in actual pesticide manufacturing effluents. At any rate, the point that initial concentration of the substrate may have an impact on the process performance as well as the rate of the process must be taken into account. Considering these points, the studies employing relatively high concentration of pesticides have been selected and evaluated in this section. Within the scope of this book the chemical oxidation processes employed to the wastewaters mentioned above were examined and their efficiencies, design criteria and costs are introduced considering their basic features. As introduced in Section 7.1.7 (see Table 7.3), chemical oxidation processes are used as primary removal technology for the treatment of cyanide and priority pollutant pesticide pollutant groups in the pesticide industry.

Table 7.4 Chemical oxidation processes examined for aqueous pesticide degradation (Ikehata and Gamal El-Din 2005b; 2006)

Process	Oxidant(s)	Other chemical(s)	Other energy source	Note
Ozonation	O_3	None	None	
High pH Ozonation	O_3	Hydroxyl ion (OH^-)	None	
O_3/UV	O_3	None	UV radiation	$\lambda = 258\text{-}260$ nm for O_3*
O_3/H_2O_2	O_3, H_2O_2	None	None	
O_3/H_2O_2/UV	O_3, H_2O_2	None	UV radiation	(see O_3/UV and H_2O_2/UV)
Fenton	H_2O_2	Ferrous ion (Fe^{2+})	None	pH<3, sludge formation
Fenton-like	H_2O_2	Ferric ion (Fe^{3+})	None	pH<3, sludge formation
Photo Fenton	H_2O_2	Fe^{2+} or Fe^{3+}	UV radiation/Visible light or solar radiation	$\lambda < 400$ nm for $Fe^{III}(OH)^{2+}$*
Photo ferrioxalate/ H_2O_2 (Fe^{3+}-chelate/ H_2O_2)	H_2O_2	Fe^{3+} , oxalate	UV/Vis (solar) radiation	$\lambda < 550$ nm for $[Fe^{III}(C_2O_4)_3]^{3-}$*
Anodic Fenton	H_2O_2	Iron electrode	Electrical current	Fenton reactions occur only in the anode half-cell
Electrochemical Fenton (Electro Fenton)	H_2O_2	Iron electrode	Electrical current	Sludge formation
Electrophoto Fenton	H_2O_2	Iron electrode	UV/Vis (solar) radiation, electrical current	
Peroxi-coagulation	Electrogenerated H_2O_2	Dioxygen and Fe^{2+} or iron electrode	Electrical current	Coagulation with iron hydroxide
H_2O_2/UV	H_2O_2	None	UV radiation	$\lambda = 250\text{-}254$ nm for H_2O_2*
TiO_2/UV	None	TiO_2	UV/Vis (solar) radiation	$\lambda < 400$ nm for TiO_2*
(Photolysis)	None	None	UV radiation	
γ-Radiolysis	None	None (water)	γ-radiation	
Sonolysis	None	None (water)	Ultrasound	

*From Oppenländer (2003).

7.2.1 Cyanide oxidation

During the production of pesticides, cyanamides, cyanates, thiocyanates, and cyanuric chloride can be used as primary materials. These cyanide-based reagents are generally used for pesticide and/or intermediate synthesis (IFC 2007). As an example, cyanuric chloride is used in the production of triazine pesticides (US EPA 1985).

Alkaline chlorination and H_2O_2 oxidation are the most commonly applied treatment methods to cyanide bearing wastewaters originating from the pesticide industry. Alkaline chlorination which has been in commercial use for over 35 years, is suitable for destroying free dissolved hydrogen cyanide and for oxidising all simple and some complex inorganic cyanides in aqueous media (Ghosh *et al.* 2006). If properly designed, maintained, and operated (proper pH and oxidation reduction potential control), the process will oxidise cyanides which are amenable to chlorination. The operating conditions, efficiency as well as typical operation and maintenance costs of alkaline chlorination are explained in detail in Subsection 4.2.1.1.

H_2O_2 is another oxidant used in the destruction of cyanide generated from the pesticide manufacturing facilities. Although, cyanide degradation with H_2O_2 for various effluents is adequate, in some cases, an extensively high dosage of H_2O_2 may be necessary to achieve convenient reaction rate and cyanide removal efficiency. In such cases, other (more powerful) oxidants may be considered to attain discharge limits, without increasing operating costs (Teixeira *et al.* 2005). The reaction mechanisms, operating costs, and feasibility of the cyanide oxidation by H_2O_2 and its modifications (copper catalysed H_2O_2 oxidation, oxidation by H_2O_2 accelerated by the addition of heterogeneous catalysts or activated carbon) have been discussed under the Subsection 4.2.1.3 and forthcoming Subsection 8.2.2.

As indicated in the previous discussions (Chapter 4), there are over ten cyanide treatment processes including ozonation, sulphur dioxide/air process, oxidation by chlorine dioxide, hexavalent ferrate oxidation, photochemical, and photocatalytic oxidation methods, electrochemical oxidation, and high temperature alkaline chlorination process that have been investigated intensively for cyanide removal. The key to successful implementation of these processes is consideration the following (Botz 1999):

- cyanide balances under both average and extreme climatic conditions;
- the range of cyanide treatment processes available and their ability to be used individually or in combination to achieve treatment objectives;
- proper testing, design, construction, maintenance, and monitoring of both water management and cyanide management facilities.

7.2.2 Priority pollutant pesticide oxidation

As indicated in Subsection 7.1.7, an ideal treatment scheme for concentrated wastewater from pesticide manufacturing is the pretreatment of the wastewaters by proper physicochemical processes to remove or reduce the most of the toxic organics before biological treatment (US EPA 1985; Cheng *et al.* 2007). Biological processes are well established and relatively cheap treatment techniques. However, these biological methods are susceptible to toxic compounds that inactivate the waste-degrading microorganisms. In such cases, a potentially useful approach is to partially pretreat the toxic waste by chemical oxidation technologies in order to produce intermediates that are more readily biodegradable (Somich *et al.* 1990; Lai *et al.* 1995a, b; Scott and Ollis 1995; Chiron *et al.* 2000). However, it should be also considered that chemical oxidation technologies are sometimes inadequate to detoxify the pesticide bearing waters/wastewaters because they may yield more toxic degradation by-products (Ikehata and Gamal El-Din 2005a). Many oxidation processes currently employed for pesticide elimination from aquatic solutions are summarised in Table 7.4. However, these processes are mostly examined under lab-scale conditions and most of the studies are conducted on synthetically prepared samples containing low level of pesticide(s) in order to simulate surface and ground water conditions. Within the scope of this book, the literature published concerning the chemical oxidation of priority pollutant pesticides in pesticide industry wastewaters and/or in synthetic aqueous solutions bearing high concentration of pesticide(s) are reviewed and discussed. Comprehensive reviews are also available elsewhere for the pesticide degradation (Burrows *et al.* 2002; Ikehata and Gamal El-Din 2005a; 2005b; 2006).

Kearney *et al.* (1987) evaluated the performance of O_3/UV process ($\lambda = 253.7$ nm, 11 W/L, in-situ O_3 generation, O_3 dose not specified) for the treatment of nine formulated herbicides (alachlor, atrazine, bentazon, butylate, cyanazine, 2,4-D, metolachlor, metribuzin, and trifluralin) and two formulated insecticides (carbofuran and malathion) at concentrations of 10, 100, and 1000 mg/L. The time required for 90 % destruction of pesticides studied was dependent on the initial concentration and extended as the concentration of pesticide increased. Relatively slow degradation of butylate and trifluralin was observed, as compared with other pesticides tested, such as alachlor, atrazine, and metolachlor. It has been concluded that all studied pesticide containing samples with concentrations of 10 and 100 mg/L were rapidly oxidised by ozone in combination with UV irradiation at 22 and 61 minutes, respectively, whereas longer reaction periods were required for pesticides at 1000 mg/L. The complete degradation of metribuzin (triazinone herbicide) was almost achieved within 2 hours of the O_3/UV treatment even at the concentrations of 10 and 100 mg/L.

Somich *et al.* (1990) studied the removal of paraquat (40 mg/L), atrazine (17 mg/L), cyanazine (30 mg/L), and metolachlor (82 mg/L) from pesticide waste rinsate obtained from a small farm by O_3/H_2O_2 process (17.3 mg applied O_3/L, 65 mg/L H_2O_2, pH 10). They reported that only 27 % of initial paraquat, 50 % of initial atrazine, 77 % of initial cyanazine and 91 % of initial metoachlor removed by O_3/H_2O_2 treatment. Nevertheless, this treatment method greatly enhanced the biodegradation of pesticides tested and diminished the phytotoxicity of rinsate.

Massey and Lavy (1997) studied treatability of imazaquin (57–893 mg/L) by ozonation and O_3/H_2O_2 processes. Imazaquin was rapidly degraded using a range of solution temperatures (1–35°C) and H_2O_2 concentrations (0–69 mM) at an ozone dose of 3.6 g O_3/h. The most rapid degradation occurred when the H_2O_2 concentration of 0.2 mM was applied. Imazaquin was also rapidly oxidised by ozone in the absence of H_2O_2. The ozonation times required for 90 % removal of the imazaquin were typically lower than 15 min. Approximately 10.5 ± 4.4 % (quinoline ring) to 17.1 ± 7.8 % (carboxylic acid) of [^{14}C] imazaquin was oxidised to $^{14}CO_2$ for pH = 5.5–8.5, reaction time = 0.5 h, and T = 20°C. Biodegradability of this imidazolinone herbicide was improved, and its phytotoxicity was reduced after the ozonation.

Kuo (2002) studied the degradation of mevinphos by direct UV photolysis (3.0 mW/cm at 90 mm distance) and O_3/UV processes. O_3/UV (400 mg O_3/L. hr) process was found very effective to reduce COD (>85 % reduction within 3 h) from simulated mevinphos rinsate containing 100 mg/L of pesticide at pH 9, whereas direct photolysis was less effective in COD reduction (30 %). In the same study, COD and toxicity removals from the aqueous solution of carbofuran with direct UV photolysis and O_3/UV processes were also investigated. More than 85 % reduction in COD was achieved using ozonation combined with UV irradiation at pH 6.5 within 3 hours. It was found that Microtox toxicity of the pesticide solutions was also reduced substantially after the O_3/UV treatment (Kuo 2002).

The degradation of carbofuran (2,3-dihydro-2,2-dimethylbenzofuran-7-yl methylcarbamate), a frequently used carbamate derivative pesticide which is considered a priority pollutant, was carried out in batch reactors by means of single oxidants such as ozone, UV radiation, and Fenton's reagent; and by the AOPs constituted by combinations of ozone plus UV radiation, UV radiation plus H_2O_2, and UV radiation plus Fenton's reagent (photo-Fenton system) by Benítez *et al.* (2002). In this study, the process performances were evaluated in terms of carbofuran removal. In the ozonation of carbofuran, increasing the temperature and ozone partial pressure had a positive effect on the removal rate of the carbofuran, while similar degradations were obtained for experiments at pH 2 and 9 which indicated a negligible contribution of the reaction with $HO^{\bullet}$

generated by the ozone self-decomposition at pH 9. The photodecomposition of carbofuran by the polychromatic UV radiation yielded a lower efficiency than the ozonation process, on the other hand, the combination of UV radiation with hydrogen peroxide increased the degradation levels for initial hydrogen peroxide concentrations of 5×10^{-4} and 5×10^{-3} M. The carbofuran oxidation by the Fenton's reagent alone exhibited a destruction power lower than single oxidants, ozone and UV radiation. For the oxidant concentrations applied in this research, the efficiency of the photo-Fenton system was found even higher than those of the conventional ozonation and O_3/UV.

Brillas *et al.* (2003) investigated the treatment of 2,4-D with ozone and ozonation catalysed with Fe^{2+} (55.8 mg/L) and/or UVA (λ_{max} = 360 nm, 8.3×10^{-7} einstein/L.min) light at pH 3.0 and 25°C. Direct ozonation (13 g O_3/L hr) gave rise to a progressive, but slow, depollution up to attained 59 % of TOC removal at 2 hours. Under UVA irradiation, 2,4-D was much more rapidly mineralised by reaction with $HO^{\bullet}$ and the solution TOC was reduced by 95 % after 2 hours of the O_3/UVA treatment. During ozonation, TOC reduction in 2,4-D aqueous solution was strongly accelerated by the addition of Fe^{2+} at the beginning, although the effect was quickly dissipated probably due to the formation of oxalate by degradation of 2,4-D that might form stable complexes with ferric ion (Fe^{3+}). However, under UVA irradiation, the enhanced degradation of 2,4-D by the O_3/Fe^{2+} process did not slow down and led to more than 95 % mineralisation in 2 hours.

Wu *et al.* (2009) investigated the treatability of organophosphorus pesticides (diazinon degraded faster than methyl parathion, and parathion) by ozonation. The aqueous ozone concentration was kept at about 1.4 mg/L throughout the experiments. The authors indicated that diazinon degraded faster than did methyl parathion or parathion at pH 5.3. Diazinon (initial concentrations of 10–30 mg/L) was completely destructed in 20 minutes, while only 80 % of either methyl parathion or parathion at an initial concentration of 20 mg/L was degraded in 30 minutes.

Lee and Carberry (1992) evaluated the Fenton process with a view to enhancing the biodegradability of pentachlorophenol (PCP) being an organochlorine fungicide and wood preservative highly resistant to microbial degradation. A very high concentration of PCP (266 mg/L) was treated by the Fenton process performed at 1 mM of H_2O_2 and 1.5 mM of Fe^{2+} for 4 h. Although, the reduction in PCP concentration was negligible (reduced to 258 mg/L), biodegradation of the PCP solution by the selected microbial consortium and a stock culture of activated sludge was substantially enhanced with Fenton pretreatment.

Alachlor is one of the most heavily used chlorinated acetoanilide herbicides. Fenton process and its modifications were employed to remove alachlor in aqueous solutions in several scientific researches. Arnold *et al.* (1996) reported nearly complete conversion of 26.4 mg/L of alachlor by Fenton process (5 mM

Fe^{2+} and H_2O_2, pH 2.5, 25°C) in a pesticide rinse water containing other pesticides, including 29 mg/L of atrazine, 0.132 mg/L of cyanazine, 59 mg/L of metolachlor, and 30 mg/L of EPTC. Complete conversion by Fenton process of other pesticides in a pesticide rinse water was reported in the same study. Huston and Pignatello (1999) studied the destruction of pesticide active ingredients and commercial formulations in acidic aqueous solution by the catalytic photo-Fenton and Fe^{3+}/H_2O_2/UV advanced oxidation processes. They reported the complete conversion of 54 mg/L of alachlor (72 % removal as TOC). 38 mg/L aldicarb (62 % removal as TOC), 53 mg/L of carbofuran (>90 % removal as TOC) and 40 mg/L of atrazine (46 % removal as TOC) by photo Fenton treatment at pH 2.8 within 120 min (λ = 300–400 nm, 0.05 mM of Fe^{3+}, 10 mM of H_2O_2).

Oturan (2000) investigated the degradation of 2,4-D by an electro-Fenton process generating H_2O_2 and reducing Fe^{3+} with continuous bubbling of O_2 and electrolysis (60 mA). More than 75 % TOC reduction (220 mg/L of 2,4-D) was achieved by the electro-Fenton treatment after 6 hours. Endosulfan is another highly toxic and persistent organochlorine insecticide. There are α- and β-isomers of this compound that are sometimes referred to as endosulfan-I and endosulfan-II, respectively. More than 80 % TOC reduction (maximum TOC = 100 mg/L) in the pesticide solutions containing endosulfan α–β was obtained using photo-Fenton (λ = 300–450 nm, 0.25 mM of Fe^{2+}, 20 mM of H_2O_2, pH 2.8). Yu (2002) reported the treatment of concentrated solutions (1 g/L) of several organophosphate insecticides including acephate, edifenphos, chlorpyrifos, and methamidophos with Fenton process at pH 2.8. After the Fenton treatment, COD in the insecticide solutions was reduced by more than 95 %. The photodegradation of alachlor, atrazine, and diuron dissolved in water at the concentrations of 50, 25, and 30 mg/L, respectively by Fe^{2+}/H_2O_2 treatment driven by solar energy at pilot-plant scale using a compound parabolic collector photoreactor was investigated by Hincapié Pérez *et al.* (2006). They studied different concentrations of Fe^{2+} (2, 10, and 55 mg/L). In the case of 55 mg/L Fe^{2+} concentration, total mineralisation of alachlor was obtained in 75 min where 60 % and 90 % TOC removals were achieved in 150 and 75 min for atrazine and diuron, respectively.

Al Momani *et al.* (2007) explored the vydine (25 mg/L) oxidation at a pilot plant using direct solar UV-light, solar UV combined with H_2O_2 or Fe^{2+} and solar photo-Fenton. Solar irradiation alone degraded low amounts of pesticide in aqueous solution. The combination of H_2O_2 or Fe^{2+} with solar irradiation was more efficient. The degradation rate of pesticide was strongly accelerated by photo-Fenton process. The degradation rate was influenced by many factors such as initial hydrogen peroxide concentration and initial iron concentration. They concluded that photo-Fenton process could be successfully used for significant pesticide degradation. More that 88 % pesticide elimination was obtained by the photo-Fenton process in only 40 min solar irradiation with; the initial reactant concentrations 20 mg/L of H_2O_2 and 20 mg/L of Fe^{2+}. In these conditions, the BOD_5 was increased from zero for pure pesticide solution to 54 mg/L and acute toxicity was decreased from 19 to 6 toxicity unit. Badawy *et al.* (2006) studied the combinations of the Fenton reaction, H_2O_2/UV, and the photo-Fenton processes in the degradation of organophosphorus containing pesticides such as fenitrothion, diazinon, and profenofos. In their study, synthetically prepared samples containing 50 mg/L of the model compounds, studied were used to simulate high-loaded organo-phosphorus pesticides. The results obtained from their study showed that TOC degradation rate of model pesticides was strongly accelerated by photo-assisted chemical oxidation processes. The photoassisted Fenton process, the combination of a homogeneous system of $H_2O_2/UV/Fe^{2+}$, exhibited the highest photochemical elimination rate in the order of profenofos > fenitrothion > diazinon. The optimum operating conditions obtained for the best degradation were: COD/H_2O_2 = 4.4/1, H_2O_2/Fe^{2+} = 100/1, and pH 3 by the oxidation of model substances by the Fenton treatment; and pH 3, COD/H_2O_2 = 2.2/1, and H_2O_2/Fe^{2+} = 50/1 by the photo-Fenton treatment.

Oller *et al.* (2007) studied the TiO_2 photocatalysis and photo-Fenton treatment of five common pesticides, namely methomyl, dimethoate, oxamyl, cymoxanil, and pyrimethanil. 50 mg/L concentration of each pesticide was used in experimental study. 35 L solar pilot plant equipped with three compound parabolic collectors for TiO_2 photocatalysis and 75 L solar pilot plant using four compound parabolic collectors were employed for oxidation studies. 90 % TOC removal was observed in 1197 and 512 minutes in case of TiO_2 photocatalysis and photo-Fenton treatments, respectively. Shorter irradiation time with two different iron concentrations (20 and 55 mg/L) resulted in 50 and 72 % TOC reductions. Photo-Fenton process realised greater pesticide degradation (more than twice) than the TiO_2 photocatalysis. Pretreatment by photo-Fenton process decreased toxicity defined as percent inhibition from 90 to 47 %. Biodegradability tests showed 70 % biodegradability could be obtained after 12 days.

In two studies, 42 mg/L Diuron and 75 mg/L Linuron were chosen as model pesticides for the photo-Fenton treatment (José-Farré *et al.* 2007a, b). Different doses of H_2O_2 (97.1, 143, and 202 mg/L) and Fe^{+2} (9.25, 13.3, and 15.9 mg/L) were used for photo-Fenton process. TOC reduction was significantly enhanced by the increase of Fe^{+2} and H_2O_2 doses. Organic acids such as acetic acid, oxalic acid, and formic acid were produced, reached a maximum, and then degraded during photo-Fenton process, higher dose of H_2O_2 and Fe^{+2} resulted in greater production and degradation for the acids rate. Quen and Raj (2006) studied the H_2O_2/UV and O_3/UV processes for the treatment of tetrahydrofuran, 1,4-dioxane and pyridine. H_2O_2/UV process resulted in greater efficiency for increasing biodegradability and destruction than

O_3/UV process for treating tetrahydrofuran solution. For dioxane solution, H_2O_2/UV process degraded all the contaminants within 60 min but did not improve biodegradability. No biodegradability enhancement was observed for all three substrates as COD during O_3/UV and H_2O_2/UV processes of pyridine.

In the study of Lafi and Al-Qodah (2006), 100 mg/L of three pesticides (deltamethrin, lambda, and cyhalothrin), with an initial COD of 6500, 6300, 6500 mg/L COD respectively were selected for O_3 and O_3/UV degradation. Over 80 and 92 % degradation observed for all three substrates as COD under O_3 and O_3/UV, respectively. Higher pH had positive effect on the degradation and COD reduction.

Li *et al.* (2009) investigated the catalytic oxidation of triazophos pesticide from synthetically prepared wastewater and actual industrial triazophos pesticide wastewater using Fenton's reagent. The COD values of the synthetic and actual industrial triazophos pesticide wastewater samples were 3242 and 3418 mg/L, respectively, and the triazophos concentration in these wastewater samples was 0.06 % by weight. The effects of reaction conditions including the dosages of $FeSO_4{\cdot}7H_2O$ and H_2O_2, the pH value of the environment, and the stirring time on COD removal on the synthesised wastewater were evaluated. COD removal efficiency of 96.3 % with a corresponding effluent COD value of 120 mg/L was achieved under optimal reaction conditions of a pH value of 4, a dosage of 2.5 g/L of $FeSO_4{\cdot}7H_2O$ and 100 mL/L of 30 % H_2O_2 solution, and a stirring time of 90 minutes. When the actual industrial wastewater was treated at a pH value of 4, COD removal efficiency of 85.4 % with a corresponding effluent COD value of 499 mg/L was reached at optimal condition of a dosage of 5.0 g/L of $FeSO_4{\cdot}7H_2O$ and 75 mL/L of 30 % H_2O_2 solution, and a stirring time of 90 min.

The two possibilities (photo-Fenton/bio and bio/photo-Fenton) of the combined system were evaluated (using different analytical tools and bioassays, mainly DOC, COD, toxicity, and biodegradability) in a pilot plant for the removal of a mixture of five commercial pesticides (Vydate®, Metomur®, Couraze®, Ditimur-40®, and Scala®) in the study of Zapata *et al.* (2010). All photo-Fenton experiments were carried out at an initial DOC of 500 mg/L (100 mg/L of DOC coming from each commercial pesticide). They concluded that the photo-Fenton treatment at pilot plant was able to reduce toxicity (from 96 to 50 % of inhibition) and increased biodegradability (from 50 to 95 %) of the wastewater and the most suitable level of oxidation for combining it with the biological treatment was after the total elimination of the active ingredients. The efficiency of the combined photo-Fenton/bio system in terms of mineralisation was 94 %, of which 35.5 % corresponded to the AOP and 58.5 % to the aerobic biological treatment. The combination bio/photo-Fenton was not successful. The efficiency of the industrial-scale combined system (photo-Fenton/bio) was 84 % (35 % corresponding to the photo-Fenton treatment and 49 % to the biological stage). In another study conducted by Zapata and co-workers

(2009) a combined solar photo-Fenton and biological treatment was proposed for the decontamination of a mixture of the same five commercial pesticides as the above work used in the previous study. In this study, photo-Fenton experiments were conducted in a solar pilot reactor consisting of four compound parabolic collectors in which the pesticide mixture was treated at an original dissolved organic carbon (DOC) concentration of 200 mg/L in the presence of Fe^{2+} or Fe^{3+} concentration of 5, 20, and 55 mg/L. Ferrous ions were marginally more active than ferric in terms of active ingredient degradation. Photo-Fenton was also far more effective (by at least two orders of magnitude) than the respective dark reaction under identical experimental conditions. Irradiation for 50–100 minutes (normalized at 30 W/m^2 light intensity) at 20 mg/L Fe^{2+} was able to completely eliminate the active ingredients, and reduced DOC by about 15–50 % and COD by 40–70 %, respectively.

In the study of Torres *et al.* (2009) the application of experimental design for the ultrasonic degradation of alachlor was investigated and the effect of electrical power (20-80 W), pH (3-10) and substrate concentration (10–50 mg/L) on the process performance was evaluated. The optimal conditions for the removal of 50 mg/L alachlor were found as pH 3, 300 kHz, and 80W of electric power. At these conditions, alachlor was completely removed after 75 minutes of irradiation time. Biodegradability test (BOD_5/COD) carried out during the course of the treatment indicated that the ultrasonic system noticeably increased the biodegradability of the initial solution.

The homogeneous catalytic oxidation of atrazine and fenitrothion in single compound-containing solutions and in mixture, in pure water, was explored by Kassinos *et al.* (2009) in dark and light-driven oxidation systems using H_2O_2 as the oxidant and Fe^{2+} as the catalyst. Three different pesticide-containing solutions were tested as follows: 30 mg/L atrazine, 30 mg/L fenitrothion, and a mixture of atrazine and fenitrothion at 15 mg/L each. It was shown that the degradation during the dark Fenton reactions took place during the first 2 hours and further prolongation of the reaction time did not lead to enhanced removals. The same held true for the light-driven reactions. With the photo-Fenton reaction, fenitrothion was completely mineralised while the TOC contained in the atrazine solution was removed by 57 %. Both compounds in their parent forms were completely eliminated from the first 5 minutes of the oxidation when being in the single-containing solutions and in their binary mixture.

The costs for chemical oxidation processes are highly dependent on the quality of the source water to be treated and effluent treatment goals (Rodriguez *et al.* 2000). In pesticide industry, application of chemical oxidation processes as a pretreatment stage to enhance biodegradability and reduce toxicity, followed by biological post-treatment is recommended (US EPA 1985; 1993). The conceptual basis of this approach, which has gained a lot of attention over the past several years, is relatively straightforward and based on the facts that (i) biological treatment is perhaps less costly and more environmentally friendly than any other destructive treatment, and (ii) complete mineralisation by chemical oxidation processes induces excessive treatment costs, since the highly oxidised end-products that are formed during chemical oxidation tend to be refractory to total oxidation by chemical means (Comninellis *et al.* 2008). These end-products, which are typically represented by short carboxylic acids, can, however, be easily degraded biologically (Khare *et al.* 2007). Optimal treatment schemes will eventually be decided upon achieving discharge limits set by national or international environmental regulations at a reasonable cost. This, in turn, will be dictated by the type of waste stream to be treated, *i.e.* its composition, concentration, biodegradability, and toxicity (Comninellis *et al.* 2008). An understanding of actual costs will also require pilot testing to determine site specific costs.

Chemical oxidation processes require different costs associated with the investment in the equipment needed, operating and maintenance costs (Beltrán 2003). The economy of these chemical oxidation processes are dependent also on the composition of the wastewater to be treated and the experimental conditions applied (water flow rate, oxidant, catalyst, and UV doses, *etc.*). The degradation rate of pollutants is also a key point for deciding if chemical oxidation processes can be applied at industrial scale with a reasonable cost (Gogate and Pandit 2004). These limitations underline the need to establish a step-by-step research methodology which takes these effects into account. Detailed studies are therefore necessary to find out how operating conditions affect the original properties of the pretreatment stream (contact time, oxidant and/or catalyst type, dose, and toxicity, temperature, *etc.*).

In the case of pesticide industry, manufactured pesticides are complicated organic molecules and their production requires the operation of sophisticated chemical processes (US EPA 1993). Due to the individuality of each pesticide manufacturing process and the impact of these differences on the chemical content and other characteristics of the wastewater generated. Although, it is not possible setting one generic wastewater discharge volume and characteristics for all pesticide manufacturing processes. Because of the differences in wastewater generation, recirculation, and recycle capabilities between different pesticides manufacturing processes, no one flow rate is considered applicable to all pesticides. For these reasons, no economic figures are given here. However, there are references reporting data on economic aspects of chemical oxidation processes that are applicable to pesticide production wastewater in literature. Some of these studies are indicated as follows.

For ozonation systems, Langlais *et al.* (1991) present an extensive study on the economics of different ozonation processes. For UV processes, Bolton *et al.* (1996) have established some figures-of-merit for the determination of electrical costs. They defined the concepts of electrical energy per order (EE/O) and electrical energy per mass (EE/M) to determine the energy requirements for reducing contaminants present at high and low concentrations, respectively. Applications of these figures have aided in the determination of capital costs for some units (AOT 1996; Bolton *et al.* 1996; US EPA 1998) and have also been used in laboratory studies (Arslan *et al.* 2000; Cater *et al.* 2000).

7.3 CONCLUDING REMARKS

The pesticide industry is an important part of the world economy. As aforementioned, the manufactured pesticides are the complicated organic molecules and their production requires the operation of sophisticated chemical processes. The pesticide manufacturing industry generates process wastewaters containing a variety of pollutants due to the individuality of each pesticide manufacturing process. It is not possible setting one generic wastewater discharge characteristics for all pesticide manufacturing processes because of the differences in wastewater generation. Chemical oxidation processes such as ozonation, ozone-based AOPs, H_2O_2/UV, Fenton, and Fenton-based AOPs *etc.* are used as a pretreatment stage to enhance biodegradability and reduce toxicity, prior to biological treatment in pesticide industry and these technologies have been found to be effective.

There are numerous studies conducted on pesticides removal by chemical oxidation technologies in the literature. However, most of the studies conducted on pesticide removal represent contaminated surface water or ground water rather than pesticide industry wastewater. Concentrations of pesticides in surface water or ground water and those in wastewater are substantially different. The treatment(s) performance data obtained under the condition simulating the treatment of pesticide-containing surface water or ground water may not be applicable directly to wastewater treatment. Another important aspect in the chemical oxidation applications to pesticide industry wastewater is to determine to extent of oxidation in a way that while the toxicity is minimised to remain at a adequate level of organic matter to support biological treatment. Beyond the technical feasibility, this point is also important in terms of economical point of view.

REFERENCES

Al Momani, F.A., Shawaqfeh, A.T. and Shawaqfeh, M.S. (2007) Solar wastewater treatment plant for aqueous solution of pesticide. *Sol. Energy* **81**, 1213–1218.

AOT Handbook (1996) Markham, Ontario: Calgon Carbon Corporation Technologies.

Arnold, S.M., Hickey, W.J., Harris, R.F. and Talaat, R.E. (1996) Integrating chemical and biological remediation of atrazine and s-triazine containing pesticide wastes. *Environ. Toxicol. Chem.* **15**, 1255–1262.

Arslan, I., Balcioglu, I.A. and Tuchkanen, T. (2000) Treatability of simulated reactive dyebath wastewater by photochemical and non-photochemical advanced oxidation processes. *J. Environ. Sci. Health A* **35**, 775–793.

Badawy, M.I., Ghaly, M.Y. and Gad-Allah, T.A. (2006). Advanced oxidation processes for the removal of organophosphorus pesticides from wastewater. *Desalination* **194**, 166–175.

Beltrán, F.J. (2003) Ozone-UV radiation-hydrogen peroxide oxidation technologies In *Chemical Degradation Methods for Wastes and Pollutants Environmental and Industrial Applications*, (ed. M.A. Tarr), pp. 1–71, Marcel Dekker Inc., (ISBN 0-8247-4307-5).

Benítez, F.J., Acero, J.L. and Real, F. (2002) Degradation of carbofuran by using ozone, UV radiation and advanced oxidation processes. *J. Hazard. Mater.* **89**, 51–65.

Bolton, J.R., Bircher, K.G., Tumas, W. and Toldman, C.A. (1996) Figures-of-merit for the technical development and application of advanced oxidation processes. *J. Adv. Oxid. Technol.* **1**, 13–17.

Botz, M. (1999) *Overview of cyanide treatment methods*, The Gold Institute, Washington D.C., U.S.A.

Brillas, E., Calpe, J.C. and Cabot, P.L. (2003) Degradation of the herbicide 2,4-dichlorophenoxyacetic acid by ozonation catalyzed with Fe^{2+} and UVA light. *Appl. Catal. B-Environ.* **46**, 381–391.

Burrows, H.D., Canle L, M., Santaballa, J.A. and Steenken S. (2002) Reaction pathways and mechanisms of photodegradation of pesticides. *J. Photoch. Photobio. B* **67**, 71–108.

Cater, S.R., Stefan, M.I., Bolton, J.R. and Safarzadeh-Amiri, A. (2000) UV/H_2O_2 treatment of methyl tert-butyl ether in contaminated waters. *Environ. Sci. Technol.* **34**, 659–662.

Cheng, H., Xu, W., Liu, J., Wang, H., He, Y. and Chen, G. (2007) Pretreatment of wastewater from triazine manufacturing by coagulation, electrolysis, and internal microelectrolysis. *J. Hazard. Mater.* **146**, 385–392.

Chiron, S., Fernandez-Alba, A., Rodriguez, A. and Garcia-Calvo, E. (2000) Pesticide chemical oxidation: State-of- the- art. *Water Res.* **34**, 366–377.

Comninellis, C., Kapalka, A., Malato, S., Parsons, S.A., Poulios, I. and Mantzavinos, D. (2008) Advanced oxidation processes for water treatment: advances and trends for R&D. *J. Chem. Technol. Biotechnol.* **83**, 769–776.

EC (2004) Regulation No 850/2004 of the European Parliament and of the Council of 29 April 2004 on persistent organic pollutants and amending, Directive 79/117/EEC.

EC (2006) Directive 2006/11/EC of the European Parliament and of the Council of 15 February 2006 on pollution caused by certain dangerous substances discharged into the aquatic environment of the Community.

Felsot, A.S. (1996) Options for cleanup and disposal of pesticide wastes generated on a small- scale. *J. Environ. Sci. Heal. B* **31**, 365–381.

Gogate, P.R. and Pandit, A.B. (2004) A review of comparative technologies for wastewater treatment. I: Oxidation technologies at ambient conditions. *Adv. Environ. Res.* **8**, 501–551.

Ghosh, R.S., Thheis, T.L., Smith, J.R. and Whong-Chong, G.M. (2006) Ambient temperature oxidation technologies for treatment of cyanide. In *Cyanide in Water and Soil* (eds R.S. Ghosh, D.A. Dzombak and G.M. Whong-Chong), pp 517–570, Taylor&Francis Group, LLC, CRC Press, London, United Kingdom, (ISBN 978-1-56670-66-7).

Hincapié Pérez, M., Peñuela, G., Maldonado, M.I., Malato, O., Fernández-Ibáñez, P., Oller, I., Gernjak, W. and Malato, S. (2006) Degradation of pesticides in water using solar advanced oxidation processes, *Appl. Catal. B-Environ.* **64**, 272–281.

Huston, P.L. and Pignatello, J.J. (1999) Degradation of selected pesticide active ingredients and commercial formulations in water by the photo-assisted Fenton reaction. *Water Res.* **33**, 1238–1246.

IFC (International Finance Corporation) (2007) *Environmental, Health, and Safety Guidelines Pesticide Manufacturing, Formulation and Packaging*. World Bank Group.

Ikehata, K. and Gamal El-Din, M. (2005a) Aqueous pesticide degradation by ozonation and ozone-based advanced oxidation processes: A review (Part II). *Ozone-Sci. Eng.* **27**, 173–202.

Ikehata, K. and Gamal El-Din, M. (2005b) Aqueus pesticide degradation by ozonation and ozone-based advanced oxidation processes: A review (Part I). *Ozone-Sci. Eng.* **27**, 83–114.

Ikehata, K. and Gamal El-Din, M. (2006) Aqueous pesticide degradation by hydrogen peroxide/ultraviolet irradiation and Fenton-type advanced oxidation processes: a review. *J. Environ. Eng. Sci.* **5**, 81–135.

José-Farré, M., García-Montaño, J., Ruiz, N., Muñoz, I., Domènech, X. and Peral, J. (2007a) Life cycle assessment of the removal of Diuron and Linuron herbicides from water using three environmentally friendly technologies. *Environ. Technol.* **28**, 819–830.

José-Farré, M., Brosillon, S., Domènech, X. and Peral, J. (2007b) Evaluation of the intermediates generated during the degradation of Diuron and Linuron herbicides by the photo-Fenton reaction. *J. Photoch. Photobio. A.* **189**, 364–373.

Kassinos, D., Varnava, N., Michael, C. and Piera, P. (2009) Homogeneous oxidation of aqueous solutions of atrazine and fenitrothion through dark and photo-Fenton reactions. *Chemosphere* **74**, 866–872.

Kearney, P.C., Muldoon, M.T. and Somich, C.J. (1987) UV-Ozonation of 11 major pesticides as a waste-disposal pretreatment. *Chemosphere* **16**, 2321–2330.

Khare, U.K., Bose, P. and Vankar, P.S. (2007) Impact of ozonation on subsequent treatment of azo dye solutions. *J. Chem. Technol. Biotechnol.* **82**, 1012–1022.

Kolpin, D.W., Barbash, J.E. and Gilliom R.J. (1998) Occurrence of pesticides in shallow groundwater of the United States: Initial results from the National Water Quality Assessment Program. *Environ. Sci. Technol.* **32**, 558–566.

Kuo, W. S. (2002) Photocatalytic oxidation of pesticide rinsate. *J. Environ. Sci. Heal. B.* **37**, 65–74.

Lafi, W.K. and Al-Qodah, Z. (2006) Combined advanced oxidation and biological treatment processes for the removal of pesticides from aqueous solutions. *J. Hazard. Mater.* **137**, 489–497.

Lai, M.S., Jensen, J.N. and Weber, A.S. (1995a) Oxidation of simazine - ozone, ultraviolet, and combined ozone ultraviolet oxidation. *Water Environ. Res.* **67**, 340–346.

Lai, M.S., Weber, A.S. and Jensen, J.N. (1995b) Oxidation of simazine - biological oxidation of simazine and its chemical oxidation byproducts. *Water Environ. Res.* **67**, 347–354.

Langlais, B., Reckhow, D.A. and Brink, D.R. (1991) *Ozone in Water Treatment: Application and Engineering*. Lewis Publ., Chelsea, MI.

Lee, S.H. and Carberry, J.B. (1992) Biodegradation of PCP enhanced by chemical oxidation pretreatment. *Water Environ. Res.* **64**, 682–690.

Li, R., Yang, C., Chen, H., Zeng, G., Yu, G. and Guo, J. (2009) Removal of triazophos pesticide from wastewater with Fenton reagent. *J. Hazard. Mater.* **167**, 1028–1032.

Massey, J.H. and Lavy, T.L. (1997) Degradation and disposal of scepter herbicide by hydrogen peroxide-catalyzed ozonation. *J. Agric. Food Chem.* **45**, 3315–3319.

McKenzie, W. (2001) Agrochemical service: the world market in 2000. *In Annual Review of the Crop Protection Association*, Crop Protection Association, Peterborough.

Meister, R.T. (1999) *Farm Chemicals Handbook '99*, Meister Publishing Company, Willoughby, Ohio, U.S.A.

Oller, I., Malato, S., Sánchez-Pérez, J.A., Maldonado, M.I. and Gassó, R. (2007) Detoxification of wastewater containing five common pesticides by solar AOPs-biological coupled system. *Catal. Today* **129**, 69–78.

Oppenländer, T. (2003) *Photochemical Purification of Water and Air*, Wiley-VCH Verlag, Weinheim, Germany.

Oturan, M.A. (2000) An ecologically effective water treatment technique using electrochemically generated hydroxyl radicals for in situ destruction of organic pollutants: application to herbicide 2,4-D. *J. Appl. Electrochem.* **30**, 475–482.

Quen, H.L. and Raj, C.B. (2006) Evaluation of UV/O_3 and UV/H_2O_2 processes for nonbiodegradable compounds: Implications for integration with biological processes for effluent treatment. *Chem. Eng. Commun.* **193**, 1263–1276.

Rodriguez, R.A., Browne, T., Roberts, R., Brown, A. and Stocking, A.J. (2000) Advanced oxidation processes. In *Treatment Technologies for Removal of Methyl Tertiary Butyl Ether (MTBE) from Drinking Water: Air Stripping, Advanced Oxidation Processes, Granular Activated Carbon, Synthetic Resin Sorbents*, 2nd edn (ed. G. Melin), pp 109-199, Center for Groundwater Restoration and Protection National Water Research Institute (NWRI-99-06).

SC (2001) Stockholm Convention on Persistent Organic Pollutants (POPs). http://www.pops.int/.

Scott, J.P. and Ollis, D.F. (1995) Integration of chemical and biological oxidation processes for water treatment: review and recommendations. *Environ. Prog.* **14**, 88–103.

Somich, C.J., Muldoon, M.T. and Kearney, P.C. (1990) On-site treatment of pesticide waste and rinsate using ozone and biologically-active soil. *Environ. Sci. Technol.* **24**, 745–749.

Teixeira, L.A.C., Andia, J.P.M. and Yokoyama, L. (2005) Oxidation of Cyanide solutions using Caro's acid. In *Proc. of 2nd Mercosur Congress on Chemical Engineering and 4th Mercosur Congress on Process Systems Engineering*, Costa verde, RJ, Brasil.

Torres, R.A., Mosteo, R., Pétrier, C. and Pulgarin, C. (2009) Experimental design approach to the optimization of ultrasonic degradation of alachlor and enhancement of treated water biodegradability. *Ultrason. Sonochem.* **16**, 425–430.

US EPA, (1976) *Development Document for Interim Final Effluent Limitations Guidelines for the Pesticide Chemicals Manufacturing Point Source Category*, EPA 440/

1-75/060d, Effluent Guidelines Division, Office of Water and Hazardous Materials, U.S. Environmental Protection Agency, Washington, D.C., United States.

US EPA, (1977) *Industrial Process Profiles for Environmental Use:Chapter 8. Pesticide Industry*, EPA 600/2-77/023h, Office of Research and Development, U.S. Environmental Protection Agency, Washington, D.C., United States.

US EPA, (1985) *Development Document for Best Available Technology, Pretreatment Technology and New Source Performance Technology in the Pesticide Chemicals Industry*, EPA 440/1-85/079, Industrial Technology Division, Office of Water Regulations and Standards, U.S. Environmental Protection Agency, Washington, D.C., United States.

US EPA, (1993) *Development Document for Effluent Limitations Guidelines, Pretreatment Standards and New Source Performance Standards for the Pesticide Chemicals Manufacturing Category*, EPA 821/R-93/016, Engineering and Analysis Division, Office of Science and Technology, U.S. Environmental Protection Agency, Washington, D.C., United States.

US EPA, (1996) *Technical Development Document for the Pesticide Formulating, Packaging, and Repackaging Effluent Limitation Guidelines, Pretreatment Standards and New Source Performance Standards*, EPA 821/R-96/019, Engineering and Analysis Division, Office of Science and Technology, U.S. Environmental Protection Agency, Washington, D.C., United States.

US EPA, (1998) *Handbook on Advanced Photochemical Oxidation Processes*, EPA 625/R-98/004, Centre for Environmental Research information National Risk Management Research Laboratory, Office of Research and Development, U.S. Environmental Protection Agency, Cincinnati, Ohio, United States.

Wong, J. M. (2006) Treatment of pesticide industry wastes. In *Waste Treatment in the Process Industries* (eds. L.K. Wang, Y.T. Hung, H.H. Lo and C. Yapijakis), pp 499–543, CRC Press, Taylor & Francis Group, (ISBN 0-8493-7233-9).

World Bank Group (1998) Pesticide formulation and pesticide manufacturing. In *Pollution Prevention and Abatement Handbook 1998: Toward Cleaner Production*, pp 363–370, World Bank Publications, (ISBN 0-8213-3638-0).

Wu, J., Lan, C. and Chan, G.Y. (2009) Organophosphorus pesticide ozonation and formation of oxone intermediates. *Chemosphere* **76**, 1308–1314.

Yu, J.J. (2002) Removal of organophosphate pesticides from wastewater by supercritical carbon dioxide extraction. *Water Res.* **36**, 1095–1101.

Zapata, A., Velegraki, T. Sánchez-Pérez, J.A., Mantzavinos, D. Maldonado, M.I. and Malato, S. (2009) Solar photo-Fenton treatment of pesticides in water: Effect of iron concentration on degradation and assessment of ecotoxicity and biodegradability. *Appl. Catal. B-Environ.* **88**, 448–454.

Zapata, A., Malato, S., Sánchez-Pérez, J.A., Oller, I. and Maldonado, M.I. (2010) Scale-up for a combined solar photo-Fenton/biological system for remediation of pesticide-contaminated water. *Catal. Today*, doi:10.1016/j.cattod.2010.01.34.

Chapter 8

Gold and silver mining

8.1 GOLD AND SILVER MINING

8.1.1 Overview

Gold and silver are noble metals that are resistant to corrosion, degradation, and oxidation in moist air. Owing to their rarity in the Earth's crust, they are classified as precious metals.

Gold and silver have been used for thousands of years as commodities and the basis for many monetary systems. Among all commodities, gold has retained a unique status as a long-term store of value. Gold also serves as an important investment vehicle and reserve asset. Therefore, most of gold and silver produced has been already used in the production of jewellery. During last few decades, gold and silver have also started to be used as essential industrial metals due to their superior electrical conductivity and resistance to corrosion and other desirable combinations of physical and chemical properties (USGS 2010). Gold performs critical functions in computers, communication equipments, spacecrafts, and jet aircraft engines. Worldwide, about 90 percent of the gold supplied to the market each year goes into manufactured products, and the remainder goes to private investors and to monetary reserves (USGS 2010). Some of the traditional

 Chemical Oxidation Applications for Industrial Wastewaters. By Olcay Tünay, Işık Kabdaşlı, Idil Arslan-Alaton and Tuğba Ölmez-Hancı. ISBN: 9781843393078. Published by IWA Publishing, London, UK.

applications of silver can be listed as silverware and table settings, photography, and coinage. Silver is also used in many industrial applications such as in mirrors, eyeglasses, electrical and electronic products, and silver-coated steel ball bearings of jet engines production. Additionally, recent researches in medicine demonstrated that silver promotes the regeneration of lost cells, and increases the rate of healing in wounds and bone (Silver 2010). As a result of the medical findings, there is a growing trend in silver-based bandage production and usage.

Recently, there is a great effort on substitutions of gold and silver with more economical alternatives in industrial applications in order to preserve their natural sources. Therefore, base metals clad with gold alloys have started to be used in electrical and electronic products, and in jewellery. In addition to this, many products have redesigned to maintain high-utility standards with lower gold contents (USGS 2010). Digital imaging, films with reduced silver content, silverless black-and-white films, and xerography have currently substituted for silver in photography. In some applications, silver batteries have also replaced with non-silver batteries (USGS 2010).

Worldwide, about 2350 tons of gold was processed by hundreds of gold producing mine in 2009. China was the world's leading gold producer with 300 tons in 2009. South Africa, United States and Australia were other important countries contributing to global gold production in 2009 (USGS 2010). World silver production in 2009 has been reported as about 21,400 tons. Peru, China and Mexico were the world's leader silver producers with 44 % of world's production in 2009 (USGS 2010).

Gold and silver ore mining are categorised as SIC 104 according to the Standard Industrial Classification. This category encompasses facilities engaged primarily in exploring for metallic minerals, developing mines, and ore mining.

8.1.2 Production processes

Gold and silver can be recovered from ores, refinery slimes, or precious metals refining processes. Several processes can be used to recover gold and silver from their ores. Silver is recovered using underground and open pit methods whereas gold is generally mined from surface lode mines (US EPA 2010). The processes employed in the recovery of gold and silver from their ores cover gravity separation, amalgamation, froth flotation, and cyanidation. These processes can be used in combinations in the facilities. The desired materials are separated from host rock by gravity separation based on density differences. Fine gold and silver in placer deposits which cannot be separated by density alone can be recovered by amalgamation based on the dissolution of gold and silver in mercury. The extracted ores must be milled for further recovery processes. After milling, sulphide ores are oxidised by chlorination, bio-oxidation, roasting, or autoclaving. Agglomeration process is performed to increase particle size of the ores crushed into small size. In this step, the crushed ores are mixed with Portland cement and/or lime, wet with a leach solution prior to leaching, and tumble for adhering fine particles to larger one. The extraction of gold and silver from their ores is accomplished using a lixiviant (leaching agent). Cyanide has been widely used in beneficiation of gold and silver mining all over the world since 1890 owing to its excellent performance with a great variety of ores, robustness and relatively low cost (US EPA 1994a; 1995; Butterman and Amey III 2005; Hilson and Monhemius 2006). On the other hand, in the last decade, the use of cyanide as a lixiviant has been a much discussed issue because of Bai Mare accident (EC 2009). Therefore, there has been special attention to the substitution of cyanidation with less hazardous techniques in gold and silver mining facilities. On the other hand, the researches undertaken to develop alternative techniques have demonstrated that a limited number of plausible lixiviants is available (Hilson and Monhemius 2006). The most promising alternative lixiviants listed in the related literature are thiourea (Groenwald 1976; 1977; Kai *et al.* 1997), thiosulphate (White 1905; Aylmore 2001; Molleman and Dreisinger 2002; Feng and van Deventer 2010), thiocyanate (White 1905; Barbosa-Filho and Monhemius 1994a; 1994b; Kholmogorov *et al.* 2002), halides (la Brooy *et al.* 1994), and coal-oil agglomeration (Calvez *et al.*1998; Kotze and Petersen 2000). In spite of the proven effectiveness of these alternative lixiviants, their full-scale operations are still limited, since their commercial operations have been hindered by several constrains such as their high detoxification costs, limited recyclability, low selectivity, and inapplicability to fine gold particle (Hilson and Monhemius 2006; EC 2009). Because of these reasons, cyanide has already been the chief lixiviant in use. In the cyanidation process, gold and silver are dissolved by the leach solution containing potassium or sodium cyanide via different contact methods at high pH in the presence of oxygen. Ores are contacted with cyanide by means of (i) heap leaching; (ii) vat leaching; or (iii) agitation leaching. Tank leaching is used for high-grade ore, while heap leaching is generally employed to recover gold from low-grade ore. After dissolving step, gold and silver are recovered from leach solution by application of the Merrill-Crowe process, activated carbon loading, or activated carbon stripping. The Merrill-Crowe process involves several sequential operations: filtration and deaeration of the leach solution, precipitation of precious metals with zinc powder, filtration of the precipitant, melting of the recovered precious metals, and casting into bars. In activated carbon loading process, the leach solution is brought into contact with activated carbon by carbon-in-column, carbon-in-pulp (CIP), or carbon-in-leach process. Gold and silver are then separated by acid leaching or electrolysis process. Activated carbon stripping is accomplished in alkaline cyanide solution at elevated temperature (40°C to 150°C) in the presence of alcohol. After desorption of the precious metals into the stripping solution, the precious metals are recovered by means of electrowinning (US EPA 1995; 2010). Carbon adsorption and electrowinning is the most common gold recovery method, the Merrill-Crowe process is the most widely used technique for gold ore containing large amounts of silver (US EPA 1995).

Gold and silver can also be recovered from refinery slimes as by-products of base metal refineries. Electrolytic cell slimes are a major source of precious metals originating from the copper industry. These cell slimes are processed to recover the precious metals. In lead refineries, precious metals are also recovered during the desilverising process which is a regular part of the refining operation.

Parting is the most basic refining process used in the separation of silver and gold. This operation can be performed by electrolytically or by acid leaching to separate silver from gold.

8.1.3 Waste sources

The extraction and beneficiation of precious metals produce significant amounts of wastes with different character. Total waste generation can change within a wide range depending on type of ore mined and processed, metal recovery methods, and waste minimisation practices employed in the facility.

Waste rock and tailings generated from the extraction step may be used as construction or foundation materials until permanent closure of mine or may be disposed of as wastes. Mine water is a waste stream being collected from underground mine workings or open pits. It may be discharged to surface water or may be recirculated for on-site uses. A slurry waste which is composed of gangue and the mercury bearing solution, is produced by amalgamation process. This slurry may be disposed of in a tailings pond or mercury may be recovered from this slurry. Sulphur dioxide is produced from ore preparation step of sulphide ores. This compound may be converted to sulphuric acid for on-site use or may be directly sent to an acid plant. Spent ore from tank leaching occurs in the form of slurry. This slurry is a mixture of gangue and process water bearing cyanide and metal-cyanide complexes. Cyanide bearing tailings may be treated prior to disposal. The slurry is disposed of in tailings impoundment together with some of the liquid component which is recirculated to the leach tank as make-up water (US EPA 1994a; 2010). At the end of heap or valley fill leaching operation, spent ore bearing residual cyanide is generated. After cyanide detoxification, this spent ore is disposed of at waste rock or spent ore disposal site. Recycling of barren cyanide solution used in the leaching operation is a common application employed in the facilities. However, the leach solution is partly discarded owing to interference of the build-up of metal impurities to beneficiation process. This spent stream, generally called as bleed solution, is discharged to tailings impoundments or land-applied after cyanide treatment. Washing water originating from cleaning operation of filter used in the Merrill-Crowe process is another waste stream discharged to tailings impoundment or piles. Carbon-in pulp and carbon-in leach processes also produce tailings in the form of slurry consisting of gangue and process water bearing cyanide and metal-cyanide complexes. Similar to that of tank leaching operation, this slurry is recirculated from tailings impoundment to the leach tank (US EPA 1994a; 2010). When activated carbon used in the recovery step is regenerated, the acid wash residue containing metal, cyanide and acid is neutralised prior to discharge to the tailings impoundment.

8.1.4 Waste characterisation

Typical waste streams generating from a mining facility can be listed as mine water, tailings impoundment water and leachate, spent barren cyanide solution (bleed stream), open pit mine run-off, run-off from ore stockpiles and waste rock stacks, leachate from disposal area and domestic wastewater (Smith and Mudder 1991). These waste streams exhibit different characteristics depending on geochemistry of the ore processed, the unit operations used for beneficiation, type of lixiviant used in the leaching processes, their reuse applications, and waste management policy of the facility. The composition of the ore processed at mines is one of important factors affecting waste stream quality as well as potential production of acid mine drainage. Gold ores may be composed of mercury, arsenic, bismuth, antimony, and thallium in the forms of oxides, carbonates, or sulphides with varying degree of solubility. Sulphide bearing minerals, such as pyrite and pyrrhotite, are unique sources of acid mine drainage. These minerals can form sulphuric acid via oxidation depending on acid neutralisation capacity of the rock, the amount and frequency of precipitation, and the design of disposal unit (US EPA 1994a). Typical pollutants present in acid mine drainage can be enumerated as metals, total suspended solids (TSS), weakly complexed metal cyanides, thiocyanate and sulphate in the form of sulphuric acid. Tailings impoundment leachates exhibit different characteristics depending on the ore, cyanide concentration used in the leaching operation and the source of water consumed such as fresh or recycled water in the process. These streams generally contain metals and weakly complexed metal cyanides. The unit operations employed at the mining facilities have also an important role on the characteristics of process waste stream. For instance, the application of the Merrill-Crowe process for gold recovery causes 1–2 orders of magnitude increase in metal concentration in waste stream (Smith and Mudder 1991). The application of activated carbon-in-pulp process results in tailings slurry containing high levels of cyanide, complexed metal cyanides, cyanate and thiocyanate together with metals, because this process does not involve recycle of a barren cyanide solution unlike the Merrill Crowe process (BCTRCR 1995). The recovery of silver from a complex ore matrix increases the cyanide requirement in the process. The increase in cyanide dose poses undesirable metal dissolution problem and concentration of cadmium, mercury, nickel or copper may reach high levels under these circumstances (Smith and Mudder 1991). The application of cyanidation instead of other leaching techniques brings about the generation of cyanide bearing wastes. Among these wastes, bleed solutions and spent barren solutions occurring temporarily or permanently closure of mine are important waste sources containing free cyanide and metal cyanide complexes of copper, iron, nickel, and zinc, as well as other impurities such as arsenic and antimony (US EPA 1994a). Open pit mine run-off, the run-off from ore stockpiles, and the active and uncovered waste rock stacks create turbid flows requiring only TSS removal. TSS concentrations in these run-offs vary with the intensity and frequency of precipitation impacting the catchments (Smith and Mudder 1991).

Depending on above mentioned factors, total cyanide concentrations vary from a few hundred mg/L to 1000 mg/L and thiocyanate concentrations may reach up a few thousand mg/L in waste streams originating from gold mining operations. The highest concentrations of copper, zinc, and iron are reported as 400 mg/L, 100 mg/L, and 40 mg/L, respectively, for these waste streams (Smith and Mudder 1991). Arsenic, manganese, selenium and antimony concentrations in the gold mining wastes are generally lower than 10 mg/L. Cadmium, chromium and lead concentrations do not exceed 0.1 mg/L. In the case of sulphide bearing ore processing, sulphate concentrations may reach up to 20,000 mg/L in acid mine drainage flows yielding extremely acidic conditions.

8.1.5 In-process control technologies

Pollution prevention techniques implemented in mining facilities target reduction of material inputs, reuse of by-products, improvement of management practice, employee awareness and education, and substitutions for toxic chemicals within the framework of improvement in process efficiency as well as minimisation of environmental impacts.

Control and mitigation of mining wastes involve several prevention implementations such as mining water control, proper design of waste rock disposal area and tailings impoundment, wetland treatment, and acid zone isolation (US EPA 1994a). Diversion systems, containment ponds, groundwater pumping systems, subsurface drainage systems, and subsurface barriers are the techniques employed for the control of the water flows as well as minimisation of acid drainage at mine sites. Waste rock disposal areas and tailings impoundments must be properly designed and sited so as to minimise uncontrolled run-offs and to control the velocity of water flowing into containment ponds (US EPA 1994a). Constructed wetland systems relying on bacterial sulphate reduction can be used for the treatment of acid mine drainage. This treatment technique is best suited to handle low rated flows. Contaminated ground or surface water generated through acid drainage can also be treated in containment ponds, tailings impoundment, or a separate holding basin by neutralisation using lime. Acid zone isolation may be accomplished by leaving rock barrier between mining operations and the potential acid-producing zone (US EPA 1994a). Segregation and removal of pyrite prior to discharging to tailings impoundments also reduce the potential for acid generation and minimise waste treatment and remediation costs.

Use of mine drainage as makeup at mine sites is a common application to reduce wastewater volumes and to minimise treatment requirements. In some cases, the entire process water requirement of the facility is supplied from mine drainage (US EPA 1982). Recycle of process water is another common practice employed for the reduction of wastewater discharges. Complete recycle of process water resulting in zero discharge can also be realised in mine facilities, especially, located in arid regions.

As mentioned earlier there has been a great effort in the substitution of cyanidation with less hazardous techniques in gold and silver mining facilities because of its serious impacts and risks on the environment. In addition to this effort, some countries such as Czech Republic have totally banned gold extraction by heap leach cyanidation (Müezzinoğlu 2003). Cyanide use for any new precious metal mine has been also banned since 1998 in Montana, U.S.A. The European Committee has been seriously studying on the state of cyanidation plants and cyanide containing ponds in Europe especially after Aurul accident happened in Bai Mare, Romania in 2000 (Müezzionoğlu 2003). In spite of these disincentive regulations and legislations, cyanidation seems to be still a prominent technique all over the world because the alternative lixiviants tested for leaching process pose several problems, in practice, such as the elevated chemical requirements (from 2 to 200 folds on the weight basis) except ammonia with the lowest gold recovery efficiency (38%), significantly lower gold beneficiation efficiency, low recyclability, and even more toxicity with higher detoxification costs (Whong-Chong *et al.* 2006; EC 2009).

Due to above mentioned factors, the recovery and reuse of cyanide in leaching process as well as cyanide treatment have been driven by both economics and environmental concerns. Cyanide recovery methods can be listed as acidification-volatilisation-neutralisation (AVR) known as the Mille Crows process, electrolysis, and ion exchange. Among these methods, AVR has been successfully applied at several mining sites. By applying this recovery method, weakly complexed metal cyanide concentrations of the barren or tailings impoundment water can be reduced to less than 10 mg/L and heavy metals are also removed (Smith and Mudder 1991).

8.1.6 Conventional end-of-pipe wastewater treatment

Several proven technologies are available for the treatment of cyanide originating from mining facilities. These techniques encompass natural degradation systems, adsorption with granular activated carbon, and chemical methods based on oxidation processes such as alkaline chlorination, copper catalyzed hydrogen peroxide, and the INCO sulphur dioxide/air process.

Impoundment ponds utilising natural degradation processes are the most widely practiced end-of-pipe treatment option in order to remove cyanide and heavy metals from slurries, bleed solutions and tailings pond overflows originating from gold mining operations. These degradation processes rely on natural physical/chemical means through prolonged exposure to natural elements. Natural degradation processes rely on several cyanide removal mechanisms: (i) neutralisation by CO_2 adsorption; (ii) volatilisation of hydrogen cyanide gas; (iii) dissociation of metal complexes with release of cyanide ion; (iv) photolysis of iron cyanide; and (v) precipitation of metal cyanide (BCTRCR 1995; EC 2009). Impoundment pond treatment is a climate-sensitive method and is required arid and sunny climate conditions. Therefore, it is the only cyanide treatment method used in dry and sunny climate *e.g.* in South Africa (EC 2009). In some cases, impoundment ponds are used as a pre-treatment step instead of end-of-pipe treatment option. Although, impoundment ponds are proved to be successful in the treatment of cyanide bearing mining wastes, the catastrophic failures of some impoundments at active or closed gold mining sites have confirmed their risks on the environment.

Activated carbon adsorption is only suited to handling tailings pond overflow with very low cyanide concentrations. As this method is still under development, its application is limited in mining sites. Chemical treatment methods based on oxidation processes have high potential in the treatment of cyanide bearing mining wastes. Since these oxidation methods are covered in the scope of this book, they are introduced and discussed in more detail in the following subsections.

8.2 CYANIDE OXIDATION PROCESSES

8.2.1 Alkaline chlorination

Alkaline chlorination is the oldest cyanide oxidation method being used on a commercial scale all over the world. Free cyanide and weakly complexed metal cyanides are successfully oxidised by alkaline chlorination process under proper operation conditions described in Subsection 4.2.1.1. On the other hand, this method is incapable of oxidising strongly complexed metal cyanides such as iron and cobalt cyanides. Nevertheless, its failure in the treatment of iron cyanide is not main obstacle as the mining wastes usually contain low ferro-cyanide concentrations and ferro-cyanide has also relatively low toxicity (BCTRCR 1995). The main drawbacks of this process for handling cyanide bearing wastes generating from the gold and silver mining facilities are listed as high capital costs, increased reagent costs owing to additional thiocyanate oxidation, and formation potential of chlorinated by-products. Due to these reasons, alkaline chlorination has been almost completely supplanted by the other cyanide treatment methods such as copper catalysed hydrogen peroxide oxidation, the sulphur dioxide/air processes, and natural degradation systems (BCTRCR 1995; Smith and Mudder 1991).

8.2.2 Copper catalysed hydrogen peroxide oxidation

Two patented copper-catalysed hydrogen peroxide processes are available for the treatment of cyanide bearing wastes originating from the gold mining sites. The first oxidation technique, known as Kastone Process, which has been developed by DuPont (DuPont 1974) utilises a specially stabilised 41 % solution of hydrogen peroxide containing a few mg/L formaldehyde and copper as a catalyst. The second process developed by Degussa Corporation also relies on utilisation of hydrogen peroxide and copper sulphate in varying concentrations (Smith and Mudder 1991).

Free cyanide and weakly complexed metal cyanides are oxidised by hydrogen peroxide in the presence of copper catalyst according to the following reactions (Smith and Mudder 1991; BCTRCR 1995):

$$CN^- + H_2O_2 \xrightarrow{Cu^{2+}} CNO^- + H_2O \tag{8.1}$$

$$M(CN)_4^{2-} + 4H_2O_2 + 2OH^- \rightarrow M(OH)_2(s) + 4CNO^- + 4H_2O \tag{8.2}$$

M represents zinc, cadmium, copper, or zinc in Eq. (8.2). During oxidation process, bivalent metal ions released from weakly complexed metal cyanide species into solution precipitate as metal hydroxides. Cupric ion used as the catalyst or present in the waste is also capable of precipitating ferro-cyanide as an insoluble bimetallic complex as follows:

$$2Cu^{2+} + Fe(CN)_6^{4-} \rightarrow Cu_2Fe(CN)_6 \tag{8.3}$$

Optimum pH range is 9.5 to 10 for the reactions given by Eqs (8.1) and (8.2). On the other hand, bimetallic iron cyanide complex precipitates at the pH values less than 9.0 (Smith and Mudder 1991). Therefore, selection of operation pH is of importance for the case of iron cyanide bearing wastes. Such a case, considering that heavy metal removal, especially abatement of copper used as the catalyst, is a more important issue than that of iron cyanide with very low toxicity, the oxidation process is preferably performed at higher pH values than 9.0. The pH adjustment is commonly made using lime slurry if needed.

Copper sulphate pentahydrate ($CuSO_4 \cdot 5H_2O$) is commonly used as a catalyst. It is initially added at the ratio of about 10–20 % of weakly complexed cyanide concentration into the reaction solution (Smith and Mudder 1991). Hydrogen peroxide dose is determined taking into account the concentrations of free cyanide and complexed cyanide species. When the process is applied to slurries, higher reagent doses are required (US EPA 1994b). The reaction time required varies depending on reaction conditions such as catalyst to cyanide ratio, initial cyanide concentrations, the distribution of cyanide species, and hydrogen peroxide doses used. The reaction rate dramatically accelerates with increasing initial cyanide concentrations as well as with elevating temperatures. Increasing hydrogen peroxide doses shorten the reaction times. Cyanate formed during oxidation process is slowly hydrolysed to ammonia. Therefore, the process developed by Degussa Corporation results in the effluent containing a mixture of cyanate and ammonia. The Kastone process also yields incomplete conversion of cyanate to ammonia and its reaction products are cyanate, ammonia, and glycolic acid amide (US EPA 1982). The process is incapable of oxidising ammonia and thiocyanate. About 10–15 percent of thiocyanate may be oxidised during the process. Therefore, when remaining ammonia and thiocyanate concentrations exceed the acceptable levels in the treated effluents in terms of the discharge standards, additional treatment may be necessary (Smith and Mudder 1991). Equipment requirement and application basis of the process are same as explained in Subsection 4.2.1.3 for the metal finishing industry. In addition to the advantages and disadvantages listed in Subsection 4.2.1.3, the major advantage of the process for handling cyanide bearing wastes originating from gold mining facilities is that the process is applicable to bleed solutions, spent barren solutions, heap leach rinse water, pulps, and tailings pond overflows. The other advantages of this process are that (i) as mentioned above all forms of cyanide including ferrocyanide can be reduced to acceptable levels; (ii) heavy metals originating in gold ores processed can also be effectively removed through precipitation; (iii) the process can be operated either in a batch or continuous operation mode; (iv) strict pH control is not required; and (v) full-scale effectiveness of the process has been proved (Smith and Mudder 1991). The main drawback of the process is that it is inapplicable to slurries originating from gold mining operations because of high hydrogen peroxide consumption resulting from auto-catalytic destruction on solid surface or from high heavy metal content. On the other hand, the results of a study performed on tailings slurry taken from Ovacık Gold Mine located in Bergama, Turkey has demonstrated that the efficiency of the copper catalysed hydrogen peroxide process in destructing cyanide from tailings slurries vary depending on the characteristic of the ore such as their heavy metal content and high cyanide removal efficiency can be achieved with reasonable hydrogen peroxide consumption in the presence of copper catalyst for slurries originating from the gold ores with low sulphide and heavy metal contents (Kitiş *et al.* 2005).

Although the theoretical ratio based on Eq 8.1 is 1.31 g hydrogen peroxide per gram of free cyanide oxidised, in practice, the consumption of hydrogen peroxide is considerably higher than the theoretical ratio. It varies between 2.0 and 8.0 g hydrogen peroxide per gram of cyanide oxidised (Botz 2001). For instance, this ratio was found as 1.7 g hydrogen peroxide per gram of weakly complexed metal cyanide oxidised in the presence of 30 mg/L copper catalyst for the treatment of tailings slurry taken from Ovacık Gold Mine (Kitiş *et al.* 2005). Although, no actual operating cost information is available for the copper catalyst hydrogen peroxide process, total operation costs are estimated as € 6.9 (on the basis of 1995 cost) per kilogram of cyanide oxidised in a study dealt with the treatment of the barren solution (BCTRCR 1995).

8.2.3 INCO sulphur dioxide/air process

The INCO process being a proprietary technology has been developed and patented by the International Nickel Company of Canada (US Patent 4,537,686; US Patent 4,615,873; US Patent 4,622,149). This process has proven effective and successful in the treatment of cyanide bearing wastes originating from the mining sites. The INCO process utilises a mixture of sulphur dioxide and air oxygen, in the presence of copper catalyst to oxidise cyanide to cyanate. This treatment technique is primarily developed to treat tailings slurry. But, free cyanide and weakly complexed metal cyanides can be also effectively treated by the INCO process. Similar to the copper catalysed hydrogen peroxide process, this process is also capable of removing the stable iron complexed cyanides via precipitation (Smith and Mudder 1991; Mudder *et al.* 2001). Sulphur dioxide and oxygen oxidise free cyanide and weakly complexed metal cyanides to cyanate in the presence of soluble copper catalyst according to the following reactions:

$$CN^- + SO_2 + O_2 + H_2O \xrightarrow{\text{catalyst}} CNO^- + H_2SO_4 \tag{8.4}$$

$$M(CN)_4^{2-} + 4SO_2 + 4O_2 + 4H_2O \xrightarrow{\text{catalyst}} 4CNO^- + 4H_2SO_4 + M^{2+} \tag{8.5}$$

M represents zinc, cadmium, copper, or zinc in Eq. (8.5). Cupric ion used as the catalyst also oxidises free cyanide ions to cyanate while reducing to cuprous ion. The cuprous ion is complexed by the remaining free cyanide as follows:

$$2Cu^{2+} + 5CN^- + 2OH^- \rightarrow 2Cu(CN)_2^- + CNO^- + H_2O \tag{8.6}$$

Cuprous cyanide complex formed in accordance with Eq. (8.6) remains in reaction solution without precipitating so as to continue active catalyst role until most of cyanide ion is oxidised (BCTRCR 1995). During the oxidation process, ferric ion being stable iron cyanide complexed form is reduced to ferrous state to form ferrous cyanide complex. This complex is precipitated as insoluble metal ferrous cyanide complex (Mudder *et al.* 2001). Ferro-cyanide removal through the INCO process can be represented according to the following reactions:

$$Fe(CN)_6^{3-} + SO_2 + 2H_2O \rightarrow Fe(CN)_6^{4-} + SO_4^{2-} + 4H^+ \tag{8.7}$$

$$Fe(CN)_6^{4-} + M^{2+} + H_2O \rightarrow M_2Fe(CN)_6{\bullet}H_2O(s) \tag{8.8}$$

In Eq. (8.8), M symbolises zinc or nickel present in the waste, or copper existing in the waste, or used as the catalyst. The metal ions released following the breakdown of weakly complexed metal cyanides are removed via hydroxide precipitation. Thiocyanate is partially oxidised to cyanate by the INCO process. Its conversion rate is limited to 15 to 25 % in the treatment of bleed solution (BCTRCR 1995). The oxidation of thiocyanate can be represented by the reaction given below:

$$SCN^- + 4SO_2 + 4O_2 + 5H_2O \rightarrow CNO^- + 5H_2SO_4 \tag{8.9}$$

Since hydrolysis of cyanate to carbon dioxide and ammonia is realised partially depending on pH and reaction time during the application of the INCO process, cyanate is the primary product present in the treated effluents. In some cases, the treated effluent may also bear the residual sulphite (BCTRCR 1995).

The optimum pH range has been reported as 8.0 to 9.0 to prevent the evolution of toxic HCN for the INCO process (Botz 2001). If the pH fluctuates outside this optimum range, the process performance is adversely affected (Smith and Mudder 1991; Mudder *et al.* 2001). Therefore, pH is controlled by the addition of an alkali agent such as lime or sodium hydroxide in order to neutralise sulphuric acid produced during the process and to maintain the pH in the optimum pH range. The addition of lime per gram of cyanide oxidised is made in the range of 2 to 4 grams in the treatment of tailings slurry and 4 to 7 grams in the treatment of barren solution and pond water (Devuyst 1997).

Liquid/gaseous SO_2, sodium bisulphite (Na_2SO_3), or sodium metabisulphite ($Na_2S_2O_5$) is used as sulphur dioxide source in the INCO process. Among them, liquid SO_2 is the most commonly preferred one because it is one of the most economical and available sulphur dioxide sources. In practice, sulphur dioxide requirement is generally higher than the stoichiometric dose (2.46 g SO_2/g CN_{tot}) owing to demand exerted by other reduced constituents in the waste streams such as metals and thiocyanate. The dose varies in the range of 3 to 6 g SO_2/g CN_{tot} for tailings slurry treatment, and 3 to 5 g SO_2/g CN_{tot} for barren solution and pond water treatments (Devuyst 1997).

Similar to the copper catalysed hydrogen peroxide process, copper sulphate pentahydrate is commonly used as homogenous catalyst in the INCO process. Its aqueous solution is added into the system to provide a copper concentration in the range of 0 to 20 mg/L for tailings slurry and barren solution treatments, and 0 to 5 mg/L for pond water treatment (Devuyst 1997).

Total cyanide concentration can be reduced to low levels varying between 0.1 and 10 mg/L by the INCO process. The process is also capable of achieving low metal concentrations less than 1.0 mg/L (Devuyst 1997).

The INCO process can be accomplished in single stage reactors equipped with suitable agitation, aeration, pH control, and chemical preparation and dosing systems. In a single stage process configuration, all treatment application is carried out in a single reactor. In some instances, multi-stage process configuration can be used to maximise metal removal (Mudder *et al.* 2001). For instance, a two-stage INCO process has been operating at Ovacık Gold Mine, Bergama Turkey since 2001 (Köksal *et al.* 2003). In this treatment system, sodium bisulphite, sodium hydroxide, copper sulphate, and air are introduced into the first reactor for cyanide oxidation, while metal removal is realised using ferric sulphate, copper sulphate and sodium at the second reactor. The treatment system has also included an independent third stage for additional antimony removal reactor (Köksal *et al.* 2003). At Ovacık Gold Mine, there is an official effluent limitation for discharge of tailings into the impoundment, but there is no permission for discharge into the environment (Akcil 2001). Therefore, total cyanide concentration is reduced from 144 mg/L to less than 1 mg/L to ensure total cyanide limitation (1 mg/L) set down by Turkish Ministry of Environment (Görür *et al.* 1999; Akcil 2002). Concentrations of heavy metals such as cadmium, zinc, lead, arsenic, antimony, chromium, and mercury in the treated effluent are typically found less than 0.1 mg/L except copper (0.42 mg/L) and iron (0.129 mg/L) (Akcil 2002).

The advantages of the INCO process are similar to those of the copper catalyst hydrogen peroxide oxidation given in the previous subsection. Although, capital and operating costs for the INCO process are reported to be comparable with other chemical methods, they can increase in the treatment of high level cyanide bearing wastes (Smith and Mudder 1991). The main disadvantage of the process is that this process can yield the treated effluents with high sulphate concentration. Furthermore, additional treatment may be required for the removal of cyanate, thiocyanate, ammonia, nitrate, sulphate and metals.

Capital and operating costs vary depending on the type of waste (slurry or solution), the size and capacity of the treatment plant, the cyanide load fed into the treatment system, and type of sulphur dioxide source used in the process (Devuyst 1997). The capital costs have been reported as €390,000 (accepting C$1.34 = €1) for a 10,000 tonnes tails per day capacity (Goode *et al.* 2001; Demopoulos and Cheng 2004) and €780,000 to €1,170,000 for a 5,000 tonnes tails per day capacity (Demopoulos and Cheng 2004). The cyanide load fed into the process is a key parameter determining operating cost. It varies from 7 euro cents per tonne of treated pond water having natural degradation to a range of €0.60 to €1.31 per tonne of CIP pulps treated (Devuyst 1997).

8.2.4 Ozonation

Although, ozone provides another alternative in treating cyanide bearing wastewaters originating from the metal finishing operations, its use in the treatment of cyanidation wastes occurring in the gold mining sites is limited because of several constraints such as inapplicability to tailings slurry and pulp, high capital and operating costs. Ozonation is only applicable to bleed solutions originating from the Merrill Crowe process. Basic reactions for the treatment of bleeding solutions, application basis of the process and the other important information about cyanide destruction with ozonation are common with those explained in Subsection 4.2.1.2 for the metal finishing industry.

8.3 CONCLUDING REMARKS

As highlighted in the related literature (Korte and Coulston 1998), many real gold mines processing large pieces are no longer available all over the world. In consequence of this exhaustion, gold has to be mostly beneficiated from their low-grade ores using the extracting techniques. One of the main obstacles in extracting gold or silver from their low-grade ores is the separation from the gangue. In order to separate gold or silver with high recovery efficiencies, cyanidation has been still the most commonly applied leaching technique, all over the world, since 1890 owing to its inherent advantages on the recovery process. Additionally, flotation process applied for the recovery of base metals such as copper, lead, and zinc, is accomplished using low concentrations of cyanide. Thus, these operations realised with either low or high cyanide concentration are addressed potential toxicity issues in regard to the ecologic system surrounding the mine sites as well as human life. This potential hazard to environment has also demonstrated itself as catastrophic failures of some tailings impoundments resulting in their mismanagement operations or their faulty designs. During the last decade, this serious environmental concern has been triggered efforts on the replacement of cyanide with less hazardous lixiviant in gold and silver mining facilities. Widespread research has been undertaken for the identification and development of less toxic lixiviants. Results obtained from these studies performed at laboratory or full scale demonstrated that leaching with thiourea, thiosulphate, thiocyanate, and halides can replace with cyanidation but there are some limitations hindering their widespread adoption in gold mining operations. For instance, in spite of their effectiveness, thiourea and thiosulphate are unstable and exhibit low recycle efficiencies. Additionally, their detoxification costs are relatively expensive. Halides are expensive and difficult to handle and control. Recovery with thiocyanate is still under research. The patented coal-oil agglomeration process seems to be replaced with mercury amalgamation as it is only suited to process free gold particles. In view of these findings, inevitably cyanide will still continue to be used as the lixiviant in gold and silver mining sites unless a plausible, benign and environmentally friendly non-cyanide alternative is found out and proved with all aspects including occupational safety, environmental safety, availability, selectivity, recyclability, detoxifiability, and feasibility at a large scale. Within this framework, rigorous adherence to the best management practices for cyanidation process in the gold and silver mining sites is crucial. By doing so, cyanide can be safely used as a lixiviant without threatening to human life and the environment until a successful and technically economically feasible non-cyanide alternative is developed since cyanide destruction techniques introduced in this chapter make it possible to treat effectively cyanide bearing wastes originating from gold and silver mining sites provided that adherence to technical and managerial basis and guidance.

REFERENCES

Akcil, A. (2002) First application of cyanidation process in Turkish gold mining and its environmental impacts. *Mineral Eng.* **15**, 695–699.

Aylmore, M.G. (2001) Treatment of a refractory gold-copper sulphide concentrate by copper ammonical thiosulfate leaching. *Minerals Eng.* **14**(6), 615–637.

Barbosa-Filho, O. and Monhemius A.J. (1994a) Iodine-thiocyanate leaching system for gold. In *Hydrometallurgy '94*, pp. 425–440, UK: Institution of Mining and Metallurgy.

Barbosa-Filho, O. and Monhemius, A.J. (1994b) Leaching of gold in thiocyanate solutions-part 1: Chemistry and thermodynamics. *Transactions Inst. Mining& Metallurgy* **103**, C105–C110.

BCTRCR (1995) *Technical Guide for the Environmental Management of Cyanide in Mining*. T.W. British Columbia Technical and Research Committee on Reclamation

Cyanide Sub-Committee, Higgs Associates Ltd, E.V.S. Consultants Ltd., ASL Laboratories Ltd., Gormerly Process Engineering.

Botz M.M. (2001) Overview of cyanide treatment methods. In *Mining Environmental Management* (ed. T. Mudder), pp. 28–30, Mining Journal Ltd., London, UK.

Butterman, W.C. and Amey III, E.B. (2005) *Mineral Commodity Profiles-gold*, US Department of the Interior, US Geological Survey, Open-file report 02-303.

Calvez, J.P.S., Kim, M.J., Wong, P.L.M. and Tran, T. (1998) Use of coal-oil agglomerates for particulate gold recovery. *Minerals Eng.* **11**(9), 803–812.

Demopoulos, G.P. and Cheng, T.C. (2004) A case study of CIP tails slurry treatment: comparison of cyanide recovery to cyanide destruction. *The European J. Mineral Processing & Environ. Proct. (ejmp&ep)*, **4**(1), 1–9.

Devuyst, E.A. (1997) The INCO SO_2/air cyanide destruction process. In *Proc. of Değerli Metallerin Madenciliğinde Atık Barajlarının Yönetimi Semineri,* Çeşme, İzmir.

DuPont (1974) *Treating Cyanide, Zinc and Cadmium Rinse Waters with KASTONE Peroxygen Compound*, E.I. DuPont de Nemours Ltd., Plating Product Bull. No. 20.

EC (2009) *Reference Document on Best Available Techniques for Management of Tailing and Waste-Rock in Mining Activities*, European Commission.

Feng, D and van Deventer, J.S.J. (2010) Oxidative pre-treatment in thiosulphate leaching of sulphide gold ores. *Int. J. Mineral Processing* **94**, 28–34.

Groenwald, T. (1976) The dissolution of gold in thiourea. *Hydrometallurgy* **1**, 277–290.

Goode, J.R., McMullen, J., Wells, J.A. and Thomas, K.G. (2001). Cyanide and the environment: Barrick Gold Corporation's perspective. In *Cyanide: Social and Economic Aspects*, (eds. C.A. Young, L.G. Tidwell and L.G. Anderson), pp. 257–270, TMS, Warrendale, PA, USA.

Groenwald, T. (1977) Potential application of thiourea in the processing of gold. *J. South African Inst. Mining & Metallurgy* **77**, 217–223.

Görür, N., Orhon, D., Tünay, O., Kabdaşlı, N. I., Canbazoğlu, M, Yazıcıgil, H., Vardar, M., Eyidoğan, H., Barka, A., Üçışık, F. and Övez, S. (1999) *Ovacık Gold Mine*, Eurogold Mining Ltd., Normandy Mining Ltd. Evaluation Report, 300 pp. Turkish National Scientific and Research Council. YDABÇAG. İstanbul. (in Turkish).

Hilson, G. and Monhemius, A.J. (2006) Alternatives to cyanide in the gold mining industry: what prospects for the future? *J. Cleaner Prod.* **14**, 1158–1167.

Kai, T., Hagiwara, T, Haseb, H. and Takahashi, T. (1997) Reduction of thiourea consumption in gold extraction by acid thiourea solutions. *Ind. Eng. Chem. Res.,* **36**, 2757–2759.

Kholmogorov, A.G., Kononova, O.N., Pashkov, G.L. and Kononov, Y.S. (2002) Thiocyanate solutions in gold technology. *Hydrometallurgy*, **64**, 43–48.

Kitiş, M., Akcil, A, Karakaya, E. and Yiğit, N.O. (2005) Destruction of cyanide by hydrogen peroxide in tailings slurries from low bearing sulphidic gold ores. *Minerals Eng.* **18**, 353–362.

Korte, F., and Coulston, F. (1998) Commentary, some considerations on the impact on ecological chemical principles in practice with emphasis on gold mining and cyanide. *Ecotoxicol. & Environ. Safety, Res. Sec. B* **41**, 119–129.

Kotze, W. and Petersen, F.W. (2000) Free gold recovery by coal-oil agglomeration. *J. South African Inst. Mining & Metallurgy* **100**(1–2), 57–62.

Köksal, E., Ormanoğlu, G. and Devuyst, E.A. (2003) Cyanide destruction: full-scale operation at Ovacik gold mine. *The European J. Mineral Processing & Environ. Proct. (ejmp&ep)* **3**(3), 270–280.

la Brooy, S.R., Linge, H.G. and Walker, G.S. (1994) Review of gold extraction from ores. *Minerals Eng.* **7**(10), 1213-1241.

Mudder, T.I., Botz, M.M. and Smith, A. (2001). *Chemistry and Treatment of cyanidation Wastes*, 2nd edn, Mining Journal Books Ltd., London, UK.

Müezzinoğlu, A. (2003) A review of environmental considerations on gold mining and production. *Critical Reviews in Environ. Sci. Technol.* **33**(1), 45–71.

Molleman, E. and Dreisinger, D. (2002) The treatment of copper-gold ores by ammonium thiosulfate leaching. *Hydrometallurgy* **66**, 1–21.

Silver (2010) Silver in hospital, A Worldwide Association of Silver Miner, Refiner, Fabricator, and Manufactures. http://www.silverinstitute.org/medical_applications.

Smith, A. and Mudder, T.I. (1991) *The Chemistry and Treatment of Cyanidation Wastes*, Mining Journal Books Ltd., London, England, (ISBN 0-900117-51-69).

US EPA (1982) *Development Document for Effluent Limitations Guidelines and Standards for the Ore Mining and Dressing Point Source Category*, EPA 440/1-82/061-b, Effluent Guidelines Division US Environmental Protection Agency, Washington D.C., United States.

US EPA (1994a) *Extraction and Beneficiation of Ores and Minerals*, Vol. 2-Gold Technical Resource Document, EPA 530-R-94-013, NTIS PB94-170-305, Office of Solid Waste, Special Waste Branch, US Environmental Protection Agency, Washington D.C., United States.

US EPA (1994b) *Technical Report, Treatment of Cyanide Heap Leaches and Tailings*, EPA 530-R-94-, NTIS PB94-, Office of Solid Waste, Special Waste Branch, US Environmental Protection Agency, Washington D.C., United States.

US EPA (1995) EPA Office of Compliance Sector Notebook Project. Profile of the Metal Mining Industry, EPA/310-R-95-008, Office of Compliance, Office of Enforcement and Compliance Assurance, US Environmental Protection Agency, Washington D.C., United States.

US EPA (2010) Identification and description of mineral processing sectors and waste streams, http://www.epa.gov/wastes/nonhaz/industrial/special/mining/minedock/id.

USGS (2010) Minerals information, Commodity statistics and information. US Geological Survey, http://minerals.usgs.gov/minerals/pubs/commodity/.

White, H.A. (1905) The solubility of gold in thiosulphates and thiocyanates. *J. Chem., Metallurgical & Mining Soc.* **6**, 109–111.

Whong-Chong, G.M., Nakles, D.V. and Dzombak, D.A. (2006) Management of cyanide industrial process wastewaters. In *Cyanide in Water and Soil*, (eds. R.S. Ghosh, D.A. Dzombak and G.M. Whong-Chong), pp. 439–451, Taylor&Francis Group, LLC, CRC Press, London, United Kingdom, (ISBN 978-1-56670-66-7).

Chapter 9

Chemicals industry

9.1 CHEMICALS INDUSTRY

9.1.1 Overview

Inorganic and organic chemicals industry involves the production of a broad range of chemicals. Inorganic chemicals industry is categorised by the European Union as five basic classes as: chlor-alkali manufacturing industry, large volume inorganic chemicals-ammonia, acids and fertilisers industries, large volume inorganic chemicals, solids and others industry and production of specialty inorganic chemicals (EC 2003). In the U.S.A., the industry is defined by SIC Code 281 as Industrial Inorganic Chemicals, and over thirty subcategories were defined mainly on the product basis (US EPA 1982; 1995). The use of chemical oxidation technologies for the control and treatment of inorganic chemicals industry wastewaters is not common except for cyanide oxidation. Therefore, in this chapter organic chemicals industry will be taken as the basis for the evaluation of the chemical oxidation technologies.

The U.S.A. approach for classification of organic chemical industry is to collect all activities involving organic chemicals manufacturing into one broad category under the title of "*Organic Chemicals, Plastics, and Synthetic Fibres*", although there are some separate categories that are related to organic chemicals manufacturing such as soap, detergents and cleaning preparations. In the European Union, organic chemicals industry has been classified into three main categories as; (i) *"Large Volume Organic Chemicals Industry"*; (ii) *"Manufacture of Organic Fine Chemicals"*; and (iii) *"Production of Polymers"*. Large volume organic chemicals industry covers a large range of chemicals and processes mostly taking refinery products and transforming them into commodity chemicals or bulk chemicals. The chemicals produced in the industry are not directly used by the consumers but used as raw materials by other industries. Production of polymers is of a similar structure. These industries are located on large and highly integrated production installations. Organic fine chemical production exhibits an enormous diversity. The products are sold mostly to other chemical companies. The size of the plants varies, however, from very small (< 10 staff) to large (> 20,000 staff) (EC 2003; 2006; 2007).

9.1.2 Production processes

Production processes of the large volume organic chemicals exhibit a large variety. The processes can be described following several steps as; raw material supply and preparation, synthesis, product separation and refinement, and product handling and storage. The main product groups are; hydrocarbons (linear or cyclic, aliphatic or aromatic, organic compounds containing oxygen (alcohols, ketones, esters), organic compounds containing sulphur (sulphides, sulphonic acids, sulphates), organic compounds containing nitrogen (amines, nitro and azo compounds), organic compounds containing halogens (halocarbons, halogenated aromatic compound), plastic materials, synthetic fibres, and synthetic rubbers. The generic processes used in the production are; oxidation, halogenations, hydrogenation, esterification, alkylation, sulphonation, dehydrogenation, hydrolysis, reforming, carbonylation, oxyacetylation, nitration, dehydration, ammonolysis, condensation, and dealkylation (EC 2003).

The polymers are also produced in a great variety. The products may be grouped as; commodity polymers (such as polyethylene, polypropylene), high performance polymers (polyimide, polysulphone), and thermosetting polymers (polyesters, epoxies). The general steps of production are; preparation, reaction and separation of products. The generic processes are; suspension polymerisation, bulk polymerisation, emulsion polymerisation, gas phase polymerisation, and solution polymerisation (EC 2007).

Production of fine chemicals addresses to a wide spectrum. Among the products, dyes and pigments, plant health products, organic explosives, organic intermediates, specialised surfactants, flavours, fragrances, pheromones, plasticisers, vitamins, optical brighteners, and flame retardants can be named. Acylation, addition, alkylation, carboxylation, condensation, diazotisation, esterification, halogenation, nitration, oxidation, reduction, substitution, sulphitation, and sulphonation are the unit processes of manufacturing the fine chemicals (EC 2006).

9.1.3 Wastewater generation and characteristics

Wastewater sources of organic chemicals manufacturing are mainly; storage tank overflows, mixing vessel overflows, mother liquors from processing products, wash-water from purification of products, vapour condensates, quench water, wastewater from the treatment of exhaust gases or flue gas treatment, wastewater streams from rinsing and cleaning, contaminated water from vacuum generation, spent neutralisation agents, spills and leakage (EC 2003; 2006; 2007). The main polluting parameters are; organic matter expressed as COD, TOC, BOD_5, and VOC; oil and grease; SS; and pH (US EPA 1987). Nitrogen, phosphorus, cyanide, heavy metals, chloride, bromide, and sulphate also exist in the wastewaters. Organic matter content in the wastewater varies in a wide range. COD values of wastewaters are between hundreds to thousands mg/L. The main aspect of water pollution of organic chemicals industry lies in the toxic pollutants. US EPA (1993) has defined standards for micropollutants based on their removability in biological treatment systems, also considered the micropollutants that pass through publicly owned wastewater treatment systems. According to US EPA (2009) toxics release inventory and discharge monitoring reports, top pollutants are; dioxin and dioxin-like compounds, chlorine, hydroquinone, polycyclic aromatic compounds, hexacholorobenzene, lead and its compounds, nitrate compounds, fluoride, and nickel (US EPA 2009). EU has also brought limitations to purgeable chlorohydrocarbons (CHCs), adsorbable organic halides (AOX), extractable organic halogens (EOX), and priority substances as well as uses the measures toxicity, persistency, bioaccumulation for the control of toxic pollutants (EC 2003; 2006; 2007).

9.1.4 In-plant control and clean technologies

In-plant control is quite important for water pollution control in the organic chemicals industry. In-plant control methodologies and applications are defined for every industry and production method. In general identification of wastewater flows, minimising water flows with technology and equipment changes and employing dry cleaning techniques, minimising pollution through process operation and control of spills, maximising wastewater re-use and recovery, and solvent recovery are general applications. Separation of flows and application of pretreatment to remove non-biodegradable organics and parameters that are toxic and exerting adverse effects on the biological treatment and on the environment, are also basic in-plant control applications. Production specific in-plant control applications and use of clean technologies have been defined in-detail by US EPA (1987) and BREF documents of EU (EC 2003; 2006; 2007). However, in-plant treatment applied to properly segregated wastewater flows is the basic element of water pollution control, and determines the performance of biological treatment that is applied as the central treatment to combined wastewater. There are a variety of in-plant treatment methods for the control of a number of parameters existing in separated flows. Solvent recovery mostly by distillation is an important in-plant control for reduction of VOC and chlorinated hydrocarbons as well as toxicity. Heavy metal removal is carried out by chemical precipitation, chromium reduction, ion exchange, and filtration. Cyanide control is realised using alkaline chlorination. The in-plant treatment technologies encompass a wide range of physical, chemical, and biological methods carried out individually or in combination to separate, recover, treat, reduce or modify the characteristics of many parameters existing in a number of process wastewaters. Chemical oxidation, adsorption, filtration, extraction, steam or air stripping, hydrolysis, emulsion breaking, wet oxidation, dehalogenation, coagulation, membrane separation, distillation, retrification, ion exchange, ultrafiltration, and resin adsorption are among the methods employed. Recently, electrochemical methods have been considered as effective and versatile processes for the control of organic chemicals manufacturing industry wastewaters. AOX elimination is of great importance as an in-plant treatment. Combinations of several stages of treatment have been successfully used for this purpose; wet oxidation, hydrolysis, stripping, distillation, adsorption, chemical oxidation, solvent extraction, membrane separation, and rectification are among the unit operations/processes being used for AOX control. On the other hand aerobic and anaerobic biological treatments are also used as in-plant treatment practices to reduce, regulate, and modify the characteristics of organic pollutants. Final in-plant control application is monitoring, and close quality control of final effluent prior to joint biological treatment.

9.1.5 Wastewater treatment

Most wastewater components of organic chemicals manufacturing wastewaters are biodegradable. Therefore, biological treatment is the core of the final treatment system of an organic chemicals manufacturing plant that either discharges into the public sewer for further biological treatment in the municipal joint treatment system, or being treated in the end-of-pipe treatment system of the plant. After the in-plant control and treatment applications, combined wastewaters are sent to end-of-pipe treatment, where preparatory treatment units precede the biological treatment. Preparatory treatment involves the use of an equalisation tank and other treatment applications such as neutralisation, plain settling or coagulation-flocculation, flotation depending on the characteristics of wastewaters. Activated sludge process is the most common biological treatment, although, aerated lagoons or biofilm systems may also be used. In some cases, a two-stage biological treatment is applied. In such cases, generally an activated sludge system is preceded by a biofilm reactor acting as a roughing filter. Activated sludge process effluent may also be further treated in a polishing pond. In some cases biological treatment includes biological nitrogen and phosphorus removal units. Phosphorus removal may also be carried out using chemical precipitation. A polishing treatment following biological treatment may also be used. Multimedia filtration and activated carbon adsorption are commonly employed polishing treatment methods.

Chemical oxidation methods are of a common use, and play critical roles at the in-plant treatment stage of organic chemicals industry wastewater control. In the next subchapter, important applications of chemical oxidation are reviewed. The emphasis is placed upon organic toxic matter control and dehalogenation by new and developing chemical oxidation applications.

9.2 CHEMICAL OXIDATION PROCESSES

9.2.1 Wet air oxidation

Wet air or thermal liquid-phase oxidation (WAO) process has a great potential for the treatment of strong organic wastewaters and toxic organic compounds which are resistant to biological treatment and to most of chemical methods. The process is based on generation of active oxygen species at high temperatures and high pressures. Supercritical water oxidation is conducted using supercritical water, which exists as a phase above the critical temperature (647.3 K) and critical pressure (22.12 MPa). Both sub-critical and supercritical oxidation processes are used with catalysts to further increase the rate and oxidation power. The processes are being used as pretreatment methods in large volume, integrated organic chemicals manufacturing plants. Levec and Pintar (2007) made a review of catalytic wet air oxidation process emphasising the developments in the catalytic wet air oxidation, catalysts, and applications to real wastewaters. The authors pointed out the great potential of the process, and reported that except for a few cases (trinitrotoluene red (TNT red) and linear alkylbenzene sulphonate (LAS)), WAO was found capable of improving biodegradability of many wastewaters including pesticide, acrylonitrile, olive oil mill, phenolic, azo dye, and water soluble polymer. Vaidya and Mahajani (2002) evaluated the use of free radical initiators and scavengers and effect of catalyst as well as mass transfer limitations on minimising the end product formation as the phenol was used as model substrate. They concluded that acetic acid formation was maximum when mass transfer limitations existed, hydroquinone accelerated the degradation, but acetic acid formation increased, t-butanol suppressed both degradation and end product formation. Ferrous sulphate used as catalyst also increased the end product formation. Belkacemi *et al.* (2000) proposed a lumped kinetics for solid-catalysed wet oxidation for the description of oxidation of various pollutants in wastewaters. Stüber *et al.* (2005) compared various chemical wet oxidation techniques for the removal of toxic non-biodegradable pollutants including phenol, o-cresol, m-xylene, o-chlorophenol, p-nitrophenol, aniline, nitrobenzene, and sulpholane. They found that activated carbon catalysed WAO readily destroyed phenol, o-cresol, m-xylene, and o-cholorophenol while other pollutants required more powerful techniques such as H_2O_2 WAO, or Fenton WAO. Qin *et al.* (2001) studied wet oxidation of p-chlorophenol using supported noble metal catalyst, and found the activity of catalysts in the order: Pt > Pd > Ru. The most significant intermediate of the oxidation was determined to be p-benzoquinone. Cybulski and Trawczynski (2004) also tested Pt and Ru catalysts for wet air oxidation of phenol. The reaction yielded a lumped intermediates amounted to 64 ± 30 g O_2/mol. Goi *et al.* (2006) studied catalytic wet oxidation of halogenated liquid wastes using CeO_2-based catalyst, and found COD and AOX removals up to 80 and 90 %, respectively. Okawa *et al.* (2005) used wet peroxide oxidation to confirm the removal of polychlorinated biphenyls in soil and mineralization of acetic and oxalic acid. Degradation of three types of PCBs by the process was found to vary between 93–97 %. Rivas *et al.* (2008) obtained 80–100 % conversion of acenaphtene, phenanthrene, anthracene, and fluoranthene using wet air oxidation, with the use of promoters hydrogen peroxide and potassium monopersulphate the removal efficiencies did not change but the reaction times required were decreased. Kulkarni *et al.* (2008) made a review of dioxin sources and current remediation technologies, and reported that decomposition yield of dioxins in fly ash was 99.7 % using supercritical water oxidation with oxidisers such as air, pure oxygen gas, and hydrogen peroxide.

9.2.2 Electrochemical processes

Electrochemical processes applied to organics, particularly, the chlorinated organic compounds aim to either oxidise or degrade (anodic and cathodic treatment) the compounds. There are significant amount of research on both aspects in the literature. Oxidative treatment is common for most of the organic matter containing wastewaters originating from industrial activities such as leather, textile, and pulp and paper, however the applications for organic chemicals manufacturing wastewaters have been decreasing due to increasing use of biological treatment and reductive dechlorination mechanism. Another reason may be that formation of chlorinated intermediates during the process. Rajkumar and Palanivelu (2004) reported that AOX were detected at high concentrations during electrochemical treatment of phenolic compounds generated from phenol-formaldehyde resin manufacturing, oil refinery, and bulk drug manufacturing industries, but AOX concentrations brought down by increasing the electrolysis time. Polcaro *et al.* (2000) studied electrochemical combustion of chlorophenols using fixed bed carbon pellets and DSA electrodes as anodes. Mascia *et al.* (2010) investigated electrochemical treatment of phenolic waters in presence of chloride with boron-doped diamond anodes, and provided a mathematical model of the process. Anandhakumar *et al.* (2009) reported anodic oxidation of chlorophenols in micelles and microemulsions on glassy carbon electrode. Berrios *et al.* (2009) also studied electrooxidation of chlorophenols at glassy carbon electrodes modified with poly Ni (II) complexes.

Electrochemical reductive dechlorination is becoming a common application. The process has many advantages as well as a number of modes of operation in terms of electrodes, electrolyte medium, and use of mediators. The studies have begun in 1970s and have increased in number and variety beginning 1990s.

A great variety of systems and system elements as well as substrates have been employed in the experimental studies concerning reductive dechlorination literature. Electrolytic cells were mostly membrane separated two-compartment, although, undivided electrolyte reactors were also used (Kulikov *et al.* 1996; Chen *et al.* 2004; Bejankiwar *et al.* 2005; Al-Abed and Fang 2006; Fang and Al-Abed 2007). Electrolytes vary from solutions with adjusted conductivity using NaCl, H_2SO_4, or buffer electrolytes such as those prepared from carbonate/hydrogen carbonate, synthetic solutions to simulate groundwater medium to real wastewaters *e.g.* pulp bleaching effluent (Chen *et al.* 2004; Berrios *et al.* 2008; Patel and Suresh 2008). Inert gas atmospheres are frequently used (Miyoshi *et al.* 2004). Most of the experiments have been conducted at ambient temperatures. The electrodes varied a great range of materials with modified surfaces using different substances *e.g.* carbon cloth modified by Pd (Kulikov *et al.* 1996; Tsyganok *et al.* 1999), glassy carbon (Berrios *et al.* 2008; 2009), composite electrodes such as Pd/MoO_x/glass carbon (Chen *et al.* 2004) and sintered RuO_2/Pt/PdO, Pt/IrO_2/RuO_2, PdO/Pt (Miyoshi *et al.* 2004), Pd/foam-Ni, Pd/polymeric pyrrole film/foam-Ni (Zhirong *et al.* 2008), stainless steel (Bejankiwar *et al.* 2005) granular graphite packed electrode (Al-Abed and Fang 2006; Fang and Al-Abed 2007), boron-doped diamond (Liu *et al.* 2009), and silver (Scialdone *et al.* 2010a). Various supporting electrolytes and mediators were used such as trifluoroacetic acid and tetraalkylammonium salts in water-MeOH medium (Tsyganok *et al.* 1999), ligandmacrolyte (phenylporphry, phthalocyanine) (Berrios *et al.* 2009), sodium dodecylsulphate micellar solution and sodium dodecylsulphate/n-hexane/n-butanol/water micro emulsions) (Anandhakumar *et al.* 2009), K_2PdCl_6 (Fang and Al-Abed 2007). The process efficiencies in terms of both dechlorination and mineralisation of the organic matter were also variable. Kulikov *et al.* (1996) obtained complete dechlorination of aromatic and aliphatic compounds. Chen *et al.* (2004) observed total dechlorinated product of phenol upon reductive chlorination of 2,4,6-trichlorophenol. Miyoshi *et al.* (2004) reported dechlorination yields varying between 53–97 % for 1,2,3-trichlorobenzene for different composite cathodes. In many of the studies, the factors affecting the conversions, intermediate products and kinetic evaluation of the reactions have been made (Chen *et al.* 2004; Bejankiwar *et al.* 2005; Al-Abed and Fang 2006; Berrios *et al.* 2008; Liu *et al.* 2009; Saez *et al.* 2009; Scialdone *et al.* 2010b). The compounds tested also exhibited a wide range of structure, pentachlorophenol (Patel and Suresh 2008), chlorophenols (Berrios *et al.* 2008; 2009), 2,4,6-trichlorophenol (Chen *et al.* 2004); 1,2,3-trichlorobenzene (Miyoshi *et al.* 2004), choloroaromatics (Tsyganok *et al.* 1999), chloroform (Zhirong *et al.* 2008), 1,2-dichloroethane (Bejankiwar *et al.* 2005), trichloroethylene (Al-Abed and Fang 2006) 2,2',5,5'-tetrachloroethylene (Shibin *et al.* 2007), 2-chlorobiphenyl (Fang and Al-Abed 2007), chlorobenzone (Liu *et al.* 2009), pentachlorophenol (Lin *et al.* 1999), perchloroethylene (Saez *et al.* 2009), and aliphatic chlorides (Scialdone *et al.* 2010a). Tsyganok *et al.* (1999) evaluated selective dechlorination of various chloroaromatics in water-MeOH medium containing trifluoroacetic acid and tetraalkylammonium salts over Pd-loaded carbon felt cathode. The reactivities of 2,4-dichlorophenyl derivatives varied in the order: 2,4-dichlorotoluene $<$ 2,4-dichlorophenol $\leq$ 2,4-dichloroanisole $\leq$ 2,4-dichlorobenzoic acid $<$ 2,4-dichlorophenoxyacetic acid $<$ 2,4-dichlorophenethyl alcohol. Miyoshi *et al.* (2004), determined selectivity of dechlorination, and found that electrodes with Ru and Pb were selective mainly for meta-position dechlorination, while those with Pt group selective mainly for ortho-position for the dechlorination of 1,2,3-trichlorobenzene using composite metal-metal oxide electrodes. Patel and Suresh (2008) evaluated the effect of pretreatments using alkaline sulphide, alkaline reduction using ferrous sulphate and alum coagulation on the raw bleach effluent on AOX removal. In some studies electrochemical process was combined with other processes such as oxidation and air-stripping (Shibin *et al.* 2007). Electrochemical reductive processes are also carried out using aqueous-solvent or solvent solutions. Scialdone *et al.* (2010a) studied reduction of aliphatic chlorides at silver cathodes in acetonitrile, water and their mixture. Rondinini and Vertova (2004) investigated the reduction of dichloromethane and trichloromethane dehalogenation by electrocatalysis on silver and silver alloys in pure aprotic and mixed aqueous-aprotic solvents. Isse *et al.* (2006) evaluated reduction of benzyl chloride using silver and glassy carbon electrodes in $CH_3CN+Et_4NClO_4$. Jalil *et al.*

(2007) studied electrochemical dechlorination of chlorobenzenes in the presence of naphthalene mediator in organic solutions. Acetonitrile solution was used to assess electrochemical dechlorination of chlorobenzenes in the presence of various arene mediators (Jalil *et al.* 2010).

9.2.3 Ozonation

Ozone is one of the most commonly used oxidants for degradation of organics. Although, direct ozonation has a moderate power, ozonation is a suitable oxidation method for partial oxidation or increasing the biodegradation of organic matter. However, ozone proved to be very effective on some substrates such as phenol, even complex organics such as 3,4-dichloro-but-1-ene which is present in synthetic rubber process wastewaters (Alborzfar *et al.* 2000). Ozone oxidation of organic matter is well documented. In addition to ozone oxidation performance on a number of organics, kinetic data were also provided. Benitez *et al.* (2000) presented rate constant for the reactions of ozone with chlorophenols in aqueous solutions. There are also studies where ozone performance was compared with other oxidants particularly AOPs. Pillai *et al.* (2009) applied direct and indirect ozone oxidation using catalyst Fe^{2+}, H_2O_2 and UV light on the wastewater obtained from terephthalic acid manufacturing where the performances of 5 combinations of advanced oxidation methods and ozonation could be comparatively evaluated. Pre-ozonation of 2,4-dichlorophenol increased the BOD_5/COD ratio from zero of the original solution to 0.25 at the point where all initial compound has been removed (Contreras *et al.* 2003). Ozonation was also investigated for applications relevant to properties of organics. Kornmüller and Wiesmann (2003) evaluated mass transfer and reaction kinetics of ozonation of polycyclic aromatic hydrocarbons in oil/water-emulsions. Ozonation was also used for soil remediation. Rivas (2006) made a review of chemical oxidation based treatments for polycyclic aromatic hydrocarbons sorbed on soils. O'Mahony *et al.* (2006) evaluated the potential of ozone for the removal of phenanthrene from different soils. Haapea and Tuhkanen (2006) studied pretreatment of aged oil contaminated with polyaromatic hydrocarbons (PAHs) using ozonation.

9.2.4 Advanced oxidation methods

Due to structural complexity and varying responses to oxidants and their mixture, the methods applied for the oxidation of organics vary in a wide range in terms of the oxidants as well as their ways of application. Following discussion of the literature is not exhaustive but aims to exhibit the variety of the advanced oxidation applications to a wide range of organics. Sunder and Hempel (1997) applied ozone and hydrogen peroxide to trichloroethene and perchloroethene in a tube reactor and nearly complete mineralization of both compounds was achieved as determined by stoichiometric release of organic chlorine as chloride ions. Shimoda *et al.* (1997) used both a tubular flow reactor and stirred tank reactor for the treatment of benzene, dichlorobenzene, trichloroethene, trichloroethane and carbon tetrachloride by H_2O_2/UV-vis photo-oxidation process, and concluded that the reaction rate controlling variable was the photon-flux and stirred tank reactor was more effective than tubular flow reactor. Hugül *et al.* (2000) studied decomposition of 2-chlorophenol, 2,4-dichlorophenol and 2,4,6-trichlorophenol in UV-catalysed oxidation with H_2O_2. High conversion efficiencies were obtained and reaction rate and degradation pathways were modelled and discussed. Kralik *et al.* (2010) developed a mathematical model describing the degradation of para-cholorophenol by UV/H_2O_2 process using two-factor three-level Box-Behnken experimental design. Contreras *et al.* (2001) presented data for the oxidation of nitrobenzene using O_3/UV and UV/Fe (III) oxidations, and compared the results with those obtained from single ozonation. Catalkaya and Kargi (2007) investigated AOX removal from a pulp mill effluent using UV, UV/H_2O_2, O_3/H_2O_2 and ozonation, and made a comparative evaluation of the results. Gimeno *et al.* (2005) studied the oxidation of phenol, 4-nitrophenol and 4-chlorophenol using O_3, UV-vis, O_3/UV-vis, TiO_2/UV-vis, O_3/UV-vis/TiO_2, and O_3/TiO_2 systems, and made comparative evaluations on parent compound, COD, and TOC removals. Gimeno *et al.* (2007) conducted experimental studies to assess and compare the effectiveness of photocatalytic oxidation of gallic acid, tyrosol and syringic acid using TiO_2 and several oxidants such as persulphate, monopersulphate, and hydrogen peroxide. Koh and Thiemann (2001) conducted photocatalytic oxidation of long-chain chlorinated paraffins by combination of UV light and H_2O_2. Chain lengths of the paraffins used were C17–C24 and their chlorine contents varied 35–62 % in aqueous medium in which acetone was added for solubilisation of paraffins in 0.1 % ratio. Half-lives of hydrocarbons C10–C13 with 62 % Cl content and hydrocarbons C17–C20 with 52 % Cl content were less than 1 hour and others ranged between 2.5–5.2 hours. Degradation products of short-chain chlorinated paraffins were long-chain alkanes. Crittenden *et al.* (1997) used surface modified TiO_2 with Pt, Ag and iron oxide for photocatalytic oxidation of model compounds trichloroethylene and p-dichlorobenzene with UV. Impregnation of TiO_2 with 1 wt % Ag and Pt using photoreduction was effective and increased the photocatalytic activity of TiO_2. Model compounds were totally mineralised and a model was developed to describe trichloroethylene destruction.

Fenton oxidation being a versatile and powerful process has also found a wide-range of application for the oxidation of organics. Pera-Titus *et al.* (2004) made a general review of degradation of chlorophenols by Fenton, photo-Fenton and

Fenton like processes. Ozone and O_3/UV systems were also evaluated in the study. Half-life times, kinetic constants, and degradation pathways of cholorophenols were reviewed and discussed. Rivas (2006) made a review of Fenton's based technologies as applied to polycyclic aromatic hydrocarbons sorbed on soils. Teel *et al.* (2001) studied and compared four different Fenton's oxidation processes; a standard Fenton's system, a modified soluble iron system with a pulse input of hydrogen peroxide, and two modified mineral-catalysed systems. The substrate used in the experiments was trichloroethylene and the evaluation emphasised hydroxyl radical and non-hydroxyl radical mechanisms. In the application of Fenton's oxidation to organics reaction intermediates are of importance from several aspects such as the toxicity, biodegradability, rate of degradation, regeneration capability of oxidised iron, and complex formation with iron. There are several studies where intermediate problem was investigated. Kwon *et al.* (1999) evaluated characteristics of p-chlorophenol oxidation by Fenton's reagent emphasising the changes of rates during the reaction and the occurrence of a slow phase that was attributed to the Fe^{2+} depletion caused by Fe-organic complex formation. Du *et al.* (2006) studied the role of intermediates in the degradation of phenolic compounds phenol, 4-chlorophenol and 4-nitrophenol by a Fenton-like process, and explained the occurrence of a fast phase following the initiation of the reaction by the role of hydroquinone-like compounds, catechol and 4-chlorocatechol formed during the reaction which were identified to reduce Fe^{3+} to Fe^{2+} quickly. Kang *et al.* (2002) proposed a kinetic model for the Fenton oxidation of phenol and monochlorophenols by accounting for the interactions of organic intermediates with the Fe species. Zimbron and Reardon (2009) also developed a model for the reaction system by investigating kinetics and intermediates of pentachlorophenol (PCP) degradation in the Fenton's reaction. The level of dechlorination was found to be 70 % at complete PCP removal. Dichloromaleic acid and tertachlorohydroquinone were the reaction intermediates total amount of which accounted for up to 5 % of PCP degraded. Several variants of the Fenton's process were also studied. Du *et al.* (2007a) investigated the role of oxygen in the degradation pathway of 4-chlorophenol by Fenton system and compared with Fenton/N_2 system in terms of process rate and intermediates formed during the reactions. The authors proposed a kinetic model for Fenton/O_2 and Fenton/N_2 for 4-chlorophenol. The model served to predict the evolution of aromatic intermediates between Fenton/O_2 and Fenton/N_2 (Du *et al.* 2007b). Utset *et al.* (2000) used aniline to test the effect of O_2 on H_2O_2 consumption in the Fenton and photo-Fenton reactions. Son *et al.* (2009) proposed a Fenton-like degradation mechanism using zero-valent iron and UV light for the degradation of 1,4-dioxane. Lee *et al.* (2009) obtained up to 99 % removal of 1,2,3,4-tetrachlorinated dibenzo-p-dioxins using Fenton-like oxidation where steel dust was used as the heterogeneous catalyst. Yalfani *et al.* (2009) studied phenol degradation by Fenton's process using in situ generated hydrogen peroxide. Ioan *et al.* (2007) compared Fenton and sono-Fenton using bis-phenol A as the model compound. Arienzo (1999) proposed a modified Fenton reagent of solid pyrite in place of soluble ferrous salt for the oxidation of 2,4,6-trinitrotoluene. Martins *et al.* (2010) evaluated Fenton's oxidation process for biodegradability enhancement. Perez-Moya *et al.* (2007) studied Fenton and photo-Fenton degradation of 2-chlorophenol by multivariate analysis and toxicity monitoring. Kiwi *et al.* (1994) investigated biodegradability of 2- and 4-nitrophenol after Fenton and photo-Fenton processes. Khan *et al.* (2009) assessed the effects of iron type in Fenton reaction on mineralization and biodegradability enhancement using trichloroethene, 2,4-dichlorophenol, 1,4-dioxane and 1,2,3-trichloropropane. Chamarro *et al.* (2001) tested formic acid, phenol, 4-chlorophenol, 2,4-dichlorophenol, and nitrobenzene for the assessment of improvement in biodegradability through Fenton's oxidation. Goi and Trapido (2002) studied Fenton and photo-Fenton oxidation of 2-nitrophenol, 4-nitrophenol, 2,5-dinitrophenol, 2,6-dinitrophenol, 2,4-dinitrophenol, 2-methyl-4,6-dinitrophenol and 4-methyl-2, 6-dinitrophenol. The results indicated that all nitrophenols studied could be converted over 90 % efficiency by Fenton and photo-Fenton processes, and the extent of nitrogen conversions to nitrate after 90 % degradation of nitrophenols were 51–67 % for Fenton and 85–90 % for photo-Fenton. Fenton's oxidation is also used for remediation of contaminated soils (Kao and Wu 2000; Silva *et al.* 2009).

9.2.5 Sonochemical degradation

Yim *et al.* (2001) studied sonochemical degradation of 1,1,1-trichloroethane, trichloroethylene and tetrachloroethylene in aqueous solution at 100 kHz under an air atmosphere. In the batch mode chlorinated hydrocarbons were readily degraded. In the continuous mode the yield of chloride ions was 70–90 % of the chlorine atoms in the parent chlorinated compound. Goskonda *et al.* (2002) investigated sonochemical degradation of chlorophenols, 4-fluorophenol and nitrotoluenes in aqueous solution at 20 kHz. In 32 hours of sonification, about 80 % chlorine of parent compound was converted to chloride ion for 4-chlorophenol and 2,4-dichlorophenol, release of nitrate accounted for 63–74 % of nitrogen content of trinitrotoluene, 2-amino-4,6-dinitrotoluen, and 4-amino-2,6-dinitrotoluen. 4-fluorophenol degradation was significantly lower than that of 4-chlorophenol. Chowdhury and Viraraghavan (2009) made a review of sonochemical degradation of chlorinated organic compounds, phenolic compounds, and organic dyes. They reported that chlorinated hydrocarbons were removed relatively easily with low intermediate/product formation. Phenol removal was quite limited. This was explained by hydrophilic nature of phenol and relatively low vapour pressure.

9.2.6 Developing methods

One of the developing treatment methods is the use of light energy without the support of semiconductor type catalysts or oxidants like hydrogen peroxide. This approach is known as UV photodegradation. Shirayama *et al.* (2001) realised photodegradation of C_2Cl_4, C_2HCl_3, $C_2H_4Cl_2$, 1,1,1-$C_2H_3Cl_3$, 1,1,2- $C_2H_3Cl_3$, $CHCl_3$, and CCl_4 under UV bandwidths of 185–254 nm in the presence and absence of dissolved oxygen in water. All model compounds were degraded. The degradation mechanism was proposed to be the formation of ozone and hydroxyl radicals in the presence of dissolved oxygen and organic radicals in the absence of dissolved oxygen. The degradation rates, with the exception of tetrachloroethylene, were found to be higher in the absence of dissolved oxygen. Miller and Olejnik (2001) studied photodegradation of benzo (a) pyrene, chrysene and fluorene in aqueous solution. The substrates were removed with varying extents. Effect of process variables, *e.g.* pH, initial concentration of reactants, and existence of dissolved oxygen, were evaluated. Rodriguez *et al.* (2002) investigated photooxidation of phenol and nitrobenzene in aqueous solution under UV radiation and using ferric iron. Both substrates were effectively removed. The removal was faster in the presence of ferric iron than only UV light. Nowakowska and Szczubialka (1999) studied photosensitised dechlorination of hexachlorobenzene (HCB) in organic solvents and in aqueous solution of poly (sodium styrenesulphonate-co-N-vinylcarbazole (PSSS-VCz)). In aqueous solution HCB was rapidly consumed and less chlorinated benzenes were formed. Photoelectrocatalysis is another new technology which uses visible light. Nissen *et al.* (2009) realised remediation of 2,4-dichlorophenol in water by photoelectrolysis using a tungsten trioxide electrode.

Use of effective catalysts for dechlorination is a new application. These catalysts work without the addition of an oxidant in aqueous, solvent or mixed media, and they are either finely divided metals or noble-metal-modified metal surfaces. Hildebrand *et al.* (2009) investigated the use of Pd/Fe_3O_4 nano-catalyst for selective dehalogenation in wastewater treatment. Lin *et al.* (2004) used noble metal (Pd, Pt, Ru and Au) deposited iron surfaces for dechlorination of trichloroethylene in aqueous solution. Morales *et al.* (2002) studied dechlorination of chlorinated phenols by catalysed or uncatalysed Fe (0) and Mg (0) particles. Metal containing catalysts are also used in low-temperature incineration of chlorinated hydrocarbons (Halasz *et al.* 1999; Paukshtis *et al.* 2010).

9.3 CONCLUDING REMARKS

Organic chemicals industry manufactures a wide range of products which leads to formation of wastewaters with varying quality. However, the common parameter in wastewaters is organic matter and most of the organic content is biodegradable. Therefore, biological treatment plays a central role in the control of water pollution and all waste streams are directed to biological treatment after appropriate pretreatment. Some of the pollutants such as heavy metals and cyanide are easily handled at the source while toxic organics are hard to treat and necessitate complex treatment application. In the conventional approach to wastewater control, although pretreatment at the source was an important element control of toxics at least equally relied on polishing techniques such as activated carbon adsorption that are placed after the central biological treatment. This conventional approach is no more valid due to indirect discharge of many organic chemicals plant to make use of the municipal joint biological treatment systems as well as the priority given to toxics control. For the same reason, the pretreatment methodology applied to the control of toxics has also changed in that the aim has become total destruction of toxic organics rather than a partial treatment by using conventional phase separation methods. This new approach and the stringent limitations imposed on toxic organics have led to application of complex pretreatment processes at the source and as an integrated part of production processes. The review of chemical oxidation technologies, in this chapter, has been realised within this context and priority has been given to effective pretreatment technologies. This is not a difficult task because most of the effective treatment technologies employed for toxics control have been based on chemical oxidation. The evaluation of the chemical oxidation processes indicated that in addition to existing and applied methods there are a number of new, effective and promising chemical oxidation applications. Wet oxidation in its various modifications is suitable for many kinds of waste components and has proved quite effective for destruction or conversion of chemicals into biodegradable forms. Electrochemical methods exhibit a great potential for both reduction and oxidation. The reductive dechlorination is expected to play an important role in the control of the chlorinated organic compounds. The new electrodes developed for oxidative electrochemical processes have also a great potential in dealing with many toxic organics. Other oxidation methods, particularly, Fenton's oxidation with its a number of combinations with other oxidation means proved to be effective on a great spectrum of organics. Sonochemical oxidation is a promising method with its power and the advantage of minimum usage of chemicals.

REFERENCES

Al-Abed, S. and Fang, Y. (2006) Influences of pH and current on electrolytic dechlorination of trichloroethylene at a granular-graphite packed electrode. *Chemosphere* **64**, 462–469.

Alborzfar, M., Escande, K. and Allen, S.J. (2000) Removal of 3,4-dichlorobut-1-ene using ozone oxidation. *Water Res.* **34**(11), 2963–2970.

Anandhakumar, S., Chandrasekaran, M. and Noel, M. (2009) Anodic oxidation of chlorophenols in micelles and microemulsions on glassy carbon electrod: The medium on electroanalysis and electrochemical detoxification. *J. Appl. Electrochem.* **40**, 303–310.

Arienzo, M., (1999) Oxidizing 2,4,6-trinirotoluene with pyrite-H_2O_2 suspensions. *Chemosphere* **39**, 1629–1638.

Baycan, N., Thomanetzi E. and Sengül, F. (2007) Influence of chloride concentration on the formation of AOX in UV oxidative system. *J. Hazard. Mater.* **143**, 171–176.

Bejankiwar, R., Lalman, J.A., Seth, R. and Biswas, N. (2005) Electrochemical degradation of 1,2-dichloroethane (DCA) in a synthetic groundwater medium using stainless-steel electrodes. *Water Res.* **39**, 4715–4724.

Belkacemi, K., Larachi, F. and Sayari, A. (2000) Lumped kinetics for solid-catalyzed oxidation: a versatile model. *J. Catal.* **193**, 224–237.

Benitez, F.J., Beltran-Heredia, J., Acero, J.L. and Rubio, F.J. (2000) Rate constants for the reactions of ozone with chlorophenols in aqueous solutions. *J. Hazard. Mater.* **B79**, 271–285.

Berrios, C., Arce, R., Rezende, M.C., Ureta-Zanrtu, M.S. and Gutierrez, C. (2008) Electrooxidation of chlorophenols at a glassy carbon electrode in a pH 11 buffer. *Electrochim. Acta* **53**, 2768–2775.

Berrios, C., Marco, J.F., Gutierrez, C. and Ureta-Zanartu, M.S., (2009) Electro-oxidation of chlorophenols at glassy carbon electrodes modified with poly Ni(II) complexes. *Electrochim. Acta* **54**, 6417–6425.

Catalkaya, E.C. and Kargi, F. (2007) Color, TOC and AOX removals from pulp mill effluent by advanced oxidation processes: A comparative study. *J. Hazard. Mater.* **B139**, 244–253.

Chamarro, E., Marco, A. and Esplugas, S. (2001) Use of Fenton reagent to improve organic chemical biodegradability. *Water Res.* **35**, 1047–1051.

Chen, G., Wang, Z. and Xia, D. (2004) Electrochemically reductive dechorination of micro amounts of 2,4,6-trichlorophenol in aqueous medium on molybdenum oxide containing supported palladium. *Electrochim. Acta* **50**, 933–937.

Chowdhury, P. and Viraraghavan, T. (2009) Sonochemical degradation of chlorinated organic compounds, phenolic compounds and organic dyes- a review. *Sci.Total Environ.* **407**, 2474–2492.

Contreras, S., Rodriguez, M., Chamarro, E. and Esplugas, S. (2001) UV- and UV/Fe(III)-enhanced of nitrobenzene in aqueous solution. *J. Photoch. Photobio. A.* **142**, 79–83.

Contreras, S., Rodriguez, M., Momani, F.A., Sans, C. and Esplugas, S. (2003) Contribution of the ozone pre-treatment to the biodegradation of aqueous solutions of 2,4-dichlorophenol. *Water Res.* **37**, 3164–3171.

Crittenden, J.C., Liu, J., Hand, D.W. and Perram, D.L. (1997) Photocatalytic oxidation of chlorinated hydrocarbons in water. *Water Res.* **31**, 429–438.

Cybulski, A. and Trawczynski, J. (2004) Catalytic wet air oxidation of phenol over platinum and ruthenium catalysts. *Appl. Catal. B-Environ.* **47**, 1–13.

Donglagic, J. and Levec, J. (1998) Does the catalytic wet oxidation yield products more amenable to biodegradation. *Appl. Catal. B-Environ.* **17**, L1–L5.

Du, Y., Zhou, M. and Lei, L. (2006) Role of the intermediates in the degradation of phenolic compounds by Fenton-like process. *J. Hazard. Mater.* **B136**, 859–865.

Du, Y., Zhou, M. and Lei, L. (2007a) The role of oxygen in the degradation of p-chlorophenol by Fenton system. *J. Hazard. Mater.* **B139**, 108–115.

Du, Y., Zhou, M. and Lei, L. (2007b) Kinetic model of 4-CP degradation by Fenton/O_2 system. *Water Res.* **41**, 1121–1133.

EC (2003) Integrated Pollution Prevention and Control (IPPC) *Reference Document on Best Available Techniques in the Large Volume Organic Chemical Industry*, European Commission.

EC (2006) *Integrated Pollution Prevention and Control Reference Document on Best Available Techniques for the Manufacture of Organic Fine Chemicals*, European Commission.

EC (2007) *Reference Document on Best Available Techniques in the Production of Polymers*, European Commission.

Fang, Y. and Al-Abed, S.R. (2007) Palladium-facilitated electrolytic dechlorination of 2-chlorobiphenyl using a granular-graphite electrode. *Chemosphere* **66**, 226–233.

Gimeno, O., Carbajo, M., Beltran, F. and Rivas, F.J. (2005) Phenol and sunstituted phenols AOPs remediation. *J. Hazard. Mater.* **B119**, 99–108.

Gimeno, O., Carbajo, M., Lopez, M.J, Melero, A., Beltran, F. and Rivas, F.J. (2007) Photocatalytic promoted oxidation of phenolic mixtures: an insight into the operating and mechanistic aspects. *Water Res.* **41**, 4672–4684.

Goi, A. and Trapido, M. (2002) Hydrogen peroxide photolysis, Fenton reagent and photo-Fenton for the degradation of nitrophenols: a comparative study. *Chemosphere* **46**, 913–922.

Goi, D., Leitenburg, C., Dolcetti, G. and Trovarelli, A. (2006) COD and AOX abatement in catalytic wet oxidation of halogenated liquid wastes using CeO_2-based catalysts. *J. Alloy. Compd.* **408–412**, 1136–1140.

Goskonda, S., Catallo, W.J. and Junk, T. (2002), Sonochemical degradation of aromatic organic pollutants. *Waste Manage.* **22**, 351–356.

Haapea, P. and Tuhkanen, T. (2006) Integrated treatment of PAH contaminated soil by soil washing, ozonation and biological treatment. *J. Hazard. Mater.* **B136**, 244–250.

Halasz, J., Hagadus, M., Kun, E. and Kirisci, I. (1999) Destruction of chlorobenzenes by catalytic oxidation over transition metal containing ZSM-5 and Y(FAU) zeolites. *Stud. Surf. Sci. Catal.* **125**, 793–800.

Hildebrand, H., Mackenzie, K. and Kopinke, F.D. (2009) Pd/Fe_3O_4 nano-catalysts for selective dehalogenation in wastewater treatment processes- Influence of water constituents. *Appl. Catal. B-Environ.* **91**, 389–396.

Hugül, M., Apak, R. and Demirci, S. (2000) Modelling kinetics of UV/hydrogen peroxide oxidation of some mono-, di-, and trichlorophenols. *J. Hazard. Mater.* **B77**, 193–208.

Ioan, I., Wilson, s., Lundanes, E. and Neculai, A. (2007) Comparison of Fenton and sono-Fenton bisphenol A degradation. *J. Hazard. Mater.* **142**, 559–563.

Isse, A.A., Gottardello, S., Maccato, C. and Gennaro, A. (2006) Silver nanoparticles deposited on glassy carbon. Electrocatalytic activity for reduction of benzyl chloride. *Electochem. Commun.* **8**, 1707–1712.

Jalil, A.A., Panjang, N.F.A., Akhbar, S., Sundang, M., Tajuddin, N. and Triwahyono, S. (2007) Complete electrochemical dechlorination of chlorobenzenes in the presence of naphthalene mediator. *J. Hazard. Mater.* **148**, 1–5.

Jalil, A.A., Triwahyono, S., Razali, N.A.M., Hairom, N.H.H., Idris, A., Muhid, M.N.M., Ismail. A., Yahaya, N.A.M., Ahmad, N.A.L. and Dzinun, H. (2010) Complete electrochemical dechlorination of chlorobenzenes in the presence of various arene mediators. *J. Hazard. Mater.* **174**, 581–585.

Kang, N., Lee, D.S. and Yoon, J. (2002) Kinetic model of Fenton oxidation of phenol and monochlorophenols. *Chemosphere* **47**, 915–924.

Kao, C.M. and Wu, M.J. (2000) Enhanced TCDD degradation by Fenton's reagent preoxidation. *J. Hazard. Mater.* **B74**, 197–211.

Khan, E., Wirojanagud, W. and Sermsai, N. (2009) Effects of iron type in Fenton reaction on mineralization and biodegradability enhancement of hazardous organic compounds. *J. Hazard. Mater.* **161**, 1024–1034.

Kiwi, J., Pulgarin, C. and Peringer, P. (1994) Effect of Fenton and photo-Fenton reactions on the degradation and biodegradability of 2 and 4-nitrophenols in water treatment. *Appl. Catal. B-Environ.* **3**, 335–350.

Koh, I.O. and Thiemann, W. (2001) Study of photochemical oxidation of standard chlorinated paraffins and identification of degradation products. *J. Photoch. Photobio. A.* **139**, 205–215.

Kornmüller, A. and Weismann, U. (2003) Ozonation of polycyclic aromatic hydrocarbons in oil/water-emulsions: mass transfer and reaction kinetics. *Water Res.* **37**, 1023–1032.

Kralik, P., Kusic, H., Koprivanac, N. and Bozic, A.L. (2010) Degradation of chlorinated hydrocarbons by UV/H_2O_2: The application of experimental design and kinetic modeling approach. *Chem. Eng. J.* **158**, 154–166.

Kulikov, S.M., Plekhanov, V.P., Tsyganok, A.I., Schlimm, C. and Heitz, E. (1996) Electrochemical reductive dechlorination of chlororganic compounds on carbon cloth and metal-modified carbon cloth cathodes. *Electrochim. Acta* **41**, 527–531.

Kulkarni, P.S., Crespo, J.G. and Afonso, C.A.M. (2008) Dioxins sources and current remediation technologies- a review. *Environ. Int.* **34**, 139–153.

Kwon, B.G., Lee, D.S.,Kang, N. and Yoon, J. (1999) Characteristics of p-chlorophenol oxidation by Fenton's reagent. *Water Res.* **33**, 2110–2118.

Lee, J.M., Kim, J.H., Chang, Y.Y. and Chang, Y.S. (2009) Steel dust catalysis for Fenton-like oxidation of polychlorinated dibenze-p-dioxins. *J. Hazard. Mater.* **163**, 222–230.

Levec, J. and Pintar, A. (2007) Catalytic wet-air oxidation processes: a review. *Catal. Today* **124**, 172–184.

Lin, C.H. and Tseng, S.K. (1999) Electrochemically reductive dechlorination of pentachlorophenol using a high overpotential zinc cathode. *Chemosphere* **39**, 2375–2389.

Lin, C.J., Lo, S.L. and Liou, Y.H. (2004) Dechlorination of trichloroethylene in aqueous solution by noble metal-modified iron. *J. Hazard. Mater.* **B116**, 219–228.

Liu, L., Zhao, G., Wu, M., Lei, Y. and Geng, R. (2009) Electrochemical degradation of chlorobenzene on bron-doped diamond and platinum electrodes. *J. Hazard. Mater.* **168**, 179–186.

Martins R.C., Rossi, A.F. and Quinta-Ferreira, R.M. (2010) Fenton's oxidation process for phenolic wastewater remediation and biodegradability enhancement. *J. Hazard. Mater.* (in press).

Mascia, M., Vasca, A., Polcaro, A.M., Palmas, S., Rodriguea, J. and Pozzo, A.D. (2010) Electrochemical treatment of phenolic waters in presence of chloride with boron-doped diamond (BDD) anodes: experimental study and mathematical model. *J. Hazard. Mater.* **174**, 314–322.

Miller, J.S. and Olejnik, D. (2001) Photolysis of polycyclic aromatic hydrocarbons in water. *Water Res.* **35**, 233–243.

Miyoshi, K., Kamegaya, Y. and Matsumara, M. (2004) Short communication electrochemical reduction of organohalogen compound by noble metal sintered electrode. *Chemosphere* **56**, 187–193.

Morales, J., Hutcheson, R. and Cheng, I.F. (2002) Dechlorination of chlorinated phenols by catalyzed and uncatalyzed Fe(0) and Mg(0) particles. *J. Hazard. Mater.* **B90**, 97–108.

Nissen, S., Alexander, B.D., Dawood, I., Tillotson, M., Wells, R.P.K., Macphee, D.E. and Killham, K. (2009) Remediation of a chlorinated aromatic hydrocarbon in water by photoelectrocatalysis. *Environ. Pollut.* **157**, 72–76.

Nowakowska, M. and Szczubialka, K. (1999) Photosensitised dechlorination of polychlorinated benzenes. 1. Carbazole-photosensitized dechlorination of hexachlorobenzene. *Chemosphere* **39**(1), 71–80.

Okawa, K., Suzuki, K., Takeshita, T. and Nakano, K. (2005) Degradation of chemical substances using wet peroxide oxidation under mild conditions. *J. Hazard. Mater.* **B17**, 68–72.

O'Mahony, M., Dobson, A.D.W., Barnes, J.D. and Singleton, I. (2006) The use of ozone in the remediation of polycylic aromatic hydrocarbon contaminated soil. *Chemosphere* **63**, 307–314.

Patel, U.D. and Suresh, S. (2008) Electrochemical treatment of pentachlorophenol in water and pulp bleaching effluent. *Sep. Purif. Technol.* **61**, 115–122.

Paukshtis, E.A., Simonova., L.G., Zagoruiko, A.N. and Balzhinimaev, B.S. (2010) Oxidative destruction of chlorinated hydrocarbons on Pt-containing fiber-glass catalysts. *Chemosphere* **79**, 199–204.

Pera-Titus, M., Garcia-Molina, V., Banos, M.A., Gimenez, J. and Esplugas, S. (2004) Degradation of chlorophenols by means of advanced oxidation processes: a general review. *Appl. Catal. B-Environ.* **47**, 219–256.

Perez-Moya, M., Graells, M., Valle, L.J., Centelles, E. and Mansilla, H.D. (2007) Fenton and photo-Fenton degradation of 2-chlorophenol: multivariate analysis and toxicity monitoring. *Catal. Today* **124**, 163–171.

Pillai, K.C., Kwon, T.O. and Moon, I.S. (2009) Degradation of wastewater from terephthalic acid manufacturing process by ozonation catalyzed with Fe^{2+}, H_2O_2 and UV light:direct versus indirect ozonation reactions. *Appl. Catal. B-Environ.* **91**, 319–328.

Polcaro, A.M., Palmas, S., Renoldi, F. and Mascia, M. (2000) Three-dimensional electrodes for the electrochemical combustion of organic pollutants. *Electochim. Acta* **46**, 389–394.

Qin, J., Karl, Q.Z. and Chuang, K.T. (2001) Catalytic wet oxidation of p-chlorophenol over supported noble metal catalysts. *Appl. Catal. B-Environ.* **29**, 115–123.

Rajkumar, D. and Palanivelu, K. (2004) Electrochemical treatment of industrial wastewater. *J. Hazard. Mater.* **B113**, 123–129.

Rivas, F.J. (2006) Polycyclic aromatic hydrocarbons sorbed on soils: a short review of chemical oxidation based treatment. *J. Hazard. Mater.* **B138**, 234–251.

Rivas, F.J., Garcia, R., Garcia-Araya, J.F. and Gimeno, O. (2008) Promoted wet air oxidation of polynuclear aromatic hydrocarbons. *J. Hazard. Mater.* **153**, 792–798.

Rodriguez, M., Abderrazik, N.B., Cntreras, S., Chamarro, E., Gimenez, J. and Esplugas, S. (2002) Iron(III) photoxidation of organic compounds in aqueous solutions. *Appl. Catal. B-Environ.* **37**, 131–137.

Rondinini, S. and Vertova, A. (2004) Electrocatalysis on silver alloys for dichloromethane and trichloromethane dehalogenation. *Electrochim. Acta* **49**, 4035–4046.

Saez, V., Vicente, M.D.E., Frias-Ferre, A.J., Bonete, P. and Gonzalez-Garcia, J. (2009) Electrochemical degradation of perchloroethylene in aqueous media: an approach to different strategies. *Water Res.* **43**, 2169–2178.

Scialdone, O., Guarisco, C., Galia, A. and Herbois, R., (2010a) Electroreduction of aliphatic chlorides at silver cathodes in water. *J. Electroanal. Chem.* **641**, 14–22.

Scialdone, O., Guarisco, C., Galia, A., Gurreri, L. and Randazzo, S. (2010b) Electrochemical abatement of chloroethanes in water: reduction, oxidation and combined processes. *Electrochim. Acta* **55**, 701–708.

Shibin, X., Schuichun, X. and Changquing, Z. (2007) Removal of organic pollutants from 2,2',5,5'-tetrachlorobenzidine (TCB) industrial wastewater by micro-electrochemical oxidation and air-stripping. *J. Hazard. Mater.* **144**, 159–163.

Silva, P.T.S., Silva, V.L., Neto, B.B. and Simonnot, M.O. (2009) Phenanthrene and pyrene oxidation in contaminated soils Fenton's reagent. *J. Hazard. Mater.* **161**, 967–973.

Shirayama, H., Tohezo, Y. and Taguchi, S. (2001) Photodegradation of chlorinated hydrocarbons in the presence and absence of dissolved oxygen in water. *Water Res.* **35**, 1941–1950.

Shimoda, S., Prengle, J.H.W. and Symons, J.M. (1997) H_2O_2/VisUV photo-oxidation process for treatment of waterborne hazardous substances-reaction mechanism, rate model, and data for tubular flow and flow stirred tank reactors. *Waste Manage.* **17**, 507–515.

Son, H.S., Im, J.K., Z. and K.D. (2009) A Fenton-like degradation mechanism for 1,4-dioxane using zero-valent iron(Fe^0) and UV light. *Water Res.* **43**, 1457–1463.

Stüber, F., Font, J., Paradowska, M., Suarez, M.E., Fortuny, A., Bengoa, C. and Fabregat, A. (2005) Chemical wet oxidation for the abatement of refractory non-biodegradable organic wastewater pollutants. *Inst. Chem. E.* **83**, 371-380.

Sunder, M. and Hempel, D.C. (1997) Oxidation of tri- and perchloroethene in aqueous solution with ozone and hydrogen peroxide in a tube reactor. *Water Res.* **31**, 33–40.

Teel, A.M., Warberg, C.R., Atkinson, D.A. and Watts, R.J. (2001) Comparison of mineral and soluble iron Fenton's catalysts for the treatment of trichloroethylene. *Water Res.* **35**, 977–984.

Tsyganok, A.I., Yamanaka, I. and Otsuka, K. (1999) Dechlorination of chloro aromatics by electrocatalytic reduction over palladium-loaded carbon felt at room temperature. *Chemosphere* **39**, 1819–1831.

US EPA (1982) *Development Document for Effluent Limitations Guidelines and Standards for the Inorganic Chemicals Manufacturing Point Source Category*, EPA-440/1-82-007, Office of Water, Effluent Guidelines Division, Office of Water Regulations and Standards, US Environmental Protection Agency, Washington D.C., United States.

US EPA (1987) *Development Document for Effluent Limitations Guidelines and Standards for the Organic Chemicals, Plastics, and Synthetic Fibers (OCPSF) Point Source Category*, EPA-440/1-87-009, Office of Water, Industrial Technology Division Office of Water Regulations and Standards, US Environmental Protection Agency, Washington D.C., United States.

US EPA (1993) *Supplement to the Development Document for Effluent Limitations Guidelines and New Source Performance Standards for the Chemicals, Plastics, and Synthetic Fibers Point Source Category*, Final, EPA-821/R-93-007, Engineering & Analysis Division, Office of Water, US Environmental Protection Agency, Washington D.C., United States.

US EPA (1995) *EPA Office of Compliance Sector Notebook Project Profile of the Inorganic Chemical Industry*, EPA/310-R-95-004, Office of Compliance, Office of Enforcement and Compliance Assurance, US Environmental Protection Agency, Washington D.C., United States.

US EPA (2009) *Technical Support Document for the Preliminary 2010 Effluent Guidelines Program Plan*, EPA/821-R-09-006, US Environmental Protection Agency, Washington D.C., United States.

Utset, B., Garcia, J., Casado, J., Domenech, X. and Peral, J. (2000) Replacement of H_2O_2 by O_2 Fenton and photo-Fenton reactions. *Chemosphere* **41**, 1187–1192.

Vaidya, P.D. and Mahajani, V.V. (2002) Insight into sub-critical wet oxidation of phenol. *Adv. Environ. Res.* **6**, 429–439.

Yim, B., Okuno, H., Nagata, Y. and Maeda, Y. (2001) Sonochemical degradation of chlorinated hydrocarbons using a batch and continuous flow system. *J. Hazard. Mater.* **B81**, 253–263.

Yalfani, M.S., Contreras, S., Medina, F. and Sueiras, J. (2009) Phenol degradation by Fenton's process using catalytic in situ generated hydrogen peroxide. *Appl. Catal. B-Environ.* **89**, 519–526.

Zhirong, S., Baohua, L., Xiangi H., Min, S., Quingnan, H. and Yongzhen, P. (2008) Electrochemical dechlorination of chloroform in neutral aqueous solution on palladium/foam-nickel and palladium/polymeric pyrrole film/foam-nickel electrodes. *J. Environ. Sci.* **20**, 268–272.

Zimbron, J.A. and Reardon, K.F. (2009) Fenton's oxidation of pentachlorophenol. *Water Res.* **43**, 1831–1840.

Chapter 10

Other industries and municipal landfill leachate

10.1 ALCOHOL DISTILLERIES

10.1.1 Industry overview, processes and wastewater character

Alcohol distillation is mainly employed for the purification of alcohol from aqueous blends. Although, there are several applications of the process three main areas of use are: distillation of alcohol following molasses fermentation; distillation of wine; and distillation for the industrial production of alcohol from agricultural raw materials. The first two applications are classified by the European Commission within "*the Food, Drink and Milk Industries Category*" (EC 2006). In the U.S.A., these productions are classified within "*Sugar and Confectionary Products*" SIC 206, and "*Beverages, Distilled and Blended Liquors*" SIC 208 (US EPA 1999). Alcohol distillation is a very common application worldwide. Although, the plants within sugar factories have large capacities, the capacities of individual plants, particularly wine distillation applications, vary in a wide range.

The distillation process is realised in a pot still either in a batch or continuous mode. The distillation of alcohol from the fermentation of molasses is carried out in column stills. The alcohol-aqueous phase, sometimes called beer, enters the first column. Crude alcohol is drawn off from the top of the column and then rectified in another column or a series of columns to obtain 95 % alcohol. At the bottom of the first column, an aqueous blend is drawn which is called stillage or commonly named as vinasse (Sheehan and Greenfield 1980; Vlyssides *et al.* 2005; EC 2006).

The amount of vinasse produced from distillation varies with the application. Zayas *et al.* (2007) gave an average figure of 16 m^3 of vinasse per tonne of anhydrous alcohol produced. Kannan *et al.* (2006) reported the range of vinasse volume as 12–14 L per litre of alcohol produced. Nava *et al.* (2008) estimated 16–124 L vinasse produced per kilogram of anhydrous alcohol. Alvarez *et al.* (2005) gave the amount of vinasse as 8–15 L per litre of alcohol produced for cherry stillage. The main pollutants of vinasse are organic matter, suspended solids, and nitrogen. However, vinasse also contains in high concentrations of potassium, calcium, chloride and sulphate. The organic matter content is variable depending on the feed material and the distillation application. Sheehan and Greenfield (1980) gave the large interval of organic content as 15–176 g/L for COD, and 7–95 g/L for BOD_5. Vlyssides *et al.* (1997) presented an average figure of 72 g/L COD for the vinasse from beet molasses. Detailed wastewater compositions for vinasse were also given in the related literature (Sheehan and Greenfield 1980; Vlyssides *et al.* 1997; Kannan *et al.* 2006). Although, a great part of the organic content of vinasse is biodegradable, there are recalcitrant organic compounds such as polyaromatic compounds, polyphenols mainly lignins, tannins, and humic acids. Thermal degradation of reducing sugars and amino compounds is mainly the source of the melanoidins giving the dark colour of vinasse. These compounds exert negative effects on the microorganisms and inhibit biological treatment processes (Kannan *et al.* 2006; Zayas *et al.* 2007; Parnaudeau *et al.* 2008, Lucas *et al.* 2010). Phenolic content varies with the type and strength of the vinasse. Carbajo *et al.* (2007) measured 268 mg/L phenolic content as gallic acid in a wine-distillery vinasse which had a total COD of 9250 mg/L. Polyphenol content of a cherry stillage was found to be 2–2.5 g/L as gallic acid against total COD of 145–180 g/L (Alvarez *et al.* 2005). Beltran *et al.* (2000) measured total phenols 735 ± 127 mg/L as gallic acid and COD as 21,715 ± 1236 mg/L in a wine-distillery vinasse.

In-plant control for alcohol distilleries is mostly following good operation practices and good housekeeping (EC 2006). If a distillery is located in a sugar mill, the vinasse may be used for cane wash. The vinasse may also be recycled as 10–20 % of the molasses dilution water (Sheehan and Greenfield 1980). Sheehan and Greenfield (1980) made a review of utilisation, treatment, and disposal of distillery wastewater. Direct land application of vinasse is a common method, but it has several limitations. Parnaudeau *et al.* (2008) investigated the mineralisation potential of vinasse depending on raw material and production process. Vinasse may also be reused for the production of biochemicals and biomass, and be utilised as a food supplement (Sheehan and Greenfield 1980). The basic treatment method for vinasse is biological treatment mostly as a combination of anaerobic and aerobic treatment. Due to inhibitory substances and pH buffering capacity, direct anaerobic treatment of vinasse may pose problems and needs to be closely controlled (Batstone and Steyer 2007). However, anaerobic processes are only the applicable treatment method for vinasse if it is not diluted. (Sheehan and Greenfield 1980; EC 2006; Yang *et al.* 2008). Activated sludge process may be used after a significant dilution (1/80 dilution factor) for vinasse treatment (Alvarez *et al.* 2005). Preliminary treatment of screening is used to eliminate coarse solids. Sometimes coagulation-flocculation or flotation may be used as a pretreatment (Sheehan and Greenfield 1980). However, the basic problem is the removal of inert and inhibitory organic material and enhancement of the biodegradability in the pretreatment of vinasse. For this purpose, most commonly applied methods are chemical oxidation processes. In the next subsection, application of chemical oxidation processes to vinasse will be briefly reviewed.

10.1.2 Oxidation processes

The ozonation has been the most widely tested process for vinasse pretreatment to enhance the biodegradability. Several studies were conducted on both synthetically prepared samples (SPS) and real wastewaters to evaluate the effect of ozonation on the vinasse and specific components of the vinasse. Carbajo *et al.* (2007) experimentally evaluated the effect of ozonation on phenolic wastewaters using SPS which was composed of syringic, pyruvic and gallic acids as well as real wastewaters, wine distillery, olive oil extraction and table olive debittering. The catalyst perovskites ($LaTi_{0.15}Cu_{0.85}O_3$) was used in the ozonation experiments. This catalyst was found effective up to 0.1 g/L concentration above which no additional benefit was provided. Ozonation of SPS provided over 90 % mineralisation. The efficiency increased with elevated temperature and increasing pH up to 11. The wine distillery wastewater ozonation also yielded over 90 % COD removal. Addition of catalyst did not increase the efficiency. Santos *et al.* (2005) made an experimental study to estimate selectivity of ozone for the removal of polyphenols from vinasse. In a 2 hour ozonation at pH around 4, the removal of phenols (measured as caffeic acid) was 83 % while COD and TOC removals were 17–18 %. The authors concluded that the reaction of ozone with phenolic compounds

4 to 20 times faster than the reaction with other organic compounds. Santos *et al.* (2003) evaluated the variation of biodegradability of vinasse by ozonation in acid and alkaline media. The ozonation in acidic condition was conducted at original pH of 4.5 of the sample. Two hour ozonation in acidic condition yielded 18 % COD and 83 % phenols (as caffeic acid) removal. In alkaline conditions (pH 8.1–8.8) COD and phenol removals were 11 and 65 %, respectively. Benitez *et al.* (2003) ozonated vinasse having 24,500 mg/L COD to observe the decrease in COD and aromatic content. At pH 4, ozonation for 9 hours provided 25 % COD and 51.4 % aromatic compound removals. In a 6 hour ozonation conducted with UV/H_2O_2 application (0.168 M H_2O_2) COD and aromatic content removals increased to 58.4 and 76.9 %, respectively. Ozonation in the discontinuous mode did not provide a significant increase in the removal efficiencies. Kinetic evaluation of the process was also made. The variation of biodegradability was assessed using an activated sludge system and measuring the kinetic parameters. COD removals in an activated sludge system with 48 h retention time were 27.7 and 39.3 % for raw and ozone-pretreated wastewaters, respectively. Beltran *et al.* (2001) tested sequential ozonation of vinasse by changing the pH. In the alkali pH, the process made use of radical oxidation while succession in the acid pH direct ozone oxidation together with the stripping of radical scavenger carbonates were carried out. The process efficiency was maximised by controlling number of cycles, pH and ozonation times at both acid and alkali phases. The process, although, provided higher organic matter removal than those of ozonation at only acid or alkali pH values, constant pH ozonation proved to be more efficient in polyphenol removal. The experiments also indicated that both successive and constant pH ozonation increased the biodegradability of the wastewater in terms of BOD_5/COD ratio. Improvement in the biodegradability by ozonation was also assessed using mixtures of vinasse and domestic wastewaters and using batch and continuous activated sludge systems which yielded similar results with those of above-mentioned studies (Beltran *et al.* 1999; 2000; Benitez *et al.* 2000). Alvarez *et al.* (2005) assessed the effect of pre-ozonation on the performance of anaerobic process for cherry stillage using an anaerobic sequencing batch reactor. Ozonation at a low dose of 25 mg ozone per gram of initial COD, proved to be satisfactory by removing more than 75 % of polyphenols and enhancing the overall performance and stability of anaerobic reactor.

In addition to ozonation, advanced oxidation processes were applied for the treatment of vinasse. Navarro *et al.* (2005) studied photocatalytic H_2O_2 oxidation of vinasse provided from a wine industry. The vinasse used had 4000 mg/L COD. Solar radiation and artificial light were used in the experiments. TiO_2 and clays containing iron were used as catalysts. Up to 58 % COD removal was obtained with the optimum doses of 2.5 mL/L H_2O_2 (30 % w/w) and 1.0 g/L TiO_2. De Heredia *et al.* (2005) used a Fenton-coagulation-flocculation process for the treatment of vinasse. Lime was used as a coagulant. The vinasse with a COD of 15,000–16,500 mg/L was used in several dilutions in the experiments. With optimal ratios H_2O_2/Fe^{2+} = 15 mol/mol and 0.15 g COD/g H_2O_2, 70 % COD removal was obtained at 3/16 diluted sample. Lucas *et al.* (2010) applied ozone-based advanced oxidation processes to winery wastewater. The wastewater COD and polyphenol (as gallic acid) concentrations were 4650 mg/L and 103 mg/L, respectively. Direct photolysis and direct ozonation were found to be ineffective. The performance of O_3/UV/H_2O_2 was better than O_3/UV. Increase of pH 4 to 10 enhanced the rate of oxidation. COD removals around 50 % were obtained. The cost of €1.31/m^3–g TOC mineralised was calculated for the O_3/UV/ H_2O_2 process at pH 4 and COD/H_2O_2 ratio of 2. Martin *et al.* (2002) studied pretreatment of vinasse by O_3, O_3/UV and O_3/UV/TiO_2 to assess anaerobic treatability. H_2O_2 and TiO_2 dosages were 4×10^{-3} M and 2 g/L, respectively. The wastewater COD was 109,200 mg/L. COD removal efficiencies of O_3, O_3/UV and O_3/UV/TiO_2 treatments were 25, 31 and 31 %, respectively. Specific rate of methane production of the anaerobic digestion was given as 3.56, 3.67, 3.56, and 4.45 mL CH_4/g VSS.h for raw vinasse, ozone-pretreated vinasse, O_3/UV-pretreated vinasse and O_3/UV/TiO_2-pretreated vinasse, respectively. Santana and Machado (2008) investigated photocatalytic degradation of the vinasse under solar radiation. The degradation tests lasted 5 consecutive days with the use of TiO_2, Nb_2O_5–TiO_2 and immobilised TiO_2 in glass beads as catalysts. TOC removals were 45–55 % for all catalysts. LC_{50} values were determined by *Artemia saline* bioassay. LC_{50} was increased from 1.54 of untreated vinasse to 1.77–2.05 after oxidation. Lucas *et al.* (2009a) used SPS prepared from grape juice and wine to evaluate the photochemical treatability of winery wastewater. Photochemical treatment by the use of solar panels and TiO_2 addition at 200 mg/L yielded only 8 % TOC reduction for the sample prepared from wine and having initial COD of 4400 mg/L. Yang *et al.* (2008) applied Fenton's oxidation to biologically treated vinasse. Biological treatment was a two stage treatment involving anaerobic and aerobic. The aerobic treatment effluent was undergone Fenton's oxidation. Using 450 mg/L $FeSO_4$ and 300 mg/L H_2O_2 the effluent COD could be reduced to 250 mg/L in 30 minutes. This application yielded 60 % COD removal.

Vlyssides *et al.* (1997) used a continuous flow electrochemical cell, made up of a stainless steel cathode and a platinum alloy covered titanium alloy anode for the pretreatment of vinasse from beet molasses. Chloride ions were added to the wastewater at 4 % NaCl ratio to increase conductivity. The temperature was kept constant at 42°C. Total COD of vinasse was 72,000 mg/L. Electrochemical oxidation of vinasse yielded 90 % COD removal producing an effluent with 8000 mg/L COD at the optimum operating conditions (pH 9.5; 30 mL/min of vinasse feed rate; 20 g/h/A/m^2). A similar study was conducted by Kannan *et al.* (2006) on vinasse sample taken from a distillery industry. The strength of vinasse was lower than that used by Vlyssides *et al.* (1997) with 18,868 mg/L COD. The vinasse sample was first screened to remove large suspended solids. The electrochemical cell was comprised of a stainless steel cathode and an aluminium anode. The applied current density was 182 A/m^2 and the temperature was kept constant at 30°C. The initial pH 4.78 of the vinasse increased up to 5.96 at 60th minute operation of electrooxidation. At 60th minute, turbidity and colour were practically totally removed while COD removal was 75 %. The experiment was repeated with the addition of 1 g/L activated carbon obtained from areca nut and 80 % COD removal was obtained. The BOD_5/COD ratio (0.61) of the vinasse increased to 0.76 after EC treatment without activated carbon addition. Nava *et al.* (2008) studied electrochemical treatment of synthetic vinasse with initial COD of 75,096 mg/L prepared using citric acid, glycerol, glucose and saccarose. The electrodes were platinum coated titanium and three-dimensional boron doped diamond. The current density was 10 m A/cm^2. A high level of mineralisation (97 % COD removal) was obtained against 168 kWh/m^3 energy consumption. Rincon *et al.* (2009) investigated electrochemical oxidation of vinasse by the addition of hydrogen peroxide using galvanized steel electrodes. In their experimental study, the highest COD removal of 58 % was obtained producing an effluent COD of 90,000 mg/L, at basic pH, 20 mA/cm^2 current density and 60,000 mg/L H_2O_2 dosage. Zayas *et al.* (2007) studied electrochemical treatment of biologically treated vinasse. The effluent from anaerobic treatment was first undergone coagulation-flocculation using $FeCl_3$ as a coagulant, and then introduced to electrochemical oxidation. Electrooxidation was realised using Ti/RuPb(40 %)O_x anode and Ti/PtPd(10 %) O_x cathode electrodes with a high chloride concentration of 26,000 mg/L in the effluent obtained from $FeCl_3$ coagulation application. Electrooxidation provided 97 % COD removal producing an effluent free from turbidity and a low COD of 39 mg/L.

Alcohol distilleries are common processes worldwide. The wastewater originating from the process is comprised totally of organic matter and it is concentrated. In-plant control alternatives for improving the wastewater quality and reduction of the volume are limited, since raw materials and the production process cannot be modified. The wastewater, although a great part of it biodegradable, contains an inert fraction which poses several problems in both aerobic and anaerobic treatment. A pretreatment is necessary before biological treatment. Chemical oxidation seems to be the most suitable alternative for the modification of the characteristics of organic matter. The problem is the selectivity of the chemical oxidation application. Ozonation is selective for the important hardly biodegradable group of polyphenols, however, improvement in the biodegradability by ozonation is marginal. Advanced oxidation applications exhibit a similar behaviour, however, total organic matter removal is higher, and they partly replace the function of biological treatment. The organic matter removal efficiencies of electrooxidation are so high that they function not as a pretreatment but rather as a final treatment method. Electrocoagulation also performs well as a polishing treatment following biological treatment.

10.2 OLIVE OIL INDUSTRY

10.2.1 Industry overview, processes and wastewater character

Olive oil industry is an important agro-alimentary branch and of crucial economic importance for many Mediterranean countries that account for almost 95 % of the olive oil production worldwide. About 9.7 million tons of olives are grown annually from which approximately 1.9 million tons of olive oil are processed in several thousands of olive oil mills. Especially in Portugal, Italy, Greece, Spain, and Turkey, the majority of olive mills are small enterprises, family-owned in most cases, and employ only few workers. During olive oil production, solid and liquid wastes such as pomace and olive mill wastewater (OMW) are produced. Pomace is evaluated economically, but OMW is discharged directly into the environment. Generation and disposal of around 30 million m^3 of OMW per year constitute a severe environmental pollution problem for the Mediterranean Region, due to unique features associated with this type of agro-wastewater, namely seasonal and localised production (typically between the months November and February) and the high organic, phenolic carbon content being resistant to biodegradation (Gotsi *et al.* 2005; Cañizares *et al.* 2007; Najjar *et al.* 2007). The characteristics of OMW generated in small-to-medium scale companies largely depend on the type of extraction process employed, *i.e.* the traditional press method or the continuous, three or two-phase centrifugation process starting from the pulp of olive fruits obtained by grinding them with stones or knife-edge spinal mills. In the traditional method, the ground olives are pressed in bags and thereafter, oil is separated from the liquid mixer by resting in a series of tanks or by using a centrifuge. In the continuous three or two-phase methods, the crushed olive fruits are pumped into a decanter and the impure oil is centrifuged. While the two-phase system results in a paste-like residue, both the traditional and the multi-phase systems produce liquid OMW. The three-phase system results in the production of approximately 1.2 kg OMW and 0.5 kg solid waste per kg of olives, while the traditional pressing process generates 0.5 kg solid waste and 0.5 kg wastewater per kg of olives (Gotsi *et al.* 2005). As aforementioned, around 12,000 oil mills currently produce over 1.9 million tons of olive oil in the EU. More than half of them are estimated to be still working with the traditional or 3-phases systems (especially small olive oil mills).

OMW is a liquid of blackish colour with an unpleasant smell that usually contains, rather in suspension, olive pulp residues, mucilage, pectic substances and small amounts of oil (0.5–2.0 %) forming a stable emulsion. OMW generates a superficial film in discharged effluent and soils causing a remarkable additional toxicity. The high organic matter content of OMW varies in the range of 80–200 g/L COD, 50–100 g/L BOD_5, and 10–100 g/L TOC. The dark colour of OMW disturbs the bright appearance of water and prevents the absorption of sunlight by photosynthetically growing organisms such as water plants and algae. Oil from OMW also forms a film layer on the water surface and prevents oxygen transport from air to water. Different organic substances found in OMW include sugars, tannins, phenolic compounds (polyphenols), polyalcohols, pectins, and lipids (D'Annibale *et al.* 1998). The toxicity and antimicrobial activity of OMW are mainly attributable to its polyphenolic content (Borja *et al.* 1997; Bisignano *et al.* 1999; De Marco *et al.* 2007). The phenolic compounds present in OMW are typically classified under three categories; namely cinnamic acid derivatives such as p-coumaric, caffeic and ferulic acids, benzoic acid derivatives including benzoic, hydroxybenzoic, gallic, protocatechuic, or vanillic acids, as well as tyrosol related compounds such as hydroxytyrosol or hydroxyphenylacetic acid (Perez *et al.* 1992; Hamdi 1993; Minh *et al.* 2007). OMW is considered as an eco-toxicologically dangerous effluent negatively affecting aquatic and terrestrial living organisms. OMW also contains inorganics such as potassium and sodium chlorides, sulphates and phosphates, as well as calcium, iron, magnesium, sodium, copper and traces of other elements and metals (Minh *et al.* 2007).

Until the early 1970's, the generalised treatment solution for the management of OMW was natural evaporation, in absence of other effective technical and economic procedures; OMW was stored in lagoons near the olive oil mills. All discharges generated during the olive oil production season were transferred to these evaporation ponds and subjected to natural drying during the rest of the year until the start of a new campaign. The biological degradation of the organic matter present in OMW generates a mixture of sludge and water of dark colour and intense smell. An additional problem is the requirement of huge volumes of lagoons and therefore treatment and storage costs become very high. However, these methods are no longer acceptable as they lead to foul odours and serious surface and groundwater contamination. The effluent OMW from this kind of treatment no longer complies with the recent EU directives and other discharge consents. Nevertheless, the problem was not only resolute but postponed in time. In order to have an idea on the high polluting effect of OMW, it can be stated that the processing of 1000 kg of olive causes a contamination of a population equivalent of 300–500 capita (Gotsi *et al.* 2005).

10.2.2 Oxidation processes

Different chemical oxidation methods have so far been applied for OMW treatment. Some examples include oxidation with the Fenton's reagent (Lin and Lo 1997; Beltran-Heredia *et al.* 2001; Rivas *et al.* 2001a; Neyens and Baeyens 2003; Vlyssides *et al.* 2003; Neyens *et al.* 2004; Vlyssides *et al.* 2004; Ahmadi *et al.* 2005; Paraskeva and Diamadopoulos 2006; Dogruel *et al.* 2009); electrocoagulation with aluminium or iron electrodes (Lin and Chang 2000; Ciardelli and Ranieri 2001; Adhoum and Monser 2004; Inan *et al.* 2004; Lai and Lin 2004; Paraskeva and Diamadopoulos 2006), ozonation (Benitez *et al.* 1999) as well as electrochemical (Israilides *et al.* 1997; Gotsi *et al.* 2005; Deligiorgis *et al.* 2008), supercritical water (Rivas *et al.* 2001b) and (catalytic) wet air oxidation (Minh *et al.* 2007; Katsoni *et al.* 2008) processes that have all been explored in pilot and/or laboratory-scale treatment experiments. The above mentioned chemical oxidation methods usually resulted in complete or high phenolics removal but only partial COD and TOC reductions (40–65 %). Regarding photochemical advanced oxidation based on solar and/or UV light irradiation, TiO_2-mediated photocatalytic (Gernjak *et al.* 2004; Chatzisymeon *et al.* 2009a) and photo-Fenton treatment of OMW (Gernjak *et al.* 2007) was reported. Below, some more detailed treatment results were given for more recent chemical oxidation applications to OMW.

Mert *et al.* (2010) investigated Fenton and Fenton-like processes to treat OMW (COD = 115 g/L; filtered COD = 34 g/L; total phenols = 5.6 g/L; oil and grease = 23 g/L; SS = 32 g/L; pH = 5.2). At the same time, the acute toxicities of raw, physicochemically pretreated, Fenton and Fenton-like treated OMW on activated sludge microorganisms using respiration inhibition test (ISO 8192 2007) are presented. Chemical pretreatment (acid cracking and coagulation-flocculation) positively affected the biodegradability and reduced the inhibitory effect on activated sludge resulting in >67 % COD and >72 % total phenol removals. Fenton and Fenton-like processes showed high COD (>80 %) and total-phenol (>85 %) treatment performance on pretreated OMW.

Lucas *et al.* (2009b) described the application of Fenton's reagent for the removal of COD from OMW (COD = 61 g/L; BOD_5 = 10 g/L; total polyphenols = 0.60 g/L; SS = 7 g/L; pH = 4.3) in a laboratory-scale batch reactor. The effect of different operational conditions, namely, hydrogen peroxide and ferrous ion concentrations, temperature and initial pH were evaluated. The oxidation reduction potential, pH and dissolved oxygen were on-line monitored during the treatability studies. Working at an initial pH equal to 3.5, a temperature of 30°C, and a molar H_2O_2/Fe^{2+} of 15 (corresponding weight ratio H_2O_2/COD = 1.75) resulted in a COD removal of 70 %. The global activation energy (E_a) could be calculated for the Fenton's reaction by using the Arrhenius equation parameters and pseudo-first order reaction coefficients obtained at varying temperatures; E_a was determined as 28.2 kJ/mol.

Chatzisymeon *et al.* (2009b) studied the electrochemical treatment of OMW (parameter values prior to filtration: COD = 47 g/L; total phenols = 8.1 g/L; TS = 50.3 g/L; pH = 4.6; conductivity = 17 mS/cm) over boron-doped diamond (BDD) electrodes. In order to evaluate the effect of the various process parameters involved in the treatment process, a factorial design methodology was implemented. Process independent variables and ranges were COD (1000–5000 mg/L), treatment time (1–4 h), current intensity (10–20 A), initial pH (4–6) and the use of 500 mg/L H_2O_2 as an additional oxidant on the responses percent COD, phenols, aromatics and colour removal. Most important process variables were determined to be COD of OMW and treatment time. The model suggests that COD degradation kinetics is zero order and satisfactorily agrees with an electrochemical model for the anodic oxidation of organics over BDD developed elsewhere. The treatability results were established as 19 % COD and 36 % phenols removal at a specific energy consumption (SEC) of 96 kWh/kg COD of the undiluted effluent (40,000 mg/L COD) being tested at 20 A for 15 h electrochemical treatment. Under the experimental conditions in question, BDD electrochemical oxidation does not appear to improve the aerobic biodegradability of OMW as assessed by the Zahn-Wellens biodegradability test.

Khoufi *et al.* (2009) developed a multi-stage process for pilot scale treatment of OMW (original values: COD = 95 g/L; BOD_5 = 19 g/L; TSS = 15 g/L; lipids = 9.8 g/L; polyphenols = 11.5 g/L; pH = 5.2) that combines electro-Fenton followed by sedimentation, anaerobic digestion and ultrafiltration having a cut-off 25 kDa. Application of the electro-Fenton process in semi-continuous mode permitted 50 % COD and 95 % phenols removal. This pretreatment was found to enhance the anaerobic activity of an up-flow anaerobic filter (300 L) significantly. In the bioreactor, a COD removal efficiency of 75 % was reached at a HRT of 4.5 days and an OLR of 10 g COD/L.d. The use of ultrafiltration technology as a post-treatment completely detoxified the anaerobic effluent and removed its high molecular weight polyphenolic carbon content. The SEC of the electro-Fenton process was calculated as 31 kWh/m^3 corresponding to a running cost of €3/m^3. Valorisation of the produced anaerobic biogas to electrical energy is possible such that a surplus energy of approximately 73.5 kWh/m^3 could be recovered after methanisation.

The objective of the study carried out by Hanafi *et al.* (2010) was to assess the electrocoagulation of 5-fold diluted OMW (COD = 20,000 mg/L; polyphenols =

260 mg/L; Cl^- = 1160 mg/L; conductivity = 3.6 mS/cm; pH = 4.2) using aluminium electrodes. In this respect, the effects of electrolysis time, current density, chloride concentration and initial pH on the removal of COD, polyphenols and colour was examined. Optimum electrocoagulation conditions were established as follows: an electrolysis time of 15 min; a NaCl concentration of 2 g/L; and an initial pH of 4.2 at a current density of 250 A/m^2. Under optimised treatment conditions, 70 % COD and phenolics removal were obtained whereas the value for specific electrode consumption was 0.085 kg Al/kg COD removed and the SEC value was determined to be 2.63 kWh/kg COD. Under these optimal experimental conditions, OMW became non-toxic for the test organism *Bacillus cereus*.

TiO_2-mediated photocatalytic treatment of OMW (properties before filtration: COD = 47 g/L; total phenols = 8.1 g/L; TS = 50.3 g/L; pH = 4.6; conductivity = 17 mS/cm) was investigated in a batch type, laboratory scale photoreactor (Chatzisymeon *et al.* 2009a). UV-A irradiation was provided by a 400 W, high pressure mercury arc lamp and Degussa P25 TiO_2 was used as the photocatalyst. The OMW effluent was subjected to filtration to remove most of its total solids (TS) content prior to heterogeneous photocatalysis. A multivariate analysis was employed to study the individual and interactive effects of various process parameters (COD, TiO_2 concentration, pH, contact time and hydrogen peroxide addition) on the responses COD and total phenols removal. The photocatalytic treatment of OMW was modelled and optimised using factorial design. Effluent decolourisation as well as process efficiency in terms of energy consumption was also evaluated. Finally, the acute toxicity of OMW samples prior to and after photocatalytic treatment was assessed. COD and total phenols removal efficiencies were positively affected mostly by contact time and secondly by influent COD, whereas all other process variables had no significant statistical importance to the responses. The untreated effluent was highly toxic to *Vibrio fischeri* resulting in an EC_{50} value of 12 % on volumetric basis (v/v). The acute toxicity of the treated OMW remained nearly unchanged for high COD values (EC_{50} = 15 % v/v), indicating that the residual COD bearing 200 mg/L total phenols contained various toxic species.

Chedeville *et al.* (2009) examined the ozonation of major pollutants found in OMW as well as actual OMW. In the first phase of the study, bimolecular rate constants of three phenolic compounds (*e.g.* gallic acid, p-hydroxybenzoic acid and p-coumaric acid) with ozone were determined by using the competition kinetic model. These rate constants, obtained at pH 5, were found to be high and ranging between 3.8×10^4 L/mol.s–2.9×10^5 L/mol.s, speaking for a diffusion controlled kinetic regime. Thus, to obtain an efficient ozonation process, it is important to adapt the ozone contactor (gas-liquid mass transfer) to this regime. In the second stage of the study, real OMW (COD = 6.4 g/L; SS = 3 g/L; pH = 5) was subjected to ozonation. In order to have a pollution load compatible with the ozoniser capacities, 6.5 L of this wastewater was mixed with 33.5 L of tap water before introduction in the ozonation pilot. The experiment was carried out at 20°C, with a liquid flow rate of 0.9 L/h and a gas flow rate of 0.3 m^3/h. Ozone concentration in the inlet gas was 23 g/m^3. Up to 80% of phenolic compounds were removed and the ozone mass transfer efficiency reached 95% during the first 15 min of ozonation.

10.3 MUNICIPAL LANDFILL LEACHATE

10.3.1 Overview

Sanitary landfilling is the most common, economical and environmentally acceptable way to dispose of municipal solid wastes all over the world. This implementation covers monitoring of the incoming waste flows, placement and compaction of solid wastes, and installation of landfill environmental monitoring and control facilities (Tchobanoglous *et al.* 1993). Several types of liquid waste and wastewater produced by landfilling activity encompass landfill leachate, landfill gas condensate, truck/equipment washwater, drained free liquids, floor washings, laboratory-derived wastewater, storm water, contaminated ground water, and wastewater from recovering pumping wells (US EPA 2000).

Landfill leachate may be explained as liquid which percolates through or emerges from solid waste. It can be collected from a liner system placed at the bottom of landfill, or through the use of slurry walls, trenches, or other containment systems. This collected leachate contains soluble, suspended, or miscible constituents removed from wastes disposed of. The leachate generated from landfills exhibits different characteristics from site to site depending on several factors: (i) the type of wastes received; (ii) operating practices encompassing shedding, daily cover and capping; (iii) compaction of wastes; (iv) depth of fill; (v) annual precipitation; and (vi) landfill age, representing degree of stabilisation (US EPA 2000). Among these factors, landfill age is one of the most important parameters used in the selection of treatment method to be applied to leachate since landfill leachate character dramatically changes by time due to a complex series of biological and chemical reactions taking place during the decomposition period of solid wastes buried in a landfill. These reactions occur at least four sequential stabilisation phases, namely, initial aerobic phase, anaerobic acid phase, initial methanogenic phase, and stable methanogenic phase (Kjeldsen *et al.* 2002). Taking into account these stabilisation phases or degrees, the landfill leachate is classified into three groups as young (new), medium age and old (mature) landfill leachates. Landfill leachate characterisation based on this classification is summarised in Table 10.1

in terms of main pollutant parameters (arranged from Renou *et al.* 2008). As can be seen from table, the ratio of BOD/COD decreases from 0.70 to 0.11 with the aging of landfills. Therefore, old landfill leachate is represented by its low BOD/COD ratio and fairly high ammonia content comprised of about 80 % of TKN whereas young landfill leachate is characterised by extremely high organic matter content.

Table 10.1 Landfill leachate characterisation (arranged from Renou *et al.* 2008)

Parameter	Unit	Young	Medium age	Old
Age	year	<5	5–10	>10
COD	mg/L	1870–70,900	1180–9500	100–3460
BOD	mg/L	90–26,800	331–1436	3–150
BOD/COD	-	0.05–0.70	0.07–0.33	0.04–0.11
TSS	mg/L	950–5000	480–784	130–1600
NH_3	mg N /L	10–11,000	743–5500	0.2–1522
pH	-	5.8–9.0	6.9–9.0	7.0–9.4

Biological processes including aerobic and anaerobic systems such as aerated lagoons, conventional activated sludge systems, anaerobic filters, up-flow anaerobic sludge blanket reactors, anaerobic digesters *etc.* are the most common treatment methods used to treat landfill leachate with high organic matter contents owing to their reliabilities, simplicities and high efficiencies. Moreover, the combinations of biological processes with physical and chemical treatment methods are also widely applied to landfill leachates to remove suspended solids, floating materials, inorganic pollutants, colloidal particles, colour, and recalcitrant or toxic compounds. These physical and chemical methods such as stripping, coagulation-flocculation, chemical precipitation, floatation, adsorption, chemical oxidation, ultrafiltration *etc.* are employed as a pre- or polishing treatment step or in order to treat a specific pollutant such as ammonia. Among them, chemical oxidation methods applied to municipal landfill leachate will be introduced in the following subsection since they are covered in the scope of this book.

10.3.2 Oxidation methods applied to landfill leachate

During last few decades, Fenton and photo-Fenton processes have been intensively studied using old and biologically treated landfill leachates so as to enhance the abatement of recalcitrant organic molecules or to convert them into more readily biodegradable forms. These processes have been performed using a wide range of ferrous iron and hydrogen peroxide doses at generally acidic pH values. Table 10.2 summarises the compiled data from the related scientific literature. The published data have demonstrated that up to 76 % COD removal efficiency could be achieved and biodegradability expressed as the ratio of BOD/COD could be improved by the application of these processes depending on landfill leachate strength.

Table 10.2 Applications of Fenton and photo-Fenton (updated upon Wang *et al.* 2003)

COD mg/L	BOD mg/L	pH[1]	Fe^{2+} mg/L	H_2O_2 mg/L	UV	f[2]	R[3]	Source
1500–2020	50–270	4.0	600–800	200	–	60	na	Gau&Ch. 1996
1150	3–5	3.0	56	2440	–	70	na	Kim *et al.* 1997
2000	87	3.5	120	1500	–	69	0.58	Kim&Huh 1997
338	<8	3.0	20	10	–	72	0.30	Wel. &He. 1998
1500	30	3.5	645	1650	–	75	na	Kang&H. 2000
1800	225	3.0	2000	1500	–	52	0.22	Kim *et al.* 2001
1800	225	4.5	1500	1200	–	45	0.27	Kim *et al.* 2001
1500	75	6.0	300	200	–	70	na	Lau *et al.*2001
1000	na	2.5	2800	2550	–	61	na	Zang *et al.* 2005
1100–1300	<0.05	3.0	3920	8160	–	61	na	Deng 2007
5700	3600	3.5	56	650	–	66	0.88	Koc.&Ko. 2009
2320–2480	180–345	3.0	560	1020	–	58	na	Mokajeri *et al.* 2010a
4284	28.4	3.6	5600	8160	–	45	na	Wu *et al.* 2010
570–585	10-20	5.9	538	777	–	72	0.21	Li *et al.* 2010
4113–9257	1440–3192	4.0	437.5	3500	–	76	0.71	Bar. *et al.* 2010
1150	3–5	3.0	56	1150	+	70	na	Kim *et al.* 1997
1150	na	3.2	72	1150	+	72	na	Kim&Vo. 1998

[1]*optimum pH;* [2]*percent of COD removal efficiency;* [3]*ratio of BOD/COD; na: not available.*

The performances of other advanced oxidation processes (AOPs) using ozone with hydrogen peroxide (O_3/H_2O_2), ozone with ultraviolet (O_3/UV), and hydrogen peroxide with ultraviolet (H_2O_2/UV) have been also tested on old and biologically treated landfill leachate. Table 10.3 outlines the compiled data from the studies performed using these processes. As seen from the table, COD removal efficiencies are comparable with those of hydrogen peroxide based AOPs (*i.e.* H_2O_2/UV and O_3/H_2O_2) and, in most cases, applications of these processes yield better COD reduction than UV photolysis of ozone.

Although, the above mentioned processes are still addressed as a category of effective means to handle the recalcitrant organic compounds in old and biologically treated landfill leachate, there are several limitations standing out in their applications. One of the major limitations in their applications of UV based processes is the reduction of effectiveness of UV lamps owing to suspended solids, dark brown or yellow colour of landfill leachate, and formation of sparingly soluble oxalate film on the lamps (Wang *et al.* 2003). In addition, high background absorbance of landfill leachate or high particle matter concentrations may fail to initiate the radical reactions. Furthermore, high energy consumption is another disadvantage of UV based processes.

Table 10.3 Applications of AOPs to landfill leachate (updated upon Wang *et al.* 2003)

COD mg/L	BOD mg/L	pH	Ozone g/L	H_2O_2 g/L	UV (W)	R[1]	f[2] (%)	Source
2000	na	na	3.5[3]	0.4[4]	–	na	95	Wable *et al.* 1993
600	na	na	3.2[3]	0.4[4]	–	na	92	Wable *et al.* 1993
1090	39	8.3	8.0	0.63	–	0.18	63	Wang *et al.* 2004
6500	500	8.1	1.2	1	–	0.5[5]	na	Wu *et al.* 2004
5300	500	8.7	0.08	2	–	0.71	48	Tizaoui *et al.* 2007
620[5]	na	9.0	2.0	1	–	na	50[5]	Hagman *et al.* 2008
2300	210	8.0	1.0[3]	–	15	na	50	Bigot *et al.*1994
1280	100	2.0	na	–	100	na	54	İnce 1998
1280	100	2.0	na	–	500	na	47	İnce 1998
6500	500	8.1	1.2	–	200	0.6[5]	na	Wu *et al.* 2004
760	na	3	-	3.4	150	na	99	Schulte *et al.* 1995
1280	100	2.0	-	na	100	na	57	İnce 1998
1280	100	2.0	-	na	500	na	59	İnce 1998
3750	na	na	-	7.9	36×4	na	65	Shu *et al.* 2006

[1] *BOD/COD;* [2]*COD removal efficiency;* [3]*g/g COD;* [4]*g/g* O_3*;* [5]*approximately; na: not available.*

In recent years, electrochemical treatment techniques have been successfully applied to raw and biologically treated landfill leachate in order to reduce organic matter and colour, and to oxidize ammonia. Various configurations of electro-reactors equipped with boron doped diamond on silicon anode and stainless steel cathode (Cabeza *et al.* 2007a; b), aluminium anode and cathode (Mohajeri *et al.* 2010b), titanium coated with SnO_2 anode and cathode (Wang *et al.* 2010), titanium coated with SnO_2 or PbO_2 anode and steel cathode (Cossu *et al.*1998), boron doped diamond anode (Anglada *et al.* 2009), boron doped diamond on silicon anode and cathode (Papastavrou *et al.* 2009), and aluminium and iron anodes and cathodes (İlhan *et al.* 2008) in the shape of either rode or sheet were used in these treatment applications. Results indicated that these techniques are very efficient in reducing organic matter and colour and oxidising ammonia from raw and pretreated landfill leachate under the optimum operation conditions in batch or continuous mode.

REFERENCES

Adhoum, N. and Monser, L. (2004) Decolourisation and removal of phenolic compounds from olive mill wastewater by electrocoagulation. *Chem. Eng. Proc.* **43**, 1281–1287.

Ahmadi, M., Vahabzadeh, F., Bonakdarpour, B., Mofarrah, E. and Mehranian, M. (2005) Application of the central composite design and response surface methodology to the advanced treatment of olive oil processing wastewater using Fenton's peroxidation. *J. Hazard. Mater.* **B123**, 187–195.

Alvarez, P.M., Beltran, F.J. and Rodriguez, E.M. (2005) Integration of ozonation and an anaerobic sequencing batch reactor (AnSBR) for the treatment of cherry stillage. *Biotchenol. Prog.* **21**, 1543–1551.

Anglada, A., Urtiaga, A. and Ortiz, I. (2009) Pilot scale performance of the electro-oxidation of landfill leachate at boron-doped diamond anodes. *Environ. Sci. Technol.* **43**, 2035–2040.

Barnes, D., Li, X. and Chen, J. (2010) Determination of suitable pretreatment method for old-intermediate landfill leachate. *Environ. Technol.* **28**, 195–203.

Batstone, D.J. and Steyer, J.P. (2007) Use of modelling to evaluate best control practice for winery-type wastewaters. *Water Sci. Technol.* **56**, 147–152.

Beltran, F.J., Garcia-Araya, J.F. and Alvarez, P.M. (1999) Wine distillery wastewater degradation 2. Improvement of aerobic biodegradation by means of an integrated chemical(ozone)-biological treatment. *J. Agric. Food. Chem.* **47**, 3919–3924.

Beltran, F.J., Garcia-Araya, J.F. and Alvarez, P.M. (2000) Continuous flow integrated chemical(ozone)-activated sludge system treating combined agro industrial-domestic wastewater. *Environ. Progress* **19**, 28–35.

Beltran, F.J., Garcia-Araya, J.F. and Alvarez, P.M. (2001) pH sequential ozonation domestic aand wine-distillery wastewaters. *Water Res.* **35**, 929–936.

Beltran-Heredia, J., Torregrosa, J., Garcia, J., Dominguez, J. and Tiermo, J. (2001) Degradation of olive oil mill wastewater by the combination of Fenton's reagent and ozonation processes with an anaerobic biological treatment. *Water Sci. Technol.* **44**, 103–108.

Benitez, F.J., Beltrán-Heredia, J., Torregrosa J. and Acero, J.L. (1999) Treatment of olive mill wastewaters by ozonation, aerobic degradation and the combination of both treatments. *J. Chem. Technol. Biotechnol.* **74**, 639–646.

Benitez, F.J., Beltran-Heredia, J., Real, F.J. and Acero, J.L. (2000) Wine viasses treatments by ozone and an activated sludge system in continuous reactors. *Bioprocess Eng.* **23**, 149–154.

Benitez, F.J., Real, F.J., Acero, J.L., Garcia, J. and Sanchez, M. (2003) Kinetics of the ozonation and aerobic biodegradation of wine vinasses in discontinuous and continuous processes. *J. Hazard. Mater.* **B101**, 203–218.

Bigot, V., Luck, F., Paillard, H. and Wagner, A. (1994) Landfill leachate treatment: comparison of three oxidation processes using ozone. In *Proc. of the IOA Regional Conference*, pp. 219–228, International Ozone Association, European-African Group, Paris, France. Zürich, 31 August – 2 September.

Bisignano, G., Tomaino, A., Lo Cascio, R., Crisafi, G., Uccella, N. and Saija, A. (1999) On the in-vitro antimicrobial activity of oleuropein and hydroxytyrosol. *J. Pharm. Pharmacol.* **51**, 971–974.

Borja, R., Alba, J. and Banks, C. J. (1997) Impact of the main phenolic compounds of olive mill wastewater (OMW) on the kinetics of acetoclastic methanogenesis. *Proc. Biochem.* **32**, 121–133.

Cabeza, A., Primo, Ó., Urtiaga A.M. and Ortiz, I. (2007a) Definition of a clean processfor the treatment of landfill leachate integration of electrooxidation and ion exchange technologies. *Sep. Sci. Technol.* **42**, 1585–1596.

Cabeza, A., Urtiaga A.M., Rivero, M.R. and Ortiz, I. (2007b) ammonium removal from landfill leachate by anodic oxidation. *J. Hazard. Mater.* **144**, 715–719.

Cañizares, P., Lobato, J., Paz, R., Rodrigo, M.A. and Sáez, C. (2007) Advanced oxidation processes for the treatment of olive-oil mills wastewater. *Chemosphere* **67**, 832–838.

Carbajo, M., Beltran, F.J., Gimeno, O., Acedo, B. and Rivass, F.J. (2007) Ozonation pf phenolic wastewaters in the presence of a perovskite type catalyst. *Appl. Catal. B-Environ.* **74**, 203–210.

Chatzisymeon, E., Xekoukoulotakis, N. P. and Mantzavinos, D. (2009a) Determination of key operating conditions for the photocatalytic treatment of olive mill wastewaters *Catal. Today* **144**, 143–148.

Chatzisymeon, E., Xekoukoulotakis, N.P., Diamadopoulos, E., Katsaounis, A. and Mantzavinos, D. (2009b) Boron-doped diamond anodic treatment of olive mill wastewaters: statistical analysis, kinetic modeling and biodegradability. *Water Res.* **43**, 3999–4009.

Chedeville, O., Debacq, M. and Porte, C. (2009). Removal of phenolic compounds present in olive mill wastewaters by ozonation. *Desalination* **249**, 865–869.

Ciardelli, G. and Ranieri, N. (2001) The treatment and reuse of wastewater in the textile industry by means of ozonation and electrofloculation. *Water Res.* **35**, 567–572.

Cossu, R., Polcaro, A.M., Lavagnolo, M.C., Mascia, M, Palmas, S. and Renoldi, F. (1998) Electrochemical treatment of landfill leachate: oxidation at Ti/PbO_2 and Ti/SnO_2 anodes. *Environ. Sci. Technol.* **32**, 3570–3573.

D'Annibale, A., Crestini, C., Vinciguerra, V., Giovannozzi, S G. (1998) The biodegradation of recalcitrant effluents from an olive mill by a white-rot fungus. *J. Biotechnol.* **61**, 209–218.

De Heredia, J.B., Dominguez, J.R. and Partido, E. (2005) Physico-chemical treatment for the depuration of wine distillery wastewaters (vinasses). *Water Sci.Technol.* **51**, 159–166.

Deligiorgis, A., Xekoukoulotakisa, N.P., Diamadopoulos, E. and Mantzavinos, D. (2008) Electrochemical oxidation of table olive processing wastewater over boron-doped diamond electrodes: treatment optimization by factorial design. *Water Res.* **42**, 1229–1237.

De Marco, E., Savarese, M., Paduano, A. and Sacchi, R. (2007) Characterization and fractionation of phenolic compounds extracted from olive mill wastewaters. *Food Chem.* **104**, 858–867.

Deng, Y. (2007) Physical and chemical removal of organics during Fenton treatment of manure municipal landfill leachate. *J. Hazard. Mater.* **146**, 334–340.

Dogruel, S., Olmez-Hanci, T., Kartal, Z., Arslan-Alaton, I. and Orhon, D. (2009) Effect of Fenton's oxidation on the particle size distribution of organic carbon in olive mill wastewater. *Water Res.* **43**, 3974–3983 (2009).

EC (2006) Integrated Pollution Prevention and Control Reference Document on Best Available Techniques in the Food, Drink and Milk Industries, European Commission.

Gau, S.H. and Chang, F.S. (1996) Improved Fenton method to remove recalcitrant organics in landfill leachate. *Water Sci. Technol.* **34**, 455–462.

Gernjak, W., Maldonado, M. I., Malato, S., Cáceres, J., Krutzler, T., Glaser, A. and Bauer, R. (2004) Pilot-plant treatment of olive mill wastewater (OMW) by solar TiO_2 photocatalysis and solar photo-Fenton. *Solar Energy* **77**, 567–72.

Gernjak, W., Krutzler, T., Bauer, R. and Malato, S. (2007) Photo-Fenton treatment of olive mill wastewater applying a combined Fenton/Flocculation pretreatment. *J. Sol. Energ. Eng.* **129,** 53–59.

Gotsi, M., Kalogerakis, N., Psillakis, E., Samaras, P. and Mantzavinos, D. (2005) Electrochemical oxidation of olive oil mill wastewaters. *Water Res.* **39**, 4177–4187.

Hagman, M., Heander, E. and Jansen, J.C. (2008) advanced oxidation of refractory organics in leachate-Potential methods and evaluation of biodegradability of remaining substrate. *Environ. Technol.* **29**, 941–946.

Hamdi, M. (1993) Thermoacidic precipitation of darkly colored polyphenols of olive mill wastewaters. *Environ. Technol.* **14**, 495–500.

Hanafi, F., Assobhei, O. and Mountadar, M. (2010) Detoxification and discoloration of Moroccan olive mill wastewater by electrocoagulation. *J. Hazard. Mater.* **174**, 807–812.

Ilhan, F., Kurt, U., Apaydın, Ö. and Gönüllü, T. (2008) Treatment of leachate by electrocoagulation using aluminum and iron electrodes. *J. Hazard. Mater.* **154**, 381–389.

Inan, H., Dimoglo, A., Simsek, H. and Karpuzcu, M. (2004) Olive oil mill wastewater treatment by means of electrocoagulation. *Sep. Purif. Technol.* **36**, 23–31.

Ince, N.H. (1998) Light-enhanced chemical oxidation for tertiary treatment of municipal landfill leachate. *Water Environ. Res.* **70**, 1161–1169.

ISO 8192 (2007) Water quality-oxygen demand inhibition assay by activated sludge, ISO 8192/TC 147/SC 5, Geneva.

Israilides, C.J., Vlyssides, A.G., Mourafeti, V. N. and Karvouni, G. (1997) Olive oil wastewater treatment with the use of an electrolysis system. *Bioresour. Technol.* **61**, 163–170.

Kannan, N., Karthikeyan, G. and Tamilselvan, N. (2006) Comparison pf treatment potential of electrocoagulation of distillery effluent with and without activated *Areca catchu* nut carbon. *J. Hazard. Mater.* **B137**, 1803–1809.

Kang, Y.W. and Hwang, K.Y. (2000) Effects of reaction conditions on the oxidation efficiency in the Fenton process. *Water Res.* **34**, 2786–2790.

Katsoni, A., Frontistis, Z., Xekoukoulotakis, N. P., Diamadopoulos, E. and Mantzavinos, D. (2008) Wet air oxidation of table olive processing wastewater: determination of key operating parameters by factorial design. *Water Res.* **42**, 3591–3600.

Khoufi, S., Aloui, F. and Sayadi, S. (2009) Pilot scale hybrid process for olive mill wastewater treatment and reuse. *Chem. Eng. Proc.* **48**, 643–650.

Kim, Y.K. and Huh, I.R. (1997) Enhancing biological treatability of landfill leachate by chemical oxidation. *Environ. Eng. Sci.* **14**, 73–79.

Kim, S.M., Geissen, S.U. and Vogelpohl, A. (1997) Landfill leachate treatment by a photoassisted Fenton reaction. *Water Sci. Technol.* **35**, 239–248.

Kim, J.S., Kim, H.Y., Won, C.H. and Kim, J.G. (2001) Treatment of leachate produced in stabilized landfills by coagulation and Fenton oxidation process. *J. Chin. Inst. Chem. Eng.* **32**, 425–429.

Kim, S.M. and Vogelpohl, A. (1998) Degradation of organic pollutants by the photo-Fenton-process. *Chem. Eng. Technol.* **21**, 187–191.

Kjeldsen, P., Barlaz, M.A., Rooker, A.P., Baun, A., Ledin, A. and Christen, T.H. (2002) Present and long-term composition of MSW landfill leachate: A review. *Critical Reviews in* Environ. *Sci. Technol.* **32**, 297–336.

Kochany, J. and Lipczynska-Kochany, E. (2009) Utilization of landfill leachate parameters for pretreatment by Fenton reaction and struvite precipitation-a comparative study, *J. Hazard. Mat.* **166**, 248–254.
Lai, C. and Lin, S. (2004) Treatment of chemical mechanical polishing wastewater by electrocoagulation: System performances and sludge settling characteristics. *Chemosphere* **54**, 235-242.
Lau, I.W.C., Wang, P. and Fang, H.H.P. (2001) Organic removal of anaerobically treated leachate by Fenton coagulation. *J. Environ. Eng.* **127**, 666–669.
Li, H., Zhou, S., Sun, Y. and Lv, J. (2010) Application of response surface methodology to the advanced treatment of biologically stabilized landfill leachate using Fenton's reagent. *Waste Man.* doi:10.1016/j.wasman.2010.03.036.
Lin, S.H. and Lo, C.C. (1997) Fenton process for treatment of desizing wastewater. *Water Res.* **31**, 2050–2056.
Lin, S. and Chang, C. (2000) Treatment of landfill leachate by combined electro-Fenton oxidation and sequencing batch reactor method. *Water Res.* **34**, 4243–4249.
Lucas, M.S., Mosteo, R., Maldonado, M.I., Malato, S. and Peres, J.A. (2009a) Solar photochemical treatment of winery wastewater in a CPC reactor. *J. Agric. Food. Chem.* **57**, 11242–11248.
Lucas, M.S. and Peres, J. A. (2009b) Removal of COD from olive mill wastewater by Fenton's reagent: kinetic study. *J. Hazard. Mater.* **168**, 1253–1259.
Lucas, M.S., Peres, J.A. and Puma, G.L. (2010) Treatment of winery wastewater by ozone-based advanced oxidation processes (O_3, O_3/UV and O_3/UV/H_2O_2) in a pilot-scale bubble column reactor and process economics. *Sep. Purif. Technol.* **72**, 235–241.
Martin, M.A., Raposo, F., Borja, R. and Martin, A. (2002) Kinetic study of the anaerobic digestion of vinasse pretreated with ozone, ozone plus ultraviolet light, and ozone plus ultraviolet light in the presence of titanium dioxide. *Process Biochem.* **37**, 699–706.
Mert, B. K., Yonar, T., Kilic, M. Y. and Kestioglu, K. (2010) Pre-treatment studies on olive oil mill effluent using physicochemical, Fenton and Fenton-like oxidation processes. *J. Hazard. Mater.* **174**, 122–128.
Minh, P., Gallezot, P. and Besson, M. (2007) Treatment of olive oil mill wastewater by catalytic wet air oxidation 3. Stability of supported ruthenium catalysts during oxidation of model pollutant p-hydroxybenzoic acid in batch and continuous reactors. *Appl. Cat. B* **75**, 71–77.
Mohajeri, S., Aziz, H.A., Isa, M.H., Bashir, M.J.K., Mohajeri, L. and Adlan, M.N. (2010a) Influence of Fenton reagent oxidation on mineralization and decolorization of municipal landfill leachate. *J. Environ. Sci. Health: A* **45**, 692–698.
Mohajeri, S., Aziz, H.A., Isa, M.H., Zahed, M.A., Bashir, M.J.K. and Adlan, M.N. (2010b) Application of the central composite design for condition optimization for semi-aerobic landfill leachate treatment using electrochemical oxidation. *Water Sci. Tecnol.* **61**, 1257–1266.
Najjar, W., Azabou, S., Sayadi, S. and Ghorbel, A. (2007) Catalytic wet peroxide photo-oxidation of phenolic olive oil mill wastewater contaminants Part I. Reactivity of tyrosol over (Al-Fe)PILC. *Appl. Catal. B-Environ.* **74**, 11–18.
Nava, J.L., Acosta, J.C. and Gonzalez, I. (2008) Electrochemical incineration of vinasse in filter-press-type FM01-C reactor using 3D BDD electrode. *Water Sci Technol.* **58**, 2413–2419.

Navarro, P., Sarasa, J., Sierra, D., Esteban, S. and Ovelleiro, J.L. (2005) Degradation of wine industry wastewater by photocatalytic advanced oxidation. *Water Sci. Technol.* **51**, 113–120.

Neyens, E. and Baeyens, J. (2003) A review of classic Fenton's peroxidation as an advance oxidation technique. *J. Hazard. Mater.* **B98**, 33–50.

Neyens, E., Baeyens, J., Dewil, R. and Deheyder, B. (2004) Advanced sludge treatment affects extracellular polymeric substances to improve activated sludge dewatering. *J. Hazard. Mater.* **B106**, 83–92.

Papastavrou, C., Matzavinos, D. and Diamadopoulos, E. (2009) A comparative treatment of stabilized landfill leachate: coagulation and activated carbon adsorption vs. electrochemical oxidation. *Environ. Technol.* **30**, 1547–1553.

Paraskeva, P. and Diamadopoulos, E. (2006) Technologies for olive mill wastewater (OMW) treatment: a review. *J. Chem. Technol. Biot.* **81**, 1475–1485.

Parnaudeau, V., Condom, N., Oliver, R., Cazevielle, P. and Recous, S. (2008) Vinasse organic matter quality and mineralization potential, as influenced by raw material, fermentation and concentration processes. *Bioresource Technol.* **99**, 1553–1562.

Perez, J., Delarubia, T., Moreno, J. and Martinez, J. (1992) Phenolic content and antibacterial activity of olive oil wastewaters. *Environ. Toxicol. Chem.* **11**, 489–495.

Renou, S., Givaudan, J.G., Poulain, S., Dirassouyan, F. and Moulin, P. (2008) Landfill leachate treatment: Review and opportunity. *J. Hazard. Mater.* **150**, 468–493.

Rincon, J.D., Martinez, F.M. and Cabrales, N.M. (2009) Using electro-flotation/oxidation for reducing chemical oxygen demand, total organic carbon and total solids in vinasses. *Rev. Ing. Invest.* **29**, 35–28 (in Spanish).

Rivas, F.J., Gimeno, O., Portela, J. R., De La Ossa, E. M. and Beltrán, F. J. (2001a) Supercritical water oxidation of olive oil mill wastewater. *Ind. Eng. Chem. Res.* **40**, 3670–3674.

Rivas, F.J., Beltrán, F. J., Gimeno, O. and Frades, J. (2001b) Treatment of olive oil mill wastewater by Fenton's reagent. *J. Agric. Food Chem.* **49**, 1873–1880.

Santana, V.S. and Machado, N.R.C.F. (2008) Photocatalytic degradation of the vinasse under solar radiation. *Catal. Today* **133–135**, 606–610.

Santos, M.M., Bocanegra, J.F., Martin, A.M. and Garcia, I.G. (2003) Ozonation of vinasse in acid and alkaline media. *J. Chem. Technol. Biotechnol.* **78**, 1121–1127.

Santos, M.M., Bocanegra, J.F., Martin, A.M. and Garcia, I.G. (2005) Estimating the selectivity of ozone in the removal of polyphenols from vinasse. *Chem. Technol. Biotechnol.* **80**, 433–438.

Schulte, P., Bayer, A., Kuhn, F., Luy, T. and Volkmer, M. (1995) H_2O_2/O_3, H_2O_2/UV and H_2O_2/Fe^{2+} processes for the oxidation of hazardous wastes. *Ozone Sci. Eng.* **17**, 119–134.

Sheehan, G.J. and Greenfield, P.F. (1980) Utilisation, treatment and disposal of distillery wastewater. *Water Res.* **14**, 257–277.

Shu, H.Y., Fan, H.J., Chang, M.C. and Hsieh, W.P. (2006) Treatment of MSW landfill leachate by a thin gap annular UV/H_2O_2 photoreactor with multi-UV lamps. *J. Hazard. Mater.* **B129**, 73–79.

Tchobanoglous, G., Theisen, H. and Vigil, S. (1993) Integrated Solid Waste Management, Engineering Principles and Management Issue, McGraw-Hill, Inc., International editions, Civil Engineering Series, Singapore, ISBN 0-07-112865-4.

Tizaoui, C., Bouselmi, L., Mansouri, L. and Ghrabi, A. (2007) Landfill leachate treatment with ozone and ozone/hydrogen peroxide systems. *J. Hazard. Mater.* **140**, 316–324.

US EPA (1999) Multimedia Environmental Compliance Guide for Food Processors, EPA-305-B-99-005, Office of Compliance, Office of Enforcement and Compliance Assurance, U.S Environmental Protection Agency, Washington D.C., United States.

US EPA (2000) Development Document for Final Effluent Limitations Guidelines and Standards for the Landfills Point Source Category, EPA-821-R-99-019, Office of Water (4303), US Environmental Protection Agency, Washington D.C., United States.

Vlyssides, A.G., Israilides, C.J., Loizidou, M., Karvouni, G. and Mourafeti, V. (1997) Electrochemical treatment of vinasse from beet molasses. *Water Sci. Technol.* **36**, 271–278.

Vlyssides, A.G., Loukakis, H.N., Israilides, C., Barampouti, E.M. and Mai, S. (2003) Detoxification of olive mill wastewater using a Fenton process. *Proceedings of the 2nd European Bioremediation Conf.*, Chania, pp. 531–534.

Vlyssides, A.G., Loukakis, H.N., Karlis, P.K., Barampouti, E.M.P. and Mai, S.T. (2004) Olive mill wastewater detoxification by applying pH related Fenton's oxidation process. *Fresen. Environ. Bull.* **13**, 501–504.

Vlyssides, A.G., Barampouti, E.M. and Mai, S. (2005) Wastewater characteristics from Greek wineries and distilleries. *Water Sci. Technol.* **51**, 53–60.

Wable, O., Jousset, M., Courant, P. and Duguet, J.P. (1993) Oxidation of landfill leachates by ozone and hydrogen peroxide: A French example. In *Proc. of the Int. Symposium on Ozone-Oxidation Methods for Water and Wastewater Treatment, Wasser, Berlin.* International Ozone Association, European-Africa Group, Paris, France. Berlin, II.6.1-II.6.11.

Wang, F., Smith, D.W. and El-Din, G. (2003) Application of advanced oxidation methods for landfill leachate treatment - A review. *J. Environ. Eng. Sci.* **2**, 413–427.

Wang, F., El-Din, G. and Smith, D.W. (2004) Oxidation of aged raw landfill leachate with O_3 only and O_3/H_2O_2: treatment efficiency and molecular size distribution analysis. *Ozone: Sci. Eng.* **26**, 287–298.

Wang, P, Lau, I.W.C. and Fang, H.H.P. (2010) Electrochemical oxidation of leachate pretreated in an upflow anaerobic sludge blanket reactor. *Environ. Technol.* **22**(4), 373–381.

Welander, U. and Herysson, T. (1998) Physical and chemical treatment of a nitrified leachate from a municipal landfill leachate. *Environ. Technol.* **19**(6), 591–599.

Wu, Y., Zhou, S., Qin, F., Ye, X. and Zheng, K. (2010) Modelling physical and oxidative removal properties of Fenton process for treatment of landfill leachate using response surface methodology (RSM). *J. Hazard. Mater.* (in press), doi:10.1016/j.hazmat.2010.04.052.

Wu, J.J., Wu, C.C., Ma, H.W. and Chang, C.C. (2004) Treatment of landfill leachate by ozone based advanced oxidation processes. *Chemosphere* **54**, 997–1003.

Yang, J., Xing, M. and Zhou, X. (2008) Degradation of recalcitrant organics from winery wastewater by Fenton's reaction. *Environ. Eng. Sci.* **25**, 1229–1234.

Zayas, T., Romero, V., Salgado, L., Meraz, M. and Morales, U. (2007) Applicability of coagulation/flocculation and electrochemical processes to the purification of biologically treated vinasse effluent. *Sep. Purif. Technol.* **57**, 270–276.

Zhang, H, Choi, H.J. and Huang C.P. (2005) Optimization of Fenton process for the treatment of landfill leachate, *J. Hazard. Mat.* **B125**, 166–174.

Index

activated carbon 127, 128
  alcohol distilleries 310
  gold and silver mining 269–270, 271–272, 274
  textile wastewater treatment 43
activated sludge treatment
  chemicals industry wastewater 289
  leather tanning wastewater 84
  pharmaceutical industry 157–158, 174
  pulp and paper wastewater 220, 227–228
  textile industry 42, 43
adsorbable organically bound halogens (AOX)
  chemicals industry 288, 291
  pulp and paper industry 211, 219, 224
  textile industry 38–39, 66
advanced oxidation processes (AOPs) 15–25
  alcohol distilleries 308–309, 311
  catalytic ozonation 17–18, 159–160, 255
  chemical 16–22
  chemicals industry 289–290, 293–296
  combined 59–60
  comparative studies 190–192, 294–296
  costs 63–65
  landfill leachate treatment 317–319
  leather tanning industry 97–100
  pesticide industry 254–255
  pharmaceutical industry 158–192
  photochemical 22–25
  supercritical water oxidation (SCWO) 19, 61–62
  textile industry 53–65
  wet air oxidation (WAO) 18–19
  see also individual processes
air oxidation
  catalytic 85–86, 87, 89–92
  of sulphide 85–92
  see also catalytic wet air oxidation (CWAO); wet air oxidation (WAO)

alcohol distilleries 305–311
biological treatment 307, 310
chemical oxidation applications 307–311
distillation process 306
in-plant controls 306–307, 310
organic matter 306, 308, 310–311
oxidation processes 307–311
ozonation 307–308
vinasse 306–307
wastewater biodegradability 307–308, 310–311
wastewater characterisation 306–307, 310
wastewater treatment 307–311
alkaline chlorination
for cyanide oxidation 121–124, 132, 252, 274–275
high temperature alkaline chlorination process (HTACP) 132
alum coagulation pre-treatment 97–98, 226
amoxicillin (AMX) 181–182, 186, 192
antibiotics 176–178, 193
amoxicillin (AMX) 181–182, 186, 192
manufacturing wastewater 159–160, 163–168, 174
AOPs *see* advanced oxidation processes (AOPs)
AOX *see* adsorbable organically bound halogens (AOX)

barren cyanide solution 270–271, 272
BAT-EC
leather tanning industry 76, 79–82
metal finishing industry 116, 117, 120–121, 132, 142
pulp and paper industry 215–216
textile industry 37–40
BAT-US
metal finishing industry 119–121, 132, 142
pesticide industry 250
pulp and paper industry 214–215
beamhouse processes 72–73, 79–80
best available techniques *see* BAT-EC; BAT-US
biodegradability
of alcohol distillery wastewater 307, 310–311
of chemicals industry wastewater 288, 293
of landfill leachate 317–318
of leather tanning wastewater 94, 99–100
of olive oil industry wastewater 313, 314
of pesticide industry wastewater 254, 255, 257–258, 259, 261
of pharmaceutical wastewater 157–159, 164–166, 170–174, 176, 178–180
of pulp and paper wastewater 217, 230
of textile industry chemicals 37–40, 42, 43, 47
of textile wastewater 57, 62–63
biological treatment
chemicals industry wastewater 288–289, 298
combined with photo-Fenton 257–258
of landfill leachate 317
leather tanning wastewater 98–100
pesticide wastewater 259–260
pharmaceutical wastewater 158, 173–174
textile industry wastewater 42, 43
of vinasse 307, 310
bleached kraft mill effluents (BKME) 219, 222, 227
bleaching
leather tanning industry 74
pulp and paper industry 206–207, 209, 210, 213–214, 220–221
wastewater characteristics 211, 212, 213
wastewater sources 209, 210
wastewater treatment 217–218, 220–224, 227
bleaching agents 38–39

carbamazepine (CBZ) 181–182, 186
carbon nanofibres (CNFs) 60–61
carbonation wastewater treatment 82
catalytic air oxidation, of sulphide 85–86, 87, 89–92
catalytic ozonation 17–18, 159–160, 255
catalytic wet air oxidation (CWAO) 19
  chemicals industry 289–290
catalytic wet peroxidation
  of pharmaceutical wastewater 183–186
  textile industry 58–59
chemical advanced oxidation processes 16–22
  catalytic ozonation 17–18, 159–160, 255
  peroxone process 16–17, 166, 167–168, 220, 224
  supercritical water oxidation (SCWO) 19, 61–62
  see also Fenton processes; sonolysis; wet air oxidation (WAO)
chemical oxidation 1–10
chemical oxidation applications 10–25
  alcohol distilleries 307–311
  chemicals industry 289–297, 298
  gold and silver mining 274–281
  landfill leachate 317–319
  leather tanning industry 85–101
  metal finishing industry 121–143
  olive oil industry 313–316
  pesticide industry 250–261
  pharmaceutical industry 157–192, 193
  pulp and paper industry 217–229, 230
  textile industry 44–66
chemical pulping 203–204
  wastewater characteristics 211, 212, 213
  wastewater treatment 218–219
chemical synthesis, of pharmaceuticals 153, 154, 155, 156
chemicals industry 285–298
  adsorbable organically bound halogens (AOX) 288, 291
  advanced oxidation processes (AOPs) 289–290, 293–296
  biological wastewater treatment 288–289, 298
  chemical oxidation applications 289–298
  dechlorination 291–293, 297
  electrochemical processes 290–293
  H_2O_2/UV treatment 294
  in-plant control 287–288
  micropollutants 287
  organic matter 287, 293
  ozonation 293
  production processes 286–287
  sonolysis 296
  wastewater biodegradability 288, 293
  wastewater characteristics 287
  wastewater flow segregation 288
  wastewater generation 287
  wastewater treatment 288–298
  wet air oxidation (WAO) 289–290
  wet peroxide oxidation 290
chlorine 11, 211, 238
  dechlorination 14–15, 291–293, 297
  see also alkaline chlorination
chlorine dioxide 12–13, 39, 128–129
chrome 81, 82
  tanning 74, 76, 80–81
coagulation 174–175, 220, 226
colour abatement *see* decolourisation
comparative studies 190–192, 294–296
complexed metal cyanides 122, 123, 125, 130
complexed metals 114–116, 119–121, 137–140
complexing agents 113
copper 127, 128–129, 275–277
costs
  advanced oxidation processes (AOPs) 63–65
  alkaline chlorination 124
  cyanide removal 124, 279–280
  electrocoagulation 45, 137, 225
  electrolytic metal recovery 142
  hexavalent chromium reduction 134

costs (*continued*)
  hydrogen peroxide 128
  INCO process 279–280
  leather tanning wastewater treatment 91–92
  ozonation 126
  pesticide wastewater treatment 259–261
  sulphide oxidation 91–92
  textile industry wastewater treatment 65–66
  ultrasound (US) treatment 63–64
CWAO *see* catalytic wet air oxidation (CWAO)
cyanidation 273–274, 281
cyanide
  barren cyanide solution 270–271, 272
  complexed metal cyanides 122, 123, 125, 130
  complexing agents 113
  destruction methods 121–132
  gold and silver mining 269, 270–281
  as a leaching agent 269–270, 281
  in metal finishing wastewater 116, 121–132
  in pesticide wastewater 245
  recovery process 273–274
  removal costs 124, 126, 128, 279–280
cyanide oxidation
  by alkaline chlorination 121–124, 132, 252, 274–275
  by chlorine dioxide 128–129
  by electrooxidation 131–132
  by hexavalent ferrate 129–130
  by high temperature alkaline chlorination process (HTACP) 132
  by hydrogen peroxide (H_2O_2) 126–128, 252, 275–277
  by ozonation 124–126, 280
  by photochemical processes 130–131
  by sulphur dioxide 128, 277–280
  copper catalysed 127, 128–129
  Degussa process 275–276
  gold and silver mining 274–280
  INCO process 128, 277–280
  Kastone Process 275, 276–277
  metal finishing industry 121–132

debarking wastewater 218
dechlorination 14–15, 291–293, 297
decolourisation
  by electrocoagulation 47–49, 225
  by electrooxidation 50–52, 53
  by ozonation 55–57, 100
  by photocatalytic oxidation 223–224
  by photoelectrocatalysis 54
  by photolysis 100
  by wet air oxidation (WAO) 61
  Fenton processes 57–59, 224
  leather tanning wastewater 100
  pulp and paper industry wastewater 218, 223–224, 225–226
  textile industry 33–34, 47–49, 50–59, 65
Degussa process of cyanide oxidation 275–276
diclofenac (DFC) 181–182, 186
dienes 246
dimensionally stable anode (DSA) 52–53
distillation process 306
  see also alcohol distilleries
dye carriers, textile industry 39
dyes and dyeing process 33, 39–40, 41, 53–54
  BAT-EC 37–40
  effluent colouration 33–34, 35
  leather tanning industry 74
  wastewater 33–34, 35, 36, 63–65
  see also decolourisation; textile industry

electro-Fenton process 96, 256, 314
electrochemical cells 3–5, 8–9
electrochemical processes 13–15
electrocoagulation 14
  colour removal 47–49, 225
  costs 45, 137, 225
  with Fenton's reagent 140
  leather tanning wastewater 93, 95–96, 98–99

metal finishing industry 137, 140
olive mill wastewater (OMW) 314–315
pharmaceutical wastewater 188–189
pulp and paper wastewater 224–225
raw wastewater 95–96
textile industry 44–49
electrode potential 4–7, 9
electroflotation 13
electrolysis 8–10
electrolytic metal recovery 140–142
electron beam irradiation 62–63
electrooxidation 13–14
of bleaching effluent 218
chrome recovery 81
cyanide removal 131–132
decolourisation 50–52, 53
of leather tanning wastewater 81, 95–96, 98–99
of olive mill wastewater (OMW) 314
of pulp and paper wastewater 225–227
sulphide oxidation 93–94
of textile industry wastewater 49–53
of vinasse 309–310, 311
electroplating wastewater 136
electroreduction 14–15, 291–293
end-of-pipe treatment
gold and silver mining 274
metal finishing industry 118–121
pesticide industry 248
pulp and paper industry 216–217
environmental impact
gold and silver mining 273, 280–281
pharmaceutical industry 192–193
pulp and paper industry 211, 229–230
textile industry 35, 37–38
enzymes 79–80

Fenton processes 18, 140
bleaching effluent 218, 221, 222
chemicals industry 294–296
colour abatement 57–59
electro-Fenton 96, 140, 256, 314
Fenton oxidation 224, 227–229, 255, 317–318
Fenton-like oxidation 218, 221, 222, 224
hybrid systems 174–176, 258–259
landfill leachate treatment 317–318
leather tanning wastewater 96
microwave enhanced 175–176
olive mill wastewater (OMW) 313–314
pesticide wastewater treatment 255–257, 258–259
pharmaceutical wastewater 170–176, 176–181
pulp and paper wastewater 218, 221, 222, 224, 227–229
textile wastewater 57–59
vinasse treatment 309
see also photo-Fenton process
fermentation 153–154, 155
ferrate 13, 94–95, 129–130
fibre furnish 202–203
wastewater treatment 218
finishing processes, textiles 32–33, 40–41
formulation
effluent 158–160
pesticides 240, 242
pharmaceuticals 154, 155, 158–160
free energy 2–4

gold and silver mining 267–281
activated carbon 269–270, 271–272, 274
chemical oxidation applications 274–281
cyanide 269, 270–281
end-of-pipe wastewater treatment 274
environmental impact 280–281
in-process controls 272–274
leachates 271–272
lixiviants 269
mine water 270, 271
pollutants 271–272
production processes 268–270
recovery and reuse processes 273–274

gold and silver mining (*continued*)
  waste sources 270–271
  wastewater characterisation 271–272
  wastewater treatment 274–281
groundwater treatment 160–161

H_2O_2/UV treatment 169–170
  chemicals industry 294
  landfill leachate 318–319
  leather tanning wastewater 97–98, 100
  metal finishing wastewater 130–131
  pesticide wastewater 255, 256–258
  pharmaceutical industry 169–170
  pulp and paper wastewater 220, 222, 224
  see also photocatalytic H_2O_2 oxidation
haloethers 245
halogenated organic compounds *see* adsorbable organically bound halogens (AOX)
halogenated organic pesticides 238
halomethanes 245
heavy metals 138, 140
hexavalent chromium 112, 114–121
hexavalent chromium reduction
  by electrochemical treatment 136–137
  by electrocoagulation 137
  by iron and iron compounds 134–136
  by reduced sulphur compounds 132–134
  costs 134
  and dissolved oxygen 134
  metal finishing industry 132–137
  metal sulphide precipitation 134–135
hexavalent ferrate oxidation, of cyanide 129–130
high temperature alkaline chlorination process (HTACP) 132
hybrid systems
  Fenton processes 174–176, 258–259
  pesticide wastewater treatment 258–259
  pharmaceutical wastewater treatment 174–176, 188–190
  pulp and paper wastewater treatment 220–221, 226
  sonolysis 20–21, 63–65, 189–190
  textile wastewater treatment 43, 62–65
hydrazine (N_2H_4) 138–140
hydrogen peroxide (H_2O_2) 12
  as a bleaching agent 38–39
  catalytic wet peroxidation 58–59, 183–186
  copper catalysed oxidation 127–128, 275–277
  cyanide oxidation 121–132, 252, 275–277
  photolysis of 23
  and sonolysis 20–21, 63–64
  sulphide oxidation 92–93
  wet peroxide oxidation 58–59, 183–186, 290
  see also Fenton processes; H_2O_2/UV treatment; peroxone treatment; photocatalytic H_2O_2 oxidation
hydroxide precipitation 45–46, 117–120, 132, 138
hydroxyl radicals 14, 20–25, 161, 164
  formation 12, 16–18, 125, 181, 221
  in photo-Fenton oxidation 23, 178, 181

ibuprofen (IBP) 180–181, 189–190
in-plant control
  alcohol distilleries 306–307, 310
  chemicals industry 287–288
  leather tanning industry 79–82
  metal finishing industry 116–118
  pesticide industry 248
  pharmaceutical industry 155–156
  pulp and paper industry 216–217
in-process controls, gold and silver mining 272–274
INCO process 128, 274, 277–280
indirect oxidation 13–14
iron and iron compounds 134–136
  see also Fenton processes

Kastone Process 127–128, 275, 276–277
kraft pulping process 204

bleached kraft mill effluents (BKME) 219, 222, 227
wastewater treatment 218, 219, 223–224

landfill leachate 316–319
biodegradability enhancement 317–318
biological treatment 317
characterisation 316–317
chemical oxidation applications 317–319
organic matter 317, 319
leachates
gold and silver mining 271–272
municipal landfill 316–319
leaching agents 269, 273, 281
leather tanning industry 71–101
beamhouse processes 72–73, 79–80
best practice 76, 79–82
bleaching 74
chemical oxidation applications 85–101
chemicals used 73–74, 76, 79–82
chrome tanning 80–81
deliming agents 80
dyes and dyeing process 74
in-plant control 79–82
micropollutants 76–77
organic matter 79, 82, 84, 85
production processes 72–74
sulphide use 79–80
surfactants 73
tanning agents 74, 80–81
tanyard processes 74, 80–82
water conservation 79
water consumption 76
leather tanning wastewater 101
activated sludge treatment 84
advanced oxidation 97–100
alum coagulation pre-treatment 97–98
biodegradability enhancement 94, 99–100
biologically pre-treated 98–100
carbonation 82
catalytic air oxidation 89–92
characterisation 74–78
colour removal 100
electro-Fenton process 96
electrocoagulation 93, 95–96, 98–99
electrooxidation 81, 95–96, 98–99
H_2O_2/UV treatment 97–98, 100
nitrogen removal 84–85
O_3/UV treatment 97
ozonation 96–97, 99–100
polishing treatment 85, 100
pre-treated 97–100
raw 95–97
sulphide oxidation 89–90, 93, 93–94
toxicity 98
treatment 82–101
treatment costs 91–92
lignin oxidation 229
lime 73, 79
lixiviants 269, 273, 281

macrolides 167–168
magnesium ammonium phosphate (MAP) precipitation 84–85
manufacturing processes
pesticide industry 237–240
pulp and paper industry 200–208
mechanical pulping 204–205, 218–219
Merrill-Crowe process 269, 270, 280
metal complexed cyanides 130
metal finishing industry 107–143
best practice 116, 117, 120–121, 132, 142
chemical oxidation applications 121–143
complexing agents 113
end-of-pipe wastewater treatment 118–121
facilities 107–108
in-plant control 116–118
metal wastes 112–113, 114–116
organic wastes 113–114, 127, 140
pollutants 112–116
production processes 108–110

metal finishing industry (*continued*)
  regulation 112–113, 114
  resource recovery and recycling 117–118, 140–142
  surfactants 113
metal finishing wastewater 142–143
  characterisation 112–116
  complexed metals 114–116, 137–140
  cyanide oxidation 121–132
  electrocoagulation 137, 140
  flow segregation 114–116, 118–121, 142
  H_2O_2/UV treatment 130–131
  heavy metals 138, 140
  hexavalent chromium reduction 132–137
  hydroxide precipitation 127–128, 132, 138
  O_3/UV treatment 131
  oxidation and reduction processes 121–143
  precious metals 114, 140–141
  sources 110–112
metal plating 9
metal sulphide precipitation 134–135
metallo-organic pesticides 236, 240
metals
  complexed metals 114–116, 137–140
  heavy metals 138, 140
  iron and iron compounds 134–136
  in pesticide wastewater 246
  precious metals 114, 140–141
  recovery 140–142, 270
  zero-valent metals 135–136, 138–140
  see also gold and silver mining; metal finishing industry; metal finishing wastewater
micropollutants 76–77, 287
microwave enhanced Fenton process 175–176
mine water 270, 271
municipal landfill leachate *see* landfill leachate

natural evaporation 312–313
Nernst equation 7
nitrogen removal 84–85
nitrogen-containing pesticides 240
nitrosamines 246

O_3/UV treatment 22–23
  landfill leachate 319
  leather tanning wastewater 97
  metal finishing wastewater 130–131
  pesticide wastewater 253, 254, 255, 257–258
  pulp and paper wastewater 221
olive mill wastewater (OMW) 311–316
  biodegradability 313, 314
  characteristics 312–313
  organic matter 312
  oxidation processes 313–316
olive oil industry 311–316
  chemical oxidation applications 313–316
  production processes 311
organic chemicals industry *see* chemicals industry
organic matter 10, 14, 18, 20
  alcohol distilleries 306, 308, 310–311
  chemicals industry 287, 293
  in landfill leachate 317, 319
  leather tanning industry 79, 82, 84, 85
  metal finishing industry 113–114, 127, 140
  olive mill wastewater (OMW) 312
  pharmaceutical industry 156, 159, 169–170
  textile industry 47–48
oxidants 2, 11–13
  ferrate 13, 94–95, 129–130
  oxygen 11–12
  potassium peroxymonosulphate 13
  see also chlorine; chlorine dioxide; hydrogen peroxide (H_2O_2); hydroxyl radicals; ozone
oxidation
  of lignin 229
  rates 11–12, 90–91

see also cyanide oxidation; redox reactions; sulphide oxidation
oxidation processes 49–53, 85–101, 307–311
  see also advanced oxidation processes (AOPs); chemical advanced oxidation processes; chemical oxidation applications
oxygen 11–12
ozonation 15, 16–18
  of bleaching effluent 222–223
  catalytic 17–18, 159–160, 255
  chemicals industry 293
  for colour removal 55–57, 100
  costs 126
  cyanide oxidation 124–126, 280
  of groundwater 160–161
  with hydrogen peroxide *see* peroxone treatment
  of leather tanning wastewater 96–97, 99–100
  of olive mill wastewater (OMW) 315–316
  of pesticide wastewater 254–255
  of pharmaceuticals 158–169
  of pre-treated wastewater 97
  of pulp and paper wastewater 221, 222–223, 224, 227–229
  of reservoir water 160–161
  sulphide oxidation 94
  textile industry 55–57
  with ultrasound (US/O_3) 21
  for vinasse 307–308
  see also O_3/UV treatment
ozone 12
  photolysis 22–23
  and sonolysis 21, 63–65

packaging processes 240, 242
paper industry *see* pulp and paper industry
paracetamol 187–188
Pauling rule 2
penicillin 190–192
peroxone treatment 16–17
  pesticide industry 254
  pharmaceutical industry 166, 167–168
  pulp and paper industry 220, 224
perozonation *see* peroxone treatment
pesticide industry 235–261
  BAT-US 250
  chemical oxidation applications 250–261
  end-of-pipe treatment 248
  formulation processes 240, 242
  in-plant control 248
  manufacturing processes 237–240
  packaging processes 240, 242
  pollution prevention and control 247–248
  priority pollutant pesticides 246–247
  repackaging process 242
  reuse and recycling techniques 247–248
  treatment technologies 248–250
  types of pesticide 235–237, 240
  waste segregation 247
pesticide industry wastewater
  advanced oxidation processes (AOPs) 254–255
  biodegradability enhancement 254, 255, 257–258, 259, 261
  biological treatment 259–260
  characterisation 243, 245–247
  chemical oxidation treatment processes 250–261
  Fenton processes 255–257, 258–259
  H_2O_2/UV treatment 255, 256–258
  hybrid treatment systems 258–259
  O_3/UV treatment 253, 254, 255, 257–258
  ozonation 254–255
  peroxone treatment 254
  solar UV irradiation 256–257
  sources 242–243
  TiO_2 photocatalysis 257
  treatment costs 259–261
  ultrasonic degradation 259

pharmaceutical industry 149–193
  chemical oxidation applications 157–192, 193
  chemical synthesis 153, 154, 155, 156
  environmental impact 192–193
  fermentation 153–154, 155
  formulation 154, 155, 158–160
  in-plant control 155–156
  natural product extraction 153, 154, 155
  organic matter 156, 159, 169–170
  pharmaceutical processes 152–154, 155
  pharmaceutical products 150–152, 154, 155, 161–163
  pollutants 154–155
  raw material inputs 154–155
  regulation 151
  research and development 152–153
  separation processes 153
  solvents 155, 193
pharmaceutical industry wastewater
  activated sludge treatment 157–158, 174
  advanced treatment 158–192
  antibiotic manufacturing 159–160, 163–168, 174
  biodegradability 157–159, 164–166, 170–174, 176, 178–180
  biological treatment 158, 173–174
  catalytic ozonation 159–160
  catalytic wet peroxidation 183–186
  characteristics 155, 156, 157, 159, 172, 174
  coagulation 174–175
  comparative treatment studies 190–192
  conventional treatment methods 156–158
  electrocoagulation 188–189
  Fenton process 170–176
  H_2O_2/UV treatment 169–170
  hybrid treatment systems 174–176, 188–190
  ozonation 158–169
  peroxone treatment 166, 167–168
  photo-Fenton treatment 176–181
  photocatalysis 181–183, 189
  pre-treatment 158–159, 166, 167, 173
  sonolysis 186–188
  toxicity 166, 167, 182–183, 187
  wet peroxide oxidation 183–186
pharmaceutical products 150–152
  formulation 154, 155
  wastewater treatment plants (WWTPs) 161–163, 180–183
photo-Fenton process 23–24
  bleaching wastewater 218, 221, 227
  chemicals industry 296
  combined with biological treatment 257–258
  landfill leachate treatment 317–318
  pesticide wastewater 255, 256–257, 258–259
  pharmaceutical wastewater 176–181
  pulp and paper wastewater 224
  solar UV irradiation 177–180, 221, 256–257, 259
  see also Fenton processes
photocatalysis 24–25
  antibiotics degradation 177–178, 181–182
  for colour removal 100, 223–224
  for complexed metals removal 140
  for cyanide destruction 130–131
  of olive mill wastewater (OMW) 315
  of pharmaceutical wastewater 181–183, 189
  of pulp and paper wastewater 219, 220, 220–221, 224–225
  sonophotocatalysis 21–22, 189–190
  of textile industry wastewater 54–55
  see also photocatalytic H_2O_2 oxidation; titanium dioxide (TiO_2) photocatalysis
photocatalytic H_2O_2 oxidation 130–131
  of pulp and paper wastewater 224, 225
  TiO_2 mediated 54–55, 100, 224, 225
  of vinasse 308–309
photoelectrocatalysis 53–54
photolysis 22–23

decolourisation 100
pesticide wastewater treatment 254
pulp mill wastewater treatment 224
see also photocatalysis; photocatalytic H_2O_2 oxidation; titanium dioxide (TiO_2) photocatalysis
polishing treatments 85, 100, 289, 311, 317
pollutants
alcohol distilleries 306
gold and silver mining 271–272
metal finishing industry 112–116
pharmaceutical industry 154–155
pulp and paper industry 209–211
textile industry 35
pollution prevention and control 117, 211–216, 247–248, 272–273
polynuclear aromatics 245–246
polyphenols 306–309
polysulphides 86–87
potassium peroxymonosulphate 13, 94
pre-treatment
alum coagulation 97–98, 226
leather tanning wastewater 97–100
ozonation 97, 307–308
pharmaceutical wastewater 158–159, 166, 167, 173
pulp and paper wastewater 217–227
of vinasse 307–308
precious metals 114–115, 140–141
see also gold and silver mining
priority pollutant pesticide oxidation 253–261
Procaine Penicillin G (PPG) 165–167
production processes
chemicals industry 286–287
gold and silver mining 268–270
leather tanning industry 72–74
metal finishing industry 108–110
olive oil industry 311
protein removal from wastewater 82, 84
pulp and paper industry 199–230
best available techniques (BAT) 214–216
bleaching 206–207, 209, 210, 213–214, 220–221
chemical oxidation applications 217–229, 230
chemical recovery 205, 208, 211
end-of-pipe wastewater treatment 216–217
environmental impact 211, 229–230
fibre furnish 202–203
in-plant control 216–217
manufacturing processes 200–208
paper mills 199–200
paper and paperboard making 207–208, 209
paperboard mills 199–200
pollutants 209, 211
pollution prevention and control 211–216
pulp mills 199–200, 208
pulp processing 205–207
pulping processes 203–205, 213
recovery and recycling 205, 208, 211
regulation 229
stock preparation 207
pulp and paper industry wastewater
activated sludge treatment 220, 227–228
adsorbable organically bound halogens (AOX) 211, 219, 224
alum coagulation 226
biodegradability 217, 230
characterisation 209, 211, 212, 213
colour removal 218, 223–224, 225–226
electrochemical sulphide oxidation 93
electrocoagulation 224–225
electrooxidation 225–227
Fenton processes 218, 221, 222, 224, 227–229
H_2O_2/UV treatment 220, 222, 224
hybrid treatment systems 220–221, 226
O_3/UV treatment 221
ozonation 221, 222–223, 224, 227–229
peroxone treatment 220, 224
photo-Fenton treatment 224

pulp and paper industry wastewater (*continued*)
  post-treatment with chemical oxidation 227–229
  pre-treatment 217–227
  pulping effluent 208, 210, 218–219
  residual COD removal 227–228
  sonolysis 218
  sources 208–209, 210
  toxicity 217
  treatment 221
  UV treatment 219, 220–222, 224–225
  wet oxidation (WO) 218–219
pulping processes 203–205, 213
  wastewater 208, 210, 218–219

raw material inputs, pharmaceutical industry 154–155
raw wastewater 95–97
reaction rates
  oxygen oxidation 11–12
  ozone oxidation 12
  redox reactions 10
  sulphide oxidation 92, 94–95
recalcitrant 17, 21, 42, 66, 187–188, 217, 220–221, 227, 306, 317–318
recovery and recycling
  by electrolysis 81
  chrome recovery 81
  of cyanide 273–274
  dyes and dyeing process 41
  electrolytic metal recovery 140–142
  gold and silver mining 273–274
  metal finishing industry 117–118, 140–142
  metals recovery 140–142, 270
  pesticide industry 247–248
  pulp and paper industry 205, 208, 211
  wastewater sources 208, 210
redox reactions 1–11
  applications 10–11
  hexavalent chromium removal 136–137
  reaction rates 10
  spontaneity 2–3, 6, 10
  thermodynamics of 2–10
reduction
  by iron and iron compounds 134–136
  of complexed metals 138–140
  electroreduction 14–15, 291–293
  see also hexavalent chromium reduction
reduction agents
  dithiocarbamate (DTC) 139
  ferrous iron 134
  hydrazine (N_2H_4) 139–140
  reduced sulphur compounds 132–134
  sodium borohydride ($NaBH_4$) 138–139
  zero-valent iron 135–136
reduction rates 133, 134, 136
reductive dechlorination 14–15, 291–293
regulation
  metal finishing industry 112–113, 114
  pharmaceutical industry 151
  pulp and paper industry 229
  textile industry 33–34
research and development, pharmaceutical industry 152–153
reservoir water treatment 160–161
reverse osmosis (RO) 161–163
run-off water 272

SCWO *see* supercritical water oxidation (SCWO)
secondary fibre pulping 205
semi-chemical pulping 204
semiconductor photocatalysis 24–25
sequestering agents 38
SHE *see* standard hydrogen electrode (SHE)
silver mining *see* gold and silver mining
sludge incineration 43
soda pulping process 204
sodium borohydride ($NaBH_4$) reduction 138–139
sodium chloride 73
sodium hypochlorite 38–39
sodium sulphate 73

sodium sulphide 73, 80
solar photo-Fenton 177–180, 221, 256–257, 259
solar titanium dioxide (TiO_2) photocatalysis 54–55, 308–309
solvents
  leather tanning industry 73
  metal finishing industry 113
  in metal finishing wastewater 114–116
  pharmaceutical industry 155, 193
sonochemical treatment *see* sonolysis
sonolysis 20–22
  costs 63–64
  enhanced 21–22, 63–65, 189–190
  hybrid systems 20–21, 63–65, 189–190
  industrial wastewater applications 63–65, 186–188, 218, 259, 296
  power ultrasound 20, 63–65, 186–188, 218
sonophotocatalysis 21–22
  ibuprofen degradation 189–190
spontaneity of redox reactions 2–3, 6, 10
standard electrode potential 5–6
standard hydrogen electrode (SHE) 5
stock preparation 207
sulphamethoxazole (SMX) 167–168, 176–178
sulphide oxidation 82
  by air oxygen 85–92
  by hydrogen peroxide (H_2O_2) 92–93
  catalytic 85–86, 87, 89–92
  costs 91–92
  electrochemical oxidation 93–94
  experimental studies 89–90, 94–95
  leather tanning wastewater 79, 89–90, 93–94
  ozonation 94
  and pH variation 87–88, 90
  polysulphides 86–87
  process mechanism 86–87
  reaction rate 87–88, 90–91, 92, 94–95
sulphite pulping 204
sulphonamides 167–168
sulphur compounds, and hexavalent chromium reduction 132–134
sulphur dioxide
  for cyanide oxidation 128, 277–280
  in gold and silver mining waste 270
  for hexavalent chromium reduction 133
sulphuric acid, leather tanning industry 73
supercritical water oxidation (SCWO) 19, 61–62
surfactants
  leather tanning industry 73
  metal finishing industry 113
  textile industry 38

tailings 270–271
  impoundment 271, 272–273, 281
  treatment 277–279
tanning agents 74, 80–81
tanyard processes 74, 80–82
textile industry 31–66
  BAT-EC 37–40
  biodegradability of chemicals used 37–40, 42–43, 47
  bleaching agents 38–39
  chemical oxidation applications 44–66
  chemicals used 37–40
  colour removal 33–34, 47–49, 50–59, 65
  dispersing agents 40, 54
  dye carriers 39
  dyestuffs 40
  environmental impact 35, 37–38
  finishing processes 32–33, 40–41
  pollutants 35
  regulation 33–34
  sequestering agents 38
  surfactants 38
  waste sources 35
  water consumption 33, 40–41
textile wastewater
  activated carbon treatment 43
  activated sludge treatment 42, 43
  adsorbable organically bound halogens (AOX) 38–39, 66

textile wastewater (*continued*)
advanced oxidation processes (AOPs) 53–65
biodegradability 57, 62–63
biological treatment 42, 43
catalytic wet peroxidation 58–59
colouration 33–34
electrocoagulation 44–49
electron beam irradiation 62–63
electrooxidation 49–53
Fenton process 57–59
hybrid treatment systems 43, 59–60, 62–65
hydroxide precipitation 45–46
organic matter 47–48
ozonation 55–57
photocatalysis 54–55
photoelectrocatalysis 53–54
supercritical water oxidation (SCWO) 61–62
toxicity 33, 39, 56–57, 61, 66
treatment costs 65–66
ultrasound (US) treatment 63–65
wet air oxidation (WAO) 43, 60–61
thermodynamics of redox reactions 2–10
titanium dioxide (TiO_2) photocatalysis 24, 223–224
metal finishing wastewater 140
olive mill wastewater (OMW) 315
pesticide wastewater treatment 257
of pharmaceuticals in WWTP effluent 180–183
pulp and paper wastewater 223–224, 225
solar UV irradiation 54–55, 308–309
sonophotocatalysis (US/UV/TiO_2) 21–22
textile industry wastewater 54–55
toxicity
leather tanning wastewater 98
pharmaceutical wastewater 166, 167, 182–183, 187
pulp and paper wastewater 217, 227
textile wastewater 33, 39, 56–57, 61, 66
toxicity dilution factor (TDF) 33, 172
Turkish Water Pollution Control Regulation (WPCR) 33

ultrafiltration 247, 314
ultrasound (US) treatment *see* sonolysis
UV irradiation *see* H_2O_2/UV treatment; O_3/UV treatment; photo-Fenton process; photocatalysis; photolysis; titanium dioxide (TiO_2) photocatalysis

vinasse 306–307
AOPs treatment 308–309, 311
biological treatment 307, 310
electrooxidation 309–310, 311
ozonation pre-treatment 307–308
volatile organics 245

WAO *see* wet air oxidation (WAO)
waste stream segregation
chemicals industry 288
metal finishing industry 114–116, 118–121, 142
pesticide industry 247
waste and wastewater sources
bleaching process 209, 210
chemical recovery 208, 210
chemicals industry 287
gold and silver mining 270–271
leather tanning industry 74–78
metal finishing industry 110–112
pesticide industry 242–243
pulp and paper industry 208–209, 210
textile industry 35
wastewater characteristics
alcohol distilleries 306–307, 310
bleaching 211, 212, 213
chemical pulping 211, 212, 213
chemicals industry 287
dyes and dyeing process 33–34, 35, 36
gold and silver mining 271–272
leather tanning industry 74–78
metal finishing industry 112–116

olive oil industry 312–313
pesticide industry 243, 245–247
pharmaceutical industry 155, 156, 157, 159, 172, 174
pulp and paper industry 209, 211
wastewater colouration 33–34
wastewater post-treatment 227–229
wastewater pre-treatment
alcohol distilleries industry 307–308
alum coagulation 97–98, 226
leather tanning industry 97–100
ozonation 97, 307–308
pharmaceutical wastewater 158–159, 166, 167, 173
pulp and paper industry 217–227
wastewater toxicity
leather tanning industry 98
pharmaceutical industry 166, 167, 182–183, 187
pulp and paper industry 217, 227
textile industry 33, 39, 56–57, 61, 66
wastewater treatment plants (WWTPs), pharmaceutical removal 161–163, 180–183
water consumption
leather tanning industry 76
textile industry 33, 40–41
wet air oxidation (WAO) 18–19
chemicals industry 289–290
colour removal 61
textile industry 43, 60–61
wet oxidation (WO) 218–219
wet peroxide oxidation 183–186, 290
WPCR *see* Turkish Water Pollution Control Regulation (WPCR)

zero-valent iron 135–136
zero-valent metals 135–136, 138–140

IWA Publishing's authorised EU representative for General Product Safety Regulations is Diane D'Arras, 15 rue Duret, 75116 Paris, France, e-mail: safety@iwap.co.uk.

Printed and bound by CPI Group (UK) Ltd, Croydon, CR0 4YY
16/06/2026
02140351-0001